ALGEBRAIC TOPOLOGY

PURE AND APPLIED MATHEMATICS

A Program of Monographs, Textbooks, and Lecture Notes

MONOGRAPHS AND TEXTBOOKS IN PURE AND APPLIED MATHEMATICS

1. K. Yano. Integral Formulas in Riemannian Geometry (1970)
2. S. Kobayashi. Hyperbolic Manifolds and Holomorphic Mappings (1970)
3. V. S. Vladimirov. Equations of Mathematical Physics (A. Jeffrey, editor; A. Littlewood, translator) (1970)
4. B. N. Pshenichnyi. Necessary Conditions for an Extremum (L. Neustadt, translation editor; K. Makowski, translator) (1971)
5. L. Narici, E. Beckenstein, and G. Bachman. Functional Analysis and Valuation Theory (1971)
6. D. S. Passman. Infinite Group Rings (1971)
7. L. Dornhoff. Group Representation Theory (in two parts). Part A: Ordinary Representation Theory. Part B: Modular Representation Theory (1971, 1972)
8. W. Boothby and G. L. Weiss (eds.). Symmetric Spaces: Short Courses Presented at Washington University (1972)
9. Y. Matsushima. Differentiable Manifolds (E. T. Kobayashi, translator) (1972)
10. L. E. Ward, Jr. Topology: An Outline for a First Course (1972)
11. A. Babakhanian. Cohomological Methods in Group Theory (1972)
12. R. Gilmer. Multiplicative Ideal Theory (1972)
13. J. Yeh. Stochastic Processes and the Wiener Integral (1973)
14. J. Barros-Neto. Introduction to the Theory of Distributions (1973)
15. R. Larsen. Functional Analysis: An Introduction (1973)
16. K. Yano and S. Ishihara. Tangent and Cotangent Bundles: Differential Geometry (1973)
17. C. Procesi. Rings with Polynomial Identities (1973)
18. R. Hermann. Geometry, Physics, and Systems (1973)
19. N. R. Wallach. Harmonic Analysis on Homogeneous Spaces (1973)
20. J. Dieudonné. Introduction to the Theory of Formal Groups (1973)
21. I. Vaisman. Cohomology and Differential Forms (1973)
22. B.-Y. Chen. Geometry of Submanifolds (1973)
23. M. Marcus. Finite Dimensional Multilinear Algebra (in two parts) (1973, 1975)
24. R. Larsen. Banach Algebras: An Introduction (1973)
25. R. O. Kujala and A. L. Vitter (eds). Value Distribution Theory: Part A; Part B. Deficit and Bezout Estimates by Wilhelm Stoll (1973)
26. K. B. Stolarsky. Algebraic Numbers and Diophantine Approximation (1974)
27. A. R. Magid. The Separable Galois Theory of Commutative Rings (1974)
28. B. R. McDonald. Finite Rings with Identity (1974)
29. I. Satake. Linear Algebra (S. Koh, T. Akiba, and S. Ihara, translators) (1975)
30. J. S. Golan. Localization of Noncommutative Rings (1975)
31. G. Klambauer. Mathematical Analysis (1975)
32. M. K. Agoston. Algebraic Topology: A First Course (1976)
33. K. R. Goodearl. Ring Theory: Nonsingular Rings and Modules (1976)

ALGEBRAIC TOPOLOGY

A FIRST COURSE

MAX K. AGOSTON

Department of Mathematics
Wesleyan University
Middletown, Connecticut

MARCEL DEKKER, INC.   New York and Basel

MARCEL DEKKER, INC.

270 Madison Avenue, New York, New York 10016

LIBRARY OF CONGRESS CATALOG CARD NUMBER: 75-18033

ISBN: 0-8247-6351-3

Current printing (last digit):
10 9 8 7 6 5 4 3 2 1

PRINTED IN THE UNITED STATES OF AMERICA

CONTENTS

Preface

1: SOME OLD TOPOLOGICAL PROBLEMS 1

1.0. Introduction 1
1.1. The Descartes-Euler Theorem 2
1.2. Coloring Maps and Graphs 9
1.3. The Jordan Curve Theorem; Knots 19
1.4. Further Early Investigations 22

2: THE GEOMETRY OF COMPLEXES 25

2.1. What is Topology 25
2.2. Simplicial Complexes and Maps; Polyhedra 31
2.3. Abstract Simplicial Complexes; Cutting and Pasting . . 46
2.4. Historical Comments 52

3: THE CLASSIFICATION OF SURFACES 55

3.0. Introduction 55
3.1. The Definition of a Surface 55
3.2. Representing Surfaces by Symbols 60
3.3. The Normal Form for Some Surfaces 78

3.4. The Classification Theorems 82
3.5. Bordered and Noncompact Surfaces 96
3.6. Historical Comments 102

4: THE HOMOLOGY GROUPS 107

4.1. Some Motivation 107
4.2. The Orientation of a Simplex 110
4.3. The Definition of the Homology Groups 115
4.4. The Homology Groups of a Cone 128
4.5. The Topological Invariance of Homology Groups 133
4.6. Historical Comments 135

5: MAPS AND HOMOTOPY 139

5.1. Simplicial Maps Again 139
5.2. Homotopy 146
5.3. Simplicial Approximations 150
5.4. The Barycentric Subdivision 155
5.5. The Simplicial Approximation Theorem; Induced Homomorphisms 164
5.6. Historical Comments 168

6: FIRST APPLICATIONS OF HOMOLOGY THEORY 169

6.1. A Quick Review 169
6.2. Local Homology Groups 174
6.3. Some Invariance Theorems 182
6.4. Homology with Arbitrary Coefficients; The Mod 2 Homology Groups 190
6.5. Pseudomanifolds; Orientability 195
6.6. Euler's Theorem Revisited 205
6.7. Retracts and the Brouwer Fixed-Point Theorem 214
6.8. Historical Comments 219

7: MAPS OF SPHERES AND MORE APPLICATIONS 225

7.1. The Degree of a Map; Vector Fields 225

7.2. The Borsuk-Ulam Theorem and the Ham Sandwich Problem 234

7.3. More on Degrees of Maps; Zeros of Polynomials 248

7.4. Local Degrees, Solvability of Equations and Some Complex Analysis 253

7.5. Extending Maps 275

7.6. The Jordan Curve Theorem and Other Separation Theorems 283

7.7. Historical Comments 289

8: CONCLUDING REMARKS 297

APPENDIX A: The Topology of $\mathbb{R}^n$ 305

APPENDIX B: Permutations and Abelian Groups 311

APPENDIX C: The Incidence Matrices 323

List of Symbols 331

Bibliography 337

Index 355

PREFACE

The main goal of this book is to serve as an introduction to algebraic topology for undergraduates. It is a sad fact that, because a thorough study of algebraic topology requires a working knowledge of other mathematical disciplines, the average mathematics student usually does not encounter this beautiful subject until he reaches graduate school. This postponement is no longer justified since undergraduates nowadays learn much more than they did in the past. Therefore, the author believes that this book will be quite accessible to seniors or advanced juniors on the undergraduate level. In addition, it should also be of interest to many beginning graduate students. Most graduate courses on algebraic topology these days develop homology theory in a rather abstract and axiomatic manner. This is certainly the efficient and elegant way to present the material and it also brings the student to current research in the quickest way possible. Furthermore, it is a natural phenomenon in mathematics and other disciplines that with the passage of time the presentation of a theory is tidied up and perhaps significant improvements are made in it. Nevertheless, the beautiful intuition which gave rise to the theory in the first place should not be forgotten. On a number of occasions the author has witnessed how the abstract approach in first-year graduate courses, by hiding the geometry involved, has left students with the feeling that they have merely learned to manipulate some definitions. Thus, the fact that this book tries to place geometrical considerations first, and emphasize the historical development, should make it a valuable reference for graduate students also.

This book grew out of some notes for a semester course on geometry taught by the author on two separate occasions to third-year college students--once at Wesleyan University and once at the University of Auckland in New Zealand. In this one semester it was possible to cover most of Chapters 1 to 5 and part of Chapter 6. Students with only a background in calculus and linear algebra were able to manage the material quite well. Actually, nothing but the most basic concepts associated to vector spaces, such as linear independence, basis, and dimension, are used in a few places. On the other hand, a background in linear algebra is helpful in understanding the group aspect of homology theory. The elementary facts of the theory of abelian groups which are used in this book are summarized in Appendix B. The only hard theorem is the Fundamental Theorem of Finitely Generated Abelian Groups, and that can be assumed. A brief discussion of the material in Appendix B by the instructor should be quite sufficient for the students. Appendix A summarizes the facts needed about $\mathbb{R}^n$ and continuous maps. Here and there, but mainly in Section 7.4, a little acquaintance with complex numbers comes in handy. It is recommended that the reader glance over both Appendix A and B before starting on the book because they introduce some basic notation which is used throughout the text.

Although the concept of an abstract topological space is an important one and indispensable to modern topology, we shall restrict ourselves entirely to subspaces of $\mathbb{R}^n$ in this book, thereby avoiding a lengthy discussion of general topology that would only slightly simplify some of the arguments while at the same time undermining the overall goal, which is to display some of the beautiful geometric insights of topology without burdening the uninitiated student with an overpowering load of technical details. One minor exception to this goal of keeping our presentation geometric is the material on abstract complexes in Section 2.3. This can be skipped because one could easily leave cutting and pasting techniques on the intuitive level. Another exception is Theorem 1 in Section 5.4. This nice result is used to define the important induced homomorphisms on homology, but the proof involves some rather abstract ideas which were difficult to avoid. Depending on the level of the class, the instructor may decide simply to sketch the proof or skip it altogether. It is included mainly for the sake of completeness.

The reader is urged to consult the excellent surveys [De-H], [Stnz], and [Ti-V] for additional historical information. The author in no way claims any completeness in this regard. The texts [Se-T 1] and [Alexf-H] are also highly recommended reading. There is no author index but one can use the bibliography instead because it gives the page numbers where the author's name appears. The birth dates of most authors in the bibliography are usually given where they first appear in the book.

Max K. Agoston

ALGEBRAIC TOPOLOGY

Chapter 1

SOME OLD TOPOLOGICAL PROBLEMS

1.0. INTRODUCTION

Algebraic topology is a young field in mathematics when compared to others such as number theory and geometry; however, it is fair to say that in recent years it has been one of the most active and has attracted the interest of many of the best mathematicians in the world. One of the nice aspects of algebraic topology, but one which makes it difficult to teach at an elementary level, is that it works against the modern trend toward specialization, and the major developments in this field would have been quite impossible without an essential interplay of many different areas of mathematics. This book discusses what is traditionally called combinatorial topology, which concerns the study of polyhedra. Although this forms only a small part of what is now called algebraic topology, it serves as a good introduction to the field as a whole; by exposing the beginner to the geometric roots he can get a better intuitive appreciation of what the subject is all about. In this chapter we shall look at some of the early history of algebraic topology and make little attempt at rigor. Our main goal here is to motivate interest in what follows.

Before the mid-nineteenth century, topology consisted mainly of a collection of isolated observations. In fact, the term "topology" itself appeared first in 1847 in *Vorstudien zur Topologie*, by J. B. Listing (1806-1882) [Lis 1], a student of C. F. Gauss (1777-1855). Before that, G. W. Leibniz (1646-1716) had introduced the term *geometria situs* for the study of certain qualitative properties of geometric figures that, today, would be considered topological (see [Leib 1] and [Leib 2]), but he did not contribute much to the subject. In the nineteenth century and early 1900s topology was usually referred to as *analysis situs*.

1.1. THE DESCARTES-EULER THEOREM

Probably the earliest significant topological observation concerned a relationship between the number of faces, edges, and vertices of a polyhedron which was already known to R. Descartes (1596-1650) [Des] around 1620. Throughout this chapter the term "polyhedron" will denote any compact 3-dimensional subspace of $\mathbb{R}^3$ that is bounded by planar faces. (The precise definition of a polyhedron, which is given in Sec. 2.2, is somewhat more general.) The surface of a polyhedron is a collection of faces, edges, and vertices. For example, the solid triangle ABC, the solid square ABCD, and the solid pentagon ABCDE are faces of the polyhedra in Fig. 1.1a, b, and f, respectively. In each of these polyhedra the line segment AB is an example of an edge. The vertices of the tetrahedron in Fig. 1.1a are A, B, C, and D.

Given a polyhedron P, let n_v, n_e, and n_f denote the number of its vertices, edges, and faces, respectively, and consider the alternating sum $n_v - n_e + n_f$. This sum is equal to 2 for the polyhedra in Fig. 1.1a-f, and it is 0 for the polyhedron in Fig. 1.1g. If it had not been for this last and more complex polyhedron it might have been natural to conjecture that $n_v - n_e + n_f$ was always equal to 2, but now we must be more careful. (Actually, we shall see that the real interest in $n_v - n_e + n_f$ arises from the very fact that it is *not* the same for all polyhedra.)

What do the first six polyhedra in Fig. 1.1 have in common and how do they differ from the last one? Let us define a polyhedron to

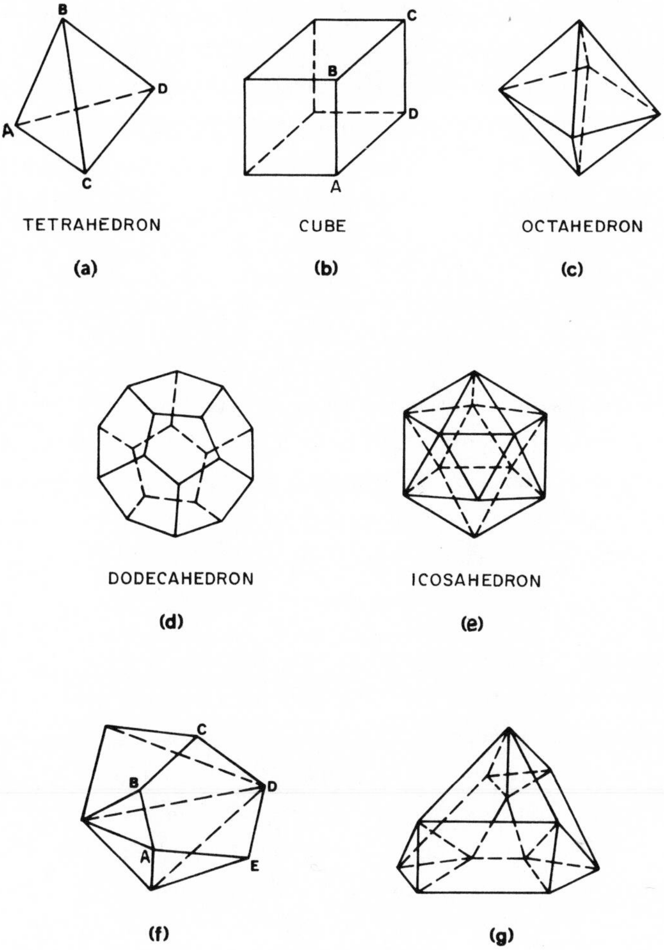

FIGURE 1.1

be *simple* if it can be deformed into a solid ball such as D^3, where, in general, the n-dimensional unit ball or disk D^n in $\mathbb{R}^n$ is defined by

$$D^n = \{(x_1, x_2, \cdots, x_n) \in \mathbb{R}^n \mid x_1^2 + x_2^2 + \cdots + x_n^2 \leq 1\}$$

What we mean by this is the following: Pretend that the polyhedron is made of rubber. If we can now stretch or squeeze it into D^3, then we shall call it simple. An example of this process is given in Fig. 1.2 where we show how a solid triangle can be deformed into a disk. It should be intuitively obvious that the polyhedra in Fig. 1.1a-f are simple. The polyhedron in Fig. 1.1g is not simple because it has a "hole," in the same way that a circle has a "hole" and cannot be deformed into a line segment without cutting it, which is not a permissible operation.

THEOREM (Descartes-Euler). $n_v - n_e + n_f = 2$ for every simple polyhedron.

The following argument is an intuitive proof of the theorem: Let P be a simple polyhedron. By hypothesis P can be deformed into D^3, and under this deformation the boundary of P will go into the boundary of D^3 which is the unit sphere S^2, where, in general, the n-dimensional unit sphere S^n in $\mathbb{R}^{n+1}$ is defined by

$$S^n = \{(x_1,x_2,\cdots,x_{n+1}) \in \mathbb{R}^{n+1} \mid x_1^2 + x_2^2 + \cdots + x_{n+1}^2 = 1\}$$

Clearly, if we delete a small disk from S^2, then the remainder can be flattened out into a plane. These remarks show that, if we delete a face from the boundary of a simple polyhedron, then the remainder of the boundary imbeds in the plane $\mathbb{R}^2$. Let n_v', n_e', n_f' denote the number of vertices, edges, and regions, respectively, that we obtain in $\mathbb{R}^2$ in this way. (We do not insist that the edges be straight-line segments.) Figure 1.3a and b shows what we would get if we apply this procedure to the cube and octahedron, respectively. Since the number of edges and vertices remain unchanged and there is now one less face, we have that $n_v' = n_v$, $n_e' = n_e$, and $n_f' = n_f - 1$. To prove the theorem it clearly suffices to prove that $n_v' - n_e' + n_f' = 1$.

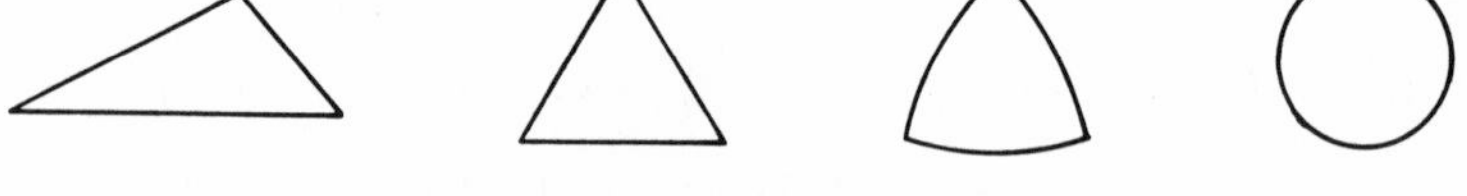

FIGURE 1.2

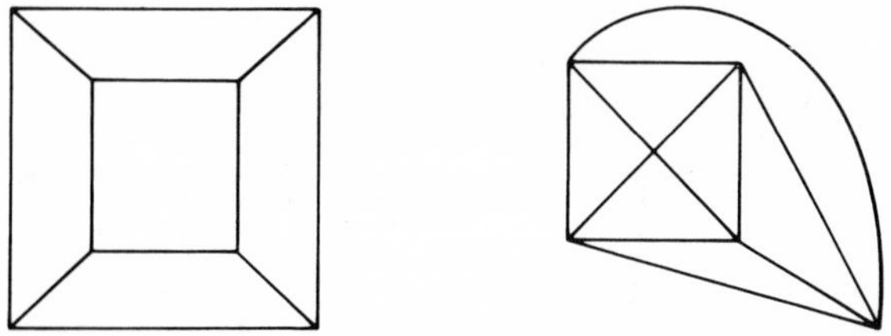

(a) (b)

FIGURE 1.3

First, we may assume that all regions are triangular, since if they are not so already, then we can obtain such a situation by successively introducing a new edge as in Fig. 1.4. This operation clearly does not alter the sum $n'_v - n'_e + n'_f$ because we also increase n'_f by 1 each time. Next, list the triangular regions in a sequence $T_1, T_2, \cdots, T_k$ such that T_i meets $\bigcup_{1 \leq j < i} T_j$ either in one or two edges (see Fig. 1.5). Obviously, for a triangle $n'_v - n'_e + n'_f = 3 - 3 + 1 = 1$. Assume that $n'_v - n'_e + n'_f = 1$ for $\bigcup_{1 \leq j < i} T_j$. It is easy to check that as we add T_i we either increase n'_v and n'_f by 1 and n'_e by 2 (see Fig. 1.5a) or we increase n'_e and n'_f by 1 and leave n'_v unchanged (see Fig. 1.5b). In any case the new sum $n'_v - n'_e + n'_f$ is still equal to 1. By induction, the theorem is proved.

In [Fre-F, p. 103-105] one can find several other methods to prove this theorem. Descartes himself had arrived at essentially that formula by considering the sum of the angles between the adjacent edges of the faces of the polyhedron (see [Des]). It is likely that he was trying to generalize an analogous formula for the case of a polygon in the plane. Although Leibniz possessed a copy of Descartes' original manuscript, it was not published and remained forgotten until its existence was pointed out in 1860. In the meantime L. Euler (1707-1783) [Eu 3,4] had independently discovered the formula

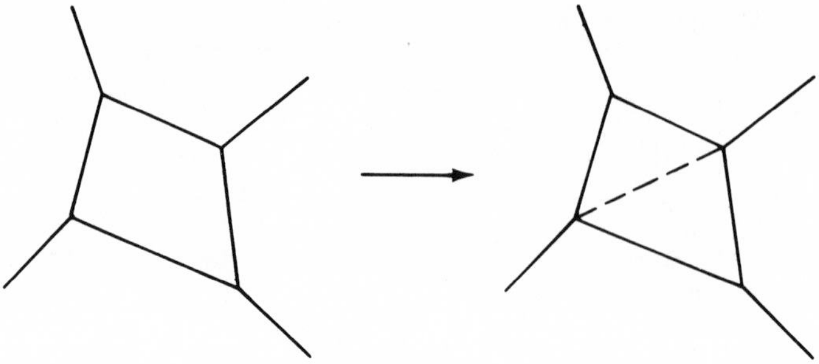

FIGURE 1.4

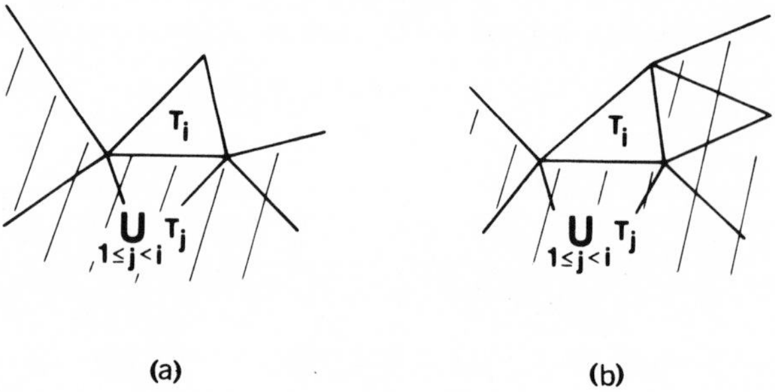

FIGURE 1.5

and published a proof of it in 1752. This explains the fact that one usually refers to the theorem as Euler's theorem. According to [Lie 2, p. 90], Archimedes (287?-212 B.C.) may have known of the Descartes-Euler formula.

The real topological significance of the above result may not be apparent for some time. Note, however, that we did not really need straight edges and planar faces and that we have somehow obtained an "intrinsic" invariant of the sphere. We can paraphrase Euler's theorem loosely as saying that no matter how we divide up a sphere into n_f regions with n_e edges and n_v vertices, the sum $n_v - n_e + n_f$ will always equal 2. This already gives us a concrete way of distinguishing between a sphere and the surface of a doughnut (see Fig. 1.1g) other than on purely intuitive grounds. Actually, neither Descartes nor Euler looked at the formula in that light. Euler and his immediate followers were concerned solely with the combinatorial aspects of the formula and it was a hundred years before its topological importance was realized.

As an application of Euler's theorem to geometry let us determine all possible regular polyhedra, that is, those polyhedra with the property that every face has the same number of edges, say h, every vertex has the same number of edges emanating from it, say k, and every edge has the same length. Now since every edge has two vertices and belongs to exactly two faces, it follows that $n_f h = 2n_e = n_v k$. Substituting this into Euler's formula we get

$$2n_e/k - n_e + 2n_e/h = 2$$

or equivalently,

$$1/n_e = 1/h + 1/k - 1/2 \qquad (*)$$

For a polyhedron we always assume that $h, k \geq 3$. On the other hand, if both h and k were larger than three, then (*) would imply that

$$0 < 1/n_e = 1/h + 1/k - 1/2 \leq 1/4 + 1/4 - 1/2 = 0$$

which is impossible. Therefore either h or k equals 3. If $h = 3$, then

$$0 < 1/n_e = 1/3 + 1/k - 1/2$$

implies that $3 \leq k \leq 5$. By symmetry, if $k = 3$, then $3 \leq h \leq 5$. Thus $(h,k,n_e) = (3,3,6), (4,3,12), (3,4,12), (5,3,30)$, and $(3,5,30)$ are the only possibilities, which are, in fact, realized by the tetrahedron, the cube, the octahedron, the dodecahedron, and the icosahedron, respectively (see Fig. 1.1a-e). Observe that we did not really use the fact that the edges of the polyhedron all have the same length, so that as long as the numbers h and k are constant we still have only five possibilities (up to stretching or contracting). These regular polyhedra were already known to Euclid (365?-300? B.C.) (see his thirteenth book in [He]) and he even made the claim that there were no others. It should be pointed out though that Euclid's definition of a "regular polyhedron" was rather imprecise and his claim is not quite correct in the generality that he states it (see [Lie 2, p. 95]). For a list of other "regular" polyhedra and additional existence or nonexistence theorems about polyhedra in general which make use of Euler's formula see [As, p. 438-455], [Stnz], [Fre-F, p. 105-107], and [Leb 3]. In [R-T, p. 200-201] it is shown that the surface of a doughnut, or torus, admits an infinite number of different regular decompositions--quite a contrast with what happens in the case of a sphere.

Problems

1.1. In proving Euler's theorem we essentially made use of the following fact: If we divide the unit disk D^2 into triangular regions (where the sides are not required to be straight lines), then it is possible to list these regions in a sequence $T_1, T_2, \cdots, T_k$ such that T_i meets $\bigcup_{1 \leq j < i} T_j$ either in one or two edges. Discuss how one might be able to prove this and what, if any, difficulties one could run into. (It was shown in 1972 that the analogous property does not hold for higher dimensional disks D^n when $n \geq 5$ so that one has to be careful about assuming even seemingly "obvious" facts in geometry.)

1.2. We have paraphrased Euler's theorem as saying that no matter how we divide up a sphere into n_f regions with n_e edges and n_v vertices, the sum $n_v - n_e + n_f$ will always equal 2. Compute $n_v - n_e + n_f$ for the decomposition of S^2 in Fig. 1.6. If we are to preserve the validity of Euler's theorem, what is bad about the decomposition? Where does the proof in this section go wrong? How can we define a notion of "permissible region" so that Euler's theorem will hold for all decompositions of S^2 into permissible regions?

1.3. Prove that no simple polyhedron can have exactly seven edges. (As usual, we assume that each face has at least three edges and each vertex belongs to at least three edges.)

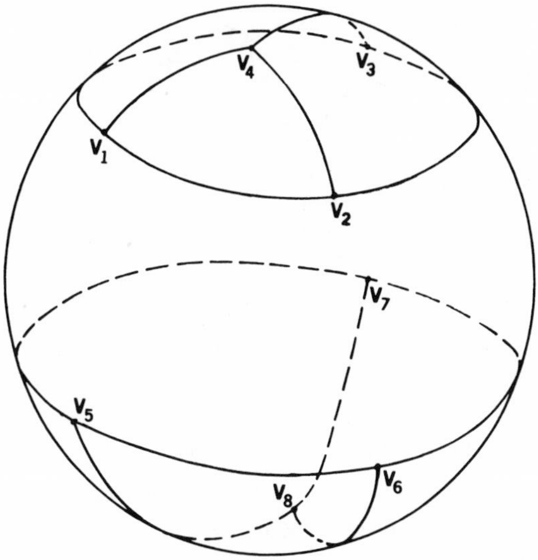

FIGURE 1.6

1.4. Draw n great circles on S^2 so that no three of them meet in a point. Into how many regions is S^2 divided? (In general, a great circle on S^n is a circle obtained by intersecting a 2-dimensional subspace or plane through the origin in $\mathbb{R}^{n+1}$ with S^n.)

1.2. COLORING MAPS AND GRAPHS

An old and famous unsolved problem which most students of mathematics hear about very early in their career is the four-color problem. The problem is to determine whether or not every map on the sphere S^2 can be colored with only four colors in such a way that no two countries with a common edge have the same color. (This problem is equivalent to coloring every map in the plane $\mathbb{R}^2$ with four colors.) For example, in Fig. 1.7 countries A and B would have to have different colors, but not A and C. The four countries C, D, E, and F also give an example of a map for which three colors would not suffice. In 1890 P. J. Heawood (1861-1955) [Hea] proved that

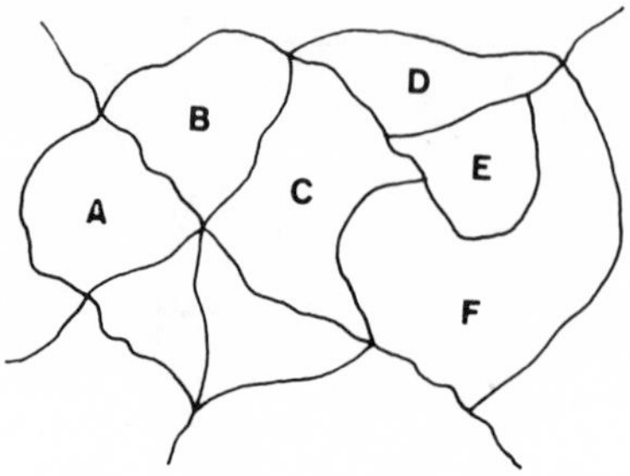

FIGURE 1.7

every map on a sphere can be colored with five colors. The conjecture that four colors would suffice originated with the mathematician Francis Guthrie (d. 1899) around 1850 (see [Fre-F, p. 78-80] or [Rin 2, p. 1-2] for the interesting history of the problem). Although many mathematicians have thought about the problem and several false proofs have been published, no one to this day has been able either to give a satisfactory proof or to construct a map which needs five colors. It has been shown, however, that any map which needs five colors would have to have at least 36 countries and hence would not be very simple. Often problems that appear to be elementary are the most difficult!

One can generalize the map-coloring problem to an arbitrary surface S (without boundary). A precise definition of a surface will be given in Sec. 3.1. The sphere and the spaces in Fig. 1.8 are examples of surfaces. Define the chromatic number of S, chr(S), to be the minimal number of colors needed to color every map on S such that no two countries with a common edge have the same color. The problem is to give a formula for chr(S) in terms of easily computed invariants of S.

We have already seen one such invariant in the discussion of Euler's theorem. In fact, subdivide S into regions having $n_v(S)$ vertices, $n_e(S)$ edges, and $n_f(S)$ faces, and such that each region can be deformed (in the sense described earlier) into the unit

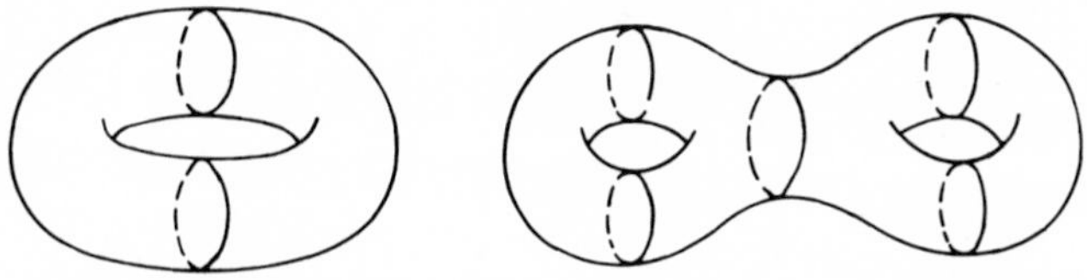

TORUS = SURFACE OF A DOUGHNUT

(a)

DOUBLE TORUS = SURFACE OF A SOLID FIGURE 8

(b)

FIGURE 1.8

disk D^2. Then it turns out that the alternating sum $\chi(S) = n_v(S) - n_e(S) + n_f(S)$ depends only on S and not on the particular subdivision of S. $\chi(S)$ is called the Euler characteristic of S. For example, Euler's theorem essentially states that $\chi(S^2) = 2$.

Another invariant of a surface is its orientability. In 1858 both A. F. Möbius (1790-1868) [Möb 2] and Listing (in a note found among his manuscripts) independently discovered one-sided surfaces (see also [Möb 3] and [Lis 2]). Consider the Möbius strip in Fig. 1.9b which is the space obtained by pasting the two ends of a strip of paper together after giving one end a 180 degree twist. At any point the strip has two sides but we can get from one side to the other merely by walking all the way around along the meridian. This is quite different from the simple cylinder in Fig. 1.9a. Here we cannot get from one side of the paper to the other simply by walking along the meridian. Therefore, we call the Möbius strip a nonorientable surface (with boundary) and the cylinder an orientable surface. In general, we say that a surface S is orientable (non-orientable) if we cannot (can) get from one side of S at a point to the other by merely walking along the surface.

One can define orientability also in terms of properties which are more suggestive of the label. For example, an orientable surface is one on which it is possible to define a notion of left and right or clockwise and counterclockwise in a consistent way. What this means is the following: Choose a point p on the surface S and

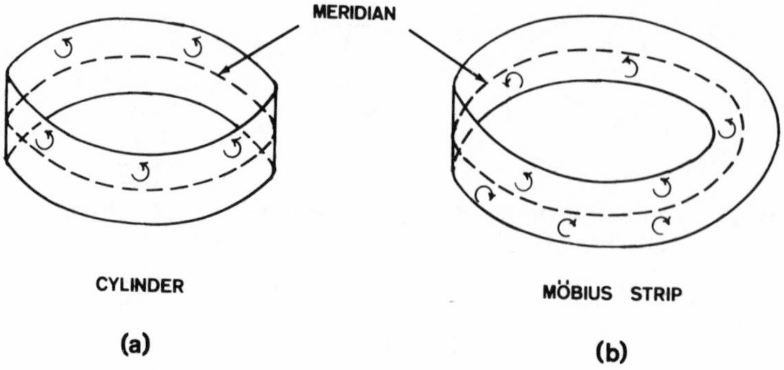

FIGURE 1.9

pretend to be standing at that point. Decide which of the two possible rotations is to be called clockwise. This is called an orientation at p. Now let q be another point of S (q may equal p) and walk to q along some path keeping track of which rotation had had been called clockwise. This will induce an orientation at q, in other words, a notion of clockwise or counterclockwise for rotations at q. The only problem is that there are many paths from p to q (nor is there always a unique shortest path) and different paths may induce different orientations. If we get the same orientation no matter which path we take, S is called orientable. In Fig. 1.9b we can see that if one walks around the meridian of the Möbius strip, one will end up with the opposite orientation from what one picked at the beginning. Therefore, the Möbius strip is nonorientable, and our new definition agrees with the old.

Orientability is an intrinsic property of surfaces. This fact was observed explicitly first by F. Klein (1849-1925) [Kle 2] in 1876. The sphere and the surfaces in Fig. 1.8 are orientable; but there are surfaces without boundary which are nonorientable and the reader is challenged to find one on his own. (*Hint:* Nonorientable surfaces without boundary cannot be drawn in $\mathbb{R}^3$. One needs a fourth dimension.)

THEOREM (Heawood conjecture). Let S be an orientable surface such that $\chi(S) < 2$. Then $\text{chr}(S) = [(7 + \sqrt{49 - 24\chi(S)})/2]$, where $[x]$ denotes the greatest integer in x.

REMARK. It can be shown (see Chapter 3) that if a surface S has $\chi(S) \geq 2$, then $\chi(S) = 2$ and S is the sphere, so that actually the sphere is the only orientable surface which is excluded from the theorem. Observe that in this case the formula in the theorem would say that $\text{chr}(S^2) = 4$, which is the four-color conjecture.

If S is a surface, let $n_S = (7 + \sqrt{49 - 24\,\chi(S)})/2$. Heawood [Hea] was able to show in 1890 that $\text{chr}(S) \leq [n_S]$. The hard part of Heawood's conjecture was to show that $[n_S] \leq \text{chr}(S)$ if S is orientable. This was finally proved by G. Ringel (1919-) and J. W. T. Youngs [Rin-Y] in 1967 who constructed a map for each such surface S which needed at least $[n_S]$ colors. Incidentally, P. Franklin (1898-) [Fr] had shown that the nonorientable surface called the Klein bottle (see Sec. 3.4 for a definition) needs only six colors to color its maps even though the "Heawood number" $[n_S]$ is 7. This explains the need for the condition of orientability in the theorem above; however, later in 1959 it was proved by Ringel [Rin 1] that the Klein bottle is the only exception.

Let us prove that $\text{chr}(S) \leq [n_S]$, which will, at the same time, explain the mysterious formula for n_S:

For any map on a surface S of Euler characteristic χ, let v, e, f denote the number of vertices, edges, faces, respectively. Let a be the average number of edges per face, that is, $af = 2e$.

LEMMA. If every vertex of S has a least three edges emanating from it, then $a \leq 6(1 - \chi/f)$.

Proof: It follows easily from the hypothesis of the lemma that $3v \leq 2e = af$. Therefore, using the fact that $\chi = v - e + f$, we get that $e \leq 3(e - v) = 3(f - \chi)$. Thus $a = 2e/f \leq 6(1 - \chi/f)$.

We are now ready to show that $\mathrm{chr}(S) \leq [n_S]$ whenever $\chi < 2$. It is not hard to see that there is no loss in generality in assuming that every vertex of the map M we are trying to color has at least three edges emanating from it.

CASE 1. $\chi = 1$. In this case, $n_S = 6$. Obviously we can color M if it has no more than 6 faces. Assume inductively that every map with $f - 1$ faces can be colored with 6 colors and suppose M has f faces. By the lemma, a is less than 6 and hence there is at least one face X with 5 or fewer edges. Contract this face to a point to obtain a map M' with $f - 1$ faces (see Fig. 1.10). By induction, M' can be colored by 6 colors and it follows easily that M can also be colored by 6 colors.

CASE 2. $\chi \leq 0$. Again it is obvious that any map with $\leq n_S$ faces can be colored by $[n_S]$ colors. Therefore assume that M has $f > n_S$ faces and that any map with fewer than f faces can be colored by $[n_S]$ colors. The definition of n_S implies that $n_S^2 - 7n_S + 6\chi = 0$, or equivalently, that $6(1 - \chi/n_S) = n_S - 1$. The lemma (and the fact that $\chi \leq 0$) now implies that $a \leq 6(1 - \chi/f) \leq 6(1- \chi/n_S) = n_S - 1$. Hence there is at least one face of M which

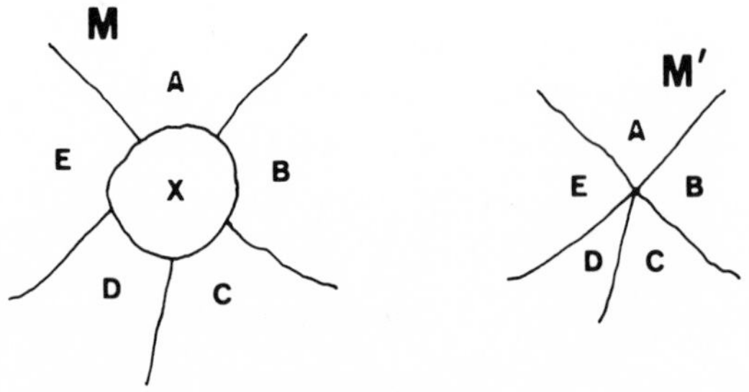

FIGURE 1.10

has no more than $[n_S] - 1$ edges. The argument can now be completed as in Case 1 by contracting this face and using induction.

We shall conclude this section with a few words about a related topic, namely graph theory. Good books on this subject are [Har] and [Ber]. A graph is a collection of points, called vertices, together with a collection of (nonintersecting) line segments, called edges, between them. Many problems can be rephrased usefully in the terminology of graph theory. For example, there is the famous Königsberg bridge problem: In the eighteenth century Königsberg had seven bridges situated as in Fig. 1.11a and the question was asked whether one could (by walking) cross all the bridges in such a way that none would be crossed more than once. Euler [Eu 1] showed in 1735 that this was impossible by considering the graph in Fig. 1.11b. Here the vertices represent land and the edges are the bridge connections. Euler actually showed much more and completely characterized those graphs which can be traversed by a path using each edge precisely once (Problem 1.10). The first important contribu-

FIGURE 1.11

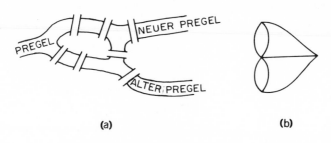

tions to the theory of linear graphs were made by G. R. Kirchhoff (1824-1887) [Kirc] in 1847.

Problems about coloring maps are usually approached via graph theory. Given a map, one represents the countries by vertices and two vertices are joined by a line if and only if the corresponding countries have a common boundary. For example, the graph in Fig. 1.12 represents the map of Fig. 1.7 consisting of countries A, B, C, D, E, and F. The map-coloring problem then translates to the following: Given a graph, what is the smallest number n with the property that it is possible to assign one of the numbers $\{1,2,\cdots,n\}$ to each vertex in such a way that no pair of vertices which are connected by an edge have been assigned the same number? The smallest such number n is called the chromatic number of the graph.

Note that not all graphs can be drawn in the plane (see Fig. 1.13). Those which can are called planar. The four-color problem is equivalent to showing that the chromatic number of every planar graph is at most four. K. Kuratowski (1896-) [Ku] showed in 1930 that any nonplanar graph must contain one of the graphs in Fig. 1.13. Incidentally, the graph in Fig. 1.13b plays a role in a popular puzzle. Assume that the top three vertices are houses and the bottom three vertices are utilities. The question is: Can each house

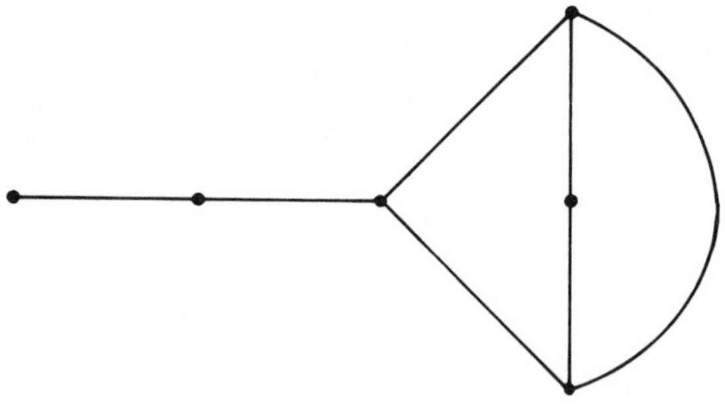

FIGURE 1.12

NONPLANAR GRAPHS

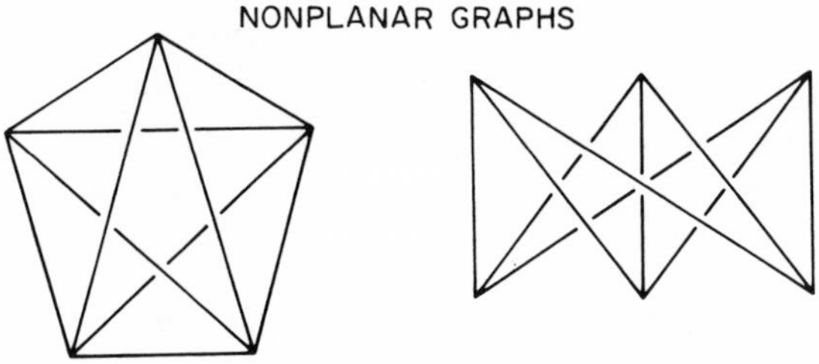

(a) (b)

FIGURE 1.13

be connected to the three utilities in such a way so that no connection passes over another? By the above remarks, the answer is "No."

Problems

1.5. Compute the Euler characteristic of the surfaces in Fig. 1.8.

1.6. Sketch a proof of the fact that the Euler characteristic of the torus does not depend on the particular subdivision of the torus into regions.

1.7. Prove that every map on the sphere can be colored with five colors.

1.8. (a) Draw a map consisting of five countries on the Möbius strip which needs five colors.
(b) Draw a map on the Möbius strip which needs six colors.
(c) Draw a map on the torus which needs seven colors.

1.9. Describe what happens when a Möbius strip is cut (a) along its center line, and (b) along a line one-third of the way in from the boundary. Try cutting along some other lines. For the many interesting configurations that can be obtained by cutting, pasting, and folding pieces of paper see [Bar],

[Lie 1, p. 117-118], and [Bal-C]. There is also an amusing limerick:

A mathematician confided
That a Möbius strip is one-sided,
  And you'll get quite a laugh
  If you cut one in half,
For it stays in one piece when divided.

1.10. Prove Euler's theorem: Given any connected graph, it is possible to arrange the edges in a path which uses each edge precisely once if and only if the number of vertices which belong to an odd number of edges is either zero or 2. (A graph is said to be connected if each pair of vertices can be joined by a path constructed from the edges of the graph.)

1.11. Prove that the graphs in Fig. 1.13 are nonplanar. The fact that the graph in Fig. 1.13a is nonplanar shows that the maximum number of points in the plane which can be connected to each other by nonintersecting lines is 4. Equivalently, we get a solution to the neighboring regions problems, namely, the maximum number of connected regions in the plane which meet each other in a line segment is 4. There is a story about a father and his five sons that has been told in this connection. When the father died he specified in his will that should his sons want to divide up the property among themselves then in order to maintain their togetherness it had to be done in such a way so that each of their individual properties were connected and adjacent to every other. The sons tried unsuccessfully for a long time until they realized that it was impossible and that their father really had wanted them not to divide the land at all. (That there exist four neighboring regions in the plane was shown by Mobius. Heawood [Hea] showed that one could find seven neighboring regions on torus.)

1.3. THE JORDAN CURVE THEOREM; KNOTS

Let us consider a circle C in the plane as in Fig. 1.14. It is intuitively obvious that C divides the plane into two regions A and B which are characterized by the property that any two points within any one of the region (such as a and b, or c and d) can be connected by a path which lies wholly within the region, whereas if we take one point from each of the regions (such as a and c) then every path between these will necessarily cut C. The bounded region A could be called the inside of C, and B, the outside. This property of the circle began to attract attention only in the latter half of the last century when, owing to developments in analysis, it was realized that continuous maps from the circle to the plane could be quite complicated and have very non-intuitive properties. For example, in 1890 G. Peano (1858-1932) [Pea] showed that it was possible to define a curve which covered an entire square in the plane. On the other hand, Peano's curve is not what is called a "simple closed curve." The latter should be thought of as the set of points in the plane traced out by a pencil which moves along a path that starts at one point, does not cross itself, and returns to its starting point. The first proof of the now famous Jordan curve theorem, that every simple closed curve in the plane divides the plane into two regions, appeared in 1893 in

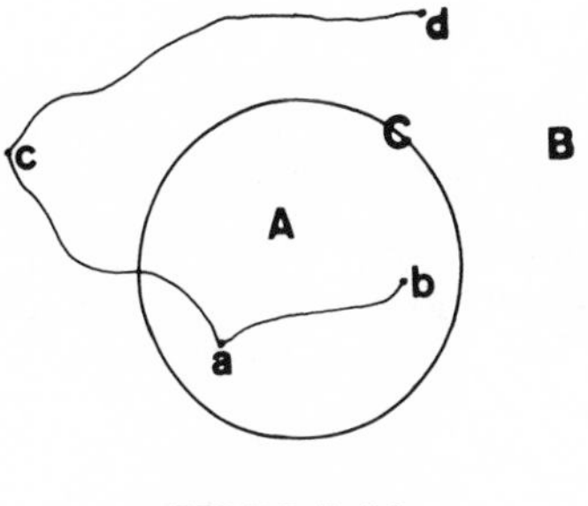

FIGURE 1.14

the important book, *Cours d'Analyse*, by C. Jordan (1838-1922) [J 3], although Jordan did not solve the problem completely (see Sec. 7.7 for more details on its history). Even though Jordan's original proof has by now been greatly simplified, it is still not easy to prove this apparently intuitively obvious fact. Clearly, the fact that there is a well-defined notion of inside and outside for simple closed curves has bearing on the map-coloring problems in the previous section and is also relevant to those who want to build or escape from mazes. (See [Bal-C] for an interesting discussion of mazes.)

If one thinks about the Jordan curve theorem for a while, it is natural to wonder if it generalizes to higher dimensions. For example, does every imbedded sphere in $\mathbb{R}^3$ divide $\mathbb{R}^3$ into two regions?

Jordan's theorem has to do with imbedded circles in $\mathbb{R}^2$. An imbedded circle in $\mathbb{R}^3$ is called a knot. The basic problem of knot theory is to determine when two knots are equivalent, that is, when they can be transformed into each other by twisting and pulling. Figure 1.15 gives an example of two inequivalent knots. (In real life one works with pieces of string. If one wants to distinguish

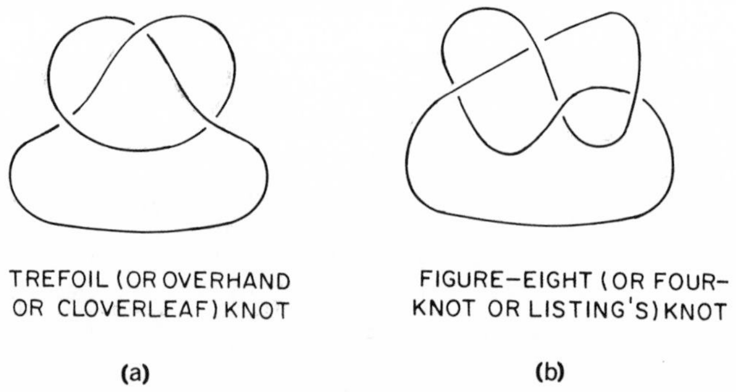

FIGURE 1.15

between different knot types in this context, then one must require that the endpoints of the string are essentially kept fixed. By bringing the ends together to form a circle we have an equivalent problem but one which is easier to handle mathematically.) Let us call a knot trivial or unknotted if it can be deformed into the standard circle S^1. A special case of the general problem is to determine when a knot is unknotted. For example, if we are given a pile of string, how can one tell if there is a knot present? Note that we cannot distinguish between knots intrinsically because they are all circles. Our problem is to distinguish between the various possible placements of the same circle in $\mathbb{R}^3$ and thus is different from the topological questions that we have asked in the previous sections where we wanted to distinguish between objects.

Gauss [G 2,3] made some brief comments on aspects of knot theory at various times between 1823 and 1840. He attempted to classify singular curves in the plane, some of which arose from the projection of a knot, and in 1833 he found a formula connecting a certain double integral with the linking number of two knots, which intuitively is the number of times two knots are intertwined (see Fig. 1.16). However, the first mathematician to make a real study of knots appears to have been Listing and part of his book [Lis 1] from 1847 is devoted to this subject. Listing's approach was to project the knot into the plane and to analyze it by considering the intersections, keeping track of the over- and under-crossings. Such a method was also used later by P. G. Tait (1831-1901) [T], T. P. Kirkman (1806-1895) [Kirk], and C. N. Little [Lit] to obtain a classification of the more elementary knots. Other early papers on knots are by O. Simony (1852-1915) [Sim 1, 2] and F. Meyer (1856-1934) [Mey 1, 2]. In 1895 W. von Dyck (1856-1934) [Dyc 3] generalized Gauss' integral to describe higher dimensional linking numbers. Consult Dyck [Dyc 1, p. 460] for a history about how analysis crept into these topological investigations through the work of several mathematicians. Linking

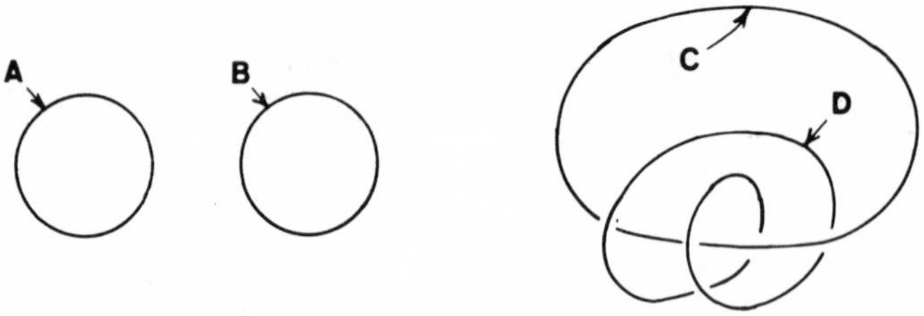

A•B = 0 AND C•D = ±2,
WHERE X•Y DENOTES THE LINKING NUMBER OF X AND Y

FIGURE 1.16

numbers "mod 2" first appeared in a note by M. H. Lebesgue (1875-1941) [Leb 2] in 1911. The first real theory of linking numbers is due to L. E. J. Brouwer (1881-1967) [Brou 4] in 1912. One can also find it discussed in [Lef 1], [Kam], [Pont 1], [Lef 3], [Se-T 1], and [Alexf-H].

The first invariants of knots which were effectively calculable were found in the 1920s by J. W. Alexander (1888-1971) [Alex 7] and by K. Reidemeister (1893-) [Rei 1,2]. These invariants (see also [Alex 8]) were obtained by studying the complement of the knot in $\mathbb{R}^3$ and its associated "knot group" which had been defined in 1910 by M. Dehn (1878-1952) [De]. However, knot theory is very difficult and many important unanswered questions remain. We have mentioned this topic only because it is another old topological problem. The interested reader can find a lengthier discussion of knots and related topics in [Lie 2], [Se-T 2], [Fox], and [Cr-F].

1.4. FURTHER EARLY INVESTIGATIONS

Soon after rediscovering Descartes' formula Euler [Eu 3, 4] launched in 1752-1753 into a detailed study of simple polyhedra and attempted to classify them in terms of the numbers N_p and M_q, where N_p is the number of faces with p edges and M_q is the number of vertices which are incident to q edges. These investigations were not immediately followed up by other mathematicians. There was a lag of some years before interest revived. Publications of A. M. Legendre (1752-1833) [Leg] from 1809, A. L. Cauchy (1789-1857) [Cau] from 1812, S. A. J. L'Huilier (1750-1840) [Lh] from 1812-1813, J. Steiner (1796-1863) [Stnr] from 1842, and E. C. Catalan (1814-1894) [Cat] from 1865 concerned themselves with this subject. Gradually it was realized though that Euler's approach to classifying polyhedra did not have much hope of success. In the middle of the nineteenth century different methods were developed by Kirkman [Kirk], A. Cayley (1821-1895) [Cay 3], and Möbius [Möb 3].

In the meantime, Euler's theorem had been generalized to other surfaces than just those of simple convex polyhedra. L. Poinsot (1777-1859) [Poins] had considered starlike polyhedra in 1809. More important was the work of L'Huilier [Lh] from 1812-1813. It had extended Euler's theorem to surfaces which are obtained by boring holes through or within a solid ball. The generalization to arbitrary orientable surfaces was accomplished by Möbius [Möb 1] in 1863. The topological classification problem of orientable surfaces (which is quite different from the kind of classification of polyhedra sought by Euler and his followers mentioned above) was finally solved in the 1850s by B. Riemann (1826-1866) [Riem 1, 3] who was motivated by his work in complex function theory. We shall return to this topic and consider it in detail in Chapter 3.

With Riemann one comes to a turning point in the history of topology. It was his work that gave topology (or *analysis situs* as it was then called) its first impetus and made mathematicians fundamentally interested in the subject.

Chapter 2

THE GEOMETRY OF COMPLEXES

2.1. WHAT IS TOPOLOGY?

A hurried reading of the last chapter could leave the impression that we have merely presented an intuitive discussion of a number of quite separate topics. In face, all these topics had something in common and it is this common element which makes them topological problems. For example, in Euler's theorem the exact shape of the faces or sides of a simple polyhedron was not important; neither was the shape of a country important to the coloring of a map. Similarly, whether or not a graph was planar did not depend on the length or straightness of its edges. In order words, we were studying properties of objects which stayed invariant under certain deformations. The problem now is to describe precisely what an allowable deformation ought to be, keeping in mind the examples from the preceding chapter. Turning this problem around slightly, what we want to do is to define what is meant by saying that two objects are topologically equivalent. From the outset, however, we can see that the deformations we are dealing with here are much more general than the ones encountered in high school (or Euclidean) geometry. There only very rigid motions (translations and rotations) are allowed in the definition of when two objects are equivalent. Equivalences are also very rigidly defined in differential geometry. Such notions as the curvature of a surface at a point, which play a significant role in differential geometry, are completely unimportant in topology.

Topology is sometimes called rubber-sheet geometry because a topologist does not distinguish between two subsets of $\mathbb{R}^n$ if, upon pretending that they are made of rubber, he can stretch one into the other without tearing or puncturing anything. Consider the four subsets of $\mathbb{R}^3$ in Fig. 2.1. The difference between the sphere and the ellipsoid is clearly minor when compared to the difference between the sphere and either the disk or the torus. The ellipsoid is just a slightly elongated or deformed sphere, but could this be the case for the disk or the torus? Observe that if one cuts along any circle on the sphere one will divide the space into two parts. One would want this property to be preserved under deformations. The ellipsoid has the property; but if one cuts the torus along the circle C, then one will not separate it. This supports our intuitive belief that the sphere and the torus should be considered to be topologically distinct spaces. An obvious difference between the sphere and the disk is that the latter has a boundary, whereas the former does not, but there is a more subtle difference. Our intuition should tell us that there is no way of flattening out a sphere without first puncturing it. To put it another way, the sphere has a "hole" but the disk does not. The "hole" in the torus is quite different from the "hole" in the sphere. At any rate, the examples in Fig. 2.1 serve as useful test cases against which one can measure the "correctness" of any definition of topological equivalence.

One important point to emphasize in the rubber-sheet geometry referred to above is that the stretching of an object is not confined to a particular $\mathbb{R}^n$ but could take place in any $\mathbb{R}^{n+k}$ for some k. For example, the trefoil knot in Fig. 1.15a which is presented to us as an object in $\mathbb{R}^3$ cannot be deformed into the standard circle S^1 in $\mathbb{R}^3$ (otherwise it would not be knotted), but it can be deformed into the circle by a deformation which takes place in $\mathbb{R}^4$. (The reader should try to visualize how this might be done!) The same is true of every other knot. The point is that the space in

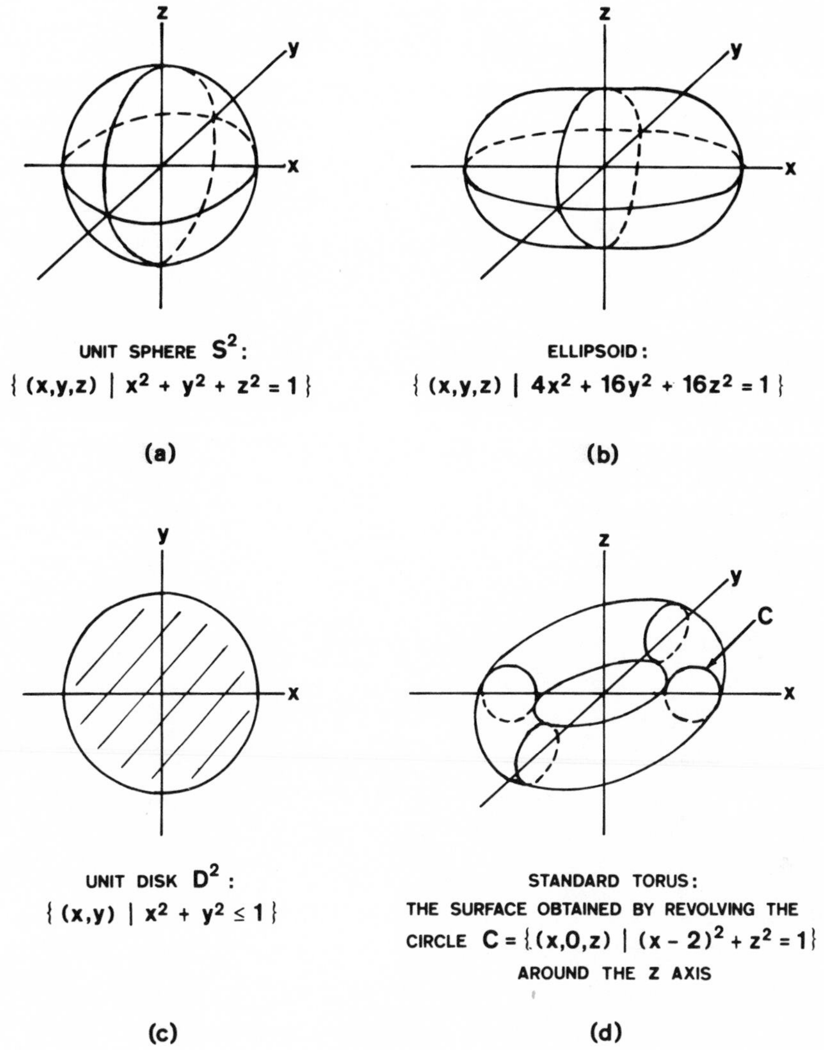

FIGURE 2.1

which an object is imbedded is unimportant and one is allowed to change it at will. In fact, it would be better to forget that it is there at all, because one is only interested in the intrinsic properties of the object. In analogy with Platonic forms one can think of equivalent spaces as being merely two different representations of some single ideal object.

REMARK. There are important problems in topology, such as knot theory, where one does want to study the various possible placements of an object in a given space. In that case one is really studying maps and their deformations, because a knot can be thought of as a continuous one-to-one map from the circle S^1 to $\mathbb{R}^3$. This subject will be taken up in Chapter 5; however, right now we are interested solely in objects and not their placements.

We have talked in vague terms long enough. Let $X, Y \subset \mathbb{R}^n$.

DEFINITION. We say that X is homeomorphic to Y, and write $X \approx Y$, if there exists a continuous bijective mapping $f : X \to Y$ which has a continuous inverse. Such a map f is called a homeomorphism.

It is clear that the relation $\approx$ is an equivalence relation among the subsets of $\mathbb{R}^n$, and we now define topology as the study of those properties of a space which are invariant under homeomorphisms. That $\approx$ does, in fact, give a precise expression to the notion of topological equivalence for which we have been searching is probably not obvious at this time. However, as an example, consider the map which sends a point p on the ellipse in Fig. 2.2 to the point p' on the circle by radial projection. This map is easily seen to be a homeomorphism because it has a continuous inverse, namely, the map which sends p' to p. One can define a similar homeomorphism between the sphere in Fig. 2.1a and the ellipsoid in Fig. 2.1b. On the other hand, we shall eventually be able to show that spaces like the sphere, the disk, and the torus are not homeomorphic to each other, justifying our claim about the appropriateness of this definition of "homeomorphism." It might be the case that one could find a continuous bijective map between almost all spaces. This would make our definition of homeomorphism useless; but, having proved no theorems, it would be premature to conclude anything to the contrary.

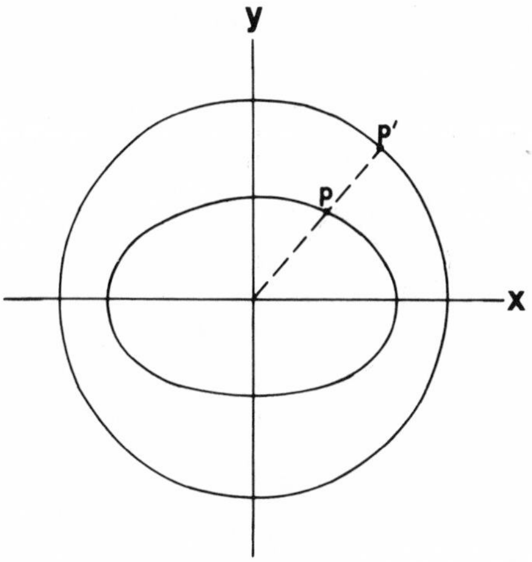

FIGURE 2.2

Since one of the basic problems in topology is to determine when two spaces are homeomorphic, it would be nice to have an algorithm whereby one could determine this. By an algorithm we mean a procedure which would settle all homeomorphism questions after a finite number of steps. It is at this point that we leave what is called "general" or "point set" topology and enter the domain of "algebraic" topology. The object of algebraic topology is to associate to each space certain algebraic invariants such that two spaces will be homeomorphic if and only if they have the same invariants. The Euler characteristic which we discussed in the last chapter serves as a perfect example of such an invariant. (Note, however, how our point of view has changed. Instead of starting with a polyhedron whose surface has a given Euler characteristic and discovering other polyhedral surfaces which have the same Euler characteristic, we are thinking of the Euler characteristic as a number which is the same for all homeomorphic spaces.) We shall also want to devise a method for computing the invariants. It would follow that, in order to study a space, it would be sufficient to study its invariants, and in this way topological or geometric questions would reduce to problems in algebra (hence the term algebraic topology) for which a great deal of theory has already been developed. The precise nature of the invariants we are talking about here will emerge only gradually because it is not possible to define them in a few words. Chapter 1 was intended to provide some motivation.

A completely trivial example of an invariant which we might associate to a space $X \subset \mathbb{R}^n$ is the number of points in X, denoted by $\|X\|$. It is easy to see that if X and Y are finite sets, then $X \approx Y$ if and only if $\|X\| = \|Y\|$. However, all interesting subsets X of $\mathbb{R}^n$ have an infinite number of points so that the invariant $\|X\|$ is not of much use since it does not distinguish between them.

Problems

2.1. (a) Prove that "homeomorphism" is an equivalence relation among the subsets of $\mathbb{R}^n$.

(b) Let $X = \{0,1,2,\cdots\}$ and $Y = \{0\} \cup \{1/n \mid n = 1,2,\cdots\}$. Define a bijective map $f : X \to Y$ by $f(0) = 0$ and $f(n) = 1/n$ for $n \geq 1$. Show that, although f is continuous, its inverse is not. In other words, if we were to drop the condition that the map have a continuous inverse in the definition of homeomorphism, then we would not have an equivalence relation.

2.2. Let $a,b,c,d \in \mathbb{R}$, $a < b$, and $c < d$. Prove that all the line segments $[a,b]$ and $[c,d]$ are homeomorphic.

2.3. Define a homeomorphism between the sphere and the ellipsoid in Fig. 2.1a and b.

2.4. Let $a,b \in \mathbb{R}$ and $a < b$. Prove that the circle S^1 is not homeomorphic to any line segment $[a,b]$.

2.5. Suppose that X and Y are finite subsets of $\mathbb{R}^n$. Prove that $X \approx Y$ if and only if $\|X\| = \|Y\|$.

2.2. SIMPLICIAL COMPLEXES AND MAPS; POLYHEDRA

Although the general problem of topology is to determine the homeomorphism type of an arbitrary space X, the methods of algebraic topology can be applied successfully only in those cases where X is faily "nice." In this section we shall give a careful definition of one particularly important class of such spaces, namely, the class of polyhedra. In the process we shall develop the basic theory of simplicial complexes which play a central role in combinatorial topology.

DEFINITION. The points $v_0, v_1, \cdots, v_k \in \mathbb{R}^n$ are said to be linearly independent if the vectors $v_1 - v_0$, $v_2 - v_0$, $\cdots$, and $v_k - v_0$ are linearly independent. (Recall that we can think of $\mathbb{R}^n$ either as a set of points or as a set of vectors which can be added and subtracted and multiplied by real numbers.)

Whether or not points are linearly independent does not depend on their ordering [Problem 2.6(a)]. In Fig. 2.3a, v_0, v_1, and v_2 are linearly independent, but not in Fig. 2.3b.

DEFINITION. The line segment between two points x and y in $\mathbb{R}^n$ is defined to be the set

$$\{ z = tx + (1 - t)y \mid t \in [0,1] \}$$

Let $A \subset \mathbb{R}^n$.

DEFINITION. We say that A is convex if $x, y \in A$ implies that the line segment from x to y is contained in A (see Fig. 2.4).

DEFINITION. The convex hull of A is the intersection of all convex sets containing A.

To justify this definition we have to make two observations:

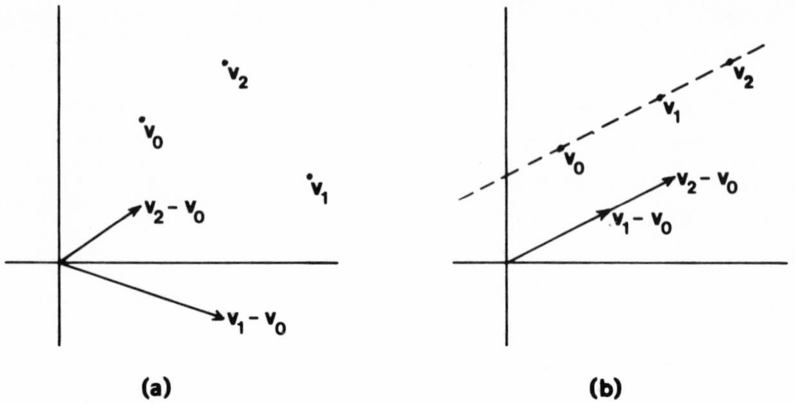

FIGURE 2.3

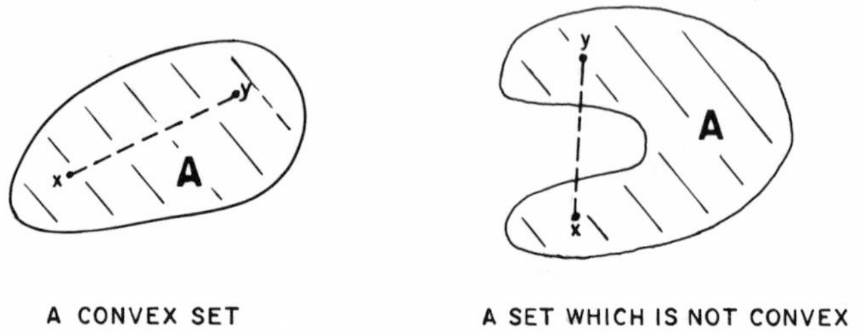

FIGURE 2.4

1. Each $\mathbb{R}^n$ is convex, so that we are not taking an empty intersection.
2. The intersection of an arbitrary number of convex sets is convex (Problem 2.7).

DEFINITION. Let $k \geq 0$. A k-dimensional simplex, or k-simplex, is the convex hull σ of $k + 1$ linearly independent points $v_0, v_1, \cdots, v_k \in \mathbb{R}^n$. We write $\sigma = v_0 v_1 \cdots v_k$. The points v_i are called the vertices of σ. Sometimes we write σ^k to emphasize the dimension of σ. If the dimension of σ is unimportant, then σ will be called simply a simplex.

Figure 2.5 gives some examples of simplices and shows that our use of the term "k-dimensional" is justified. Note that $\mathbb{R}^2$ does not contain any 3-dimensional simplex σ^3. In general, $\mathbb{R}^n$ contains at most an n-dimensional simplex because we cannot find j linearly independent points in $\mathbb{R}^n$ for $j > n + 1$. Also, a simplex depends only on the set of vertices and not on their ordering. For example, $v_0v_1 = v_1v_0$.

DEFINITION. Let $\sigma = v_0v_1 \cdots v_k$ be a k-simplex and let $\{w_0, w_1, \cdots, w_\ell\}$ be a nonempty subset of $\{v_0, v_1, \cdots, v_k\}$, where $w_i \neq w_j$ if $i \neq j$. Then $\tau = w_0w_1 \cdots w_\ell$ is called an ℓ-dimensional face of σ and we shall write $\tau < \sigma$.

Problem 2.8 justifies the above terminology. In Fig. 2.5c, σ^2 has the following faces: one 2-dimensional face, σ^2; three 1-dimensional faces, v_0v_1, v_1v_2, and v_0v_2; and three 0-dimensional faces, v_0, v_1, and v_2.

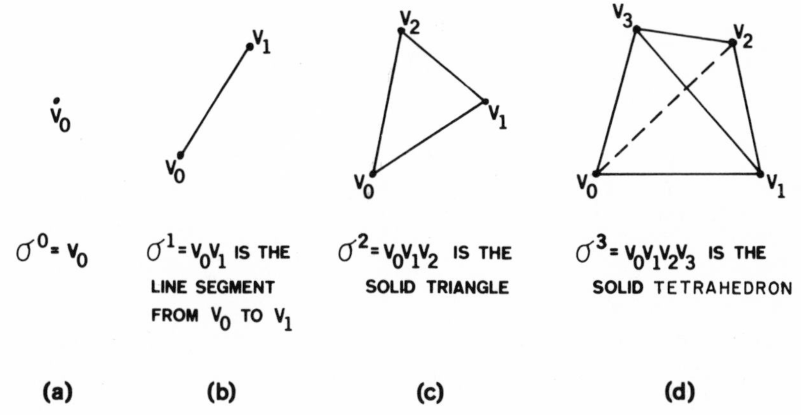

FIGURE 2.5

One of the technical advantages of simplices over other regularly shaped regions is that their points have a nice characterization.

THEOREM 1. Every point w of a simplex $v_0v_1\cdots v_k$ can be written uniquely in the form

$$w = \sum_{i=0}^{k} \lambda_i v_i, \quad \lambda_i \in [0,1] \quad \text{and} \quad \sum_{i=0}^{k} \lambda_i = 1$$

The λ_i are called the barycentric coordinates of w. Furthermore, the dimension and the vertices of a simplex are uniquely determined, that is, if $v_0v_1\cdots v_k = v_0'v_1'\cdots v_\ell'$, then $k = \ell$ and $v_i = v_i'$ after a renumbering of the v_i'.

Proof: Let

$$S = \{ \sum_{i=0}^{k} \lambda_i v_i \mid \lambda_i \in [0,1] \text{ and } \sum_{i=0}^{k} \lambda_i = 1 \}$$

We show first that S is convex. Let $w = \sum_{i=0}^{k} \lambda_i v_i$, $w' = \sum_{i=0}^{k} \mu_i v_i \in S$ and let $t \in [0,1]$. Then

$$tw + (1 - t)w' = t(\sum_{i=0}^{k} \lambda_i v_i) + (1 - t)(\sum_{i=0}^{k} \mu_i v_i)$$

$$= \sum_{i=0}^{k} (t\lambda_i + (1 - t)\mu_i)v_i$$

But

$$\sum_{i=0}^{k} (t\lambda_i + (1 - t)\mu_i) = t(\sum_{i=0}^{k} \lambda_i) + (1 - t)(\sum_{i=0}^{k} \mu_i)$$

$$= t\cdot 1 + (1 - t)\cdot 1$$

$$= 1$$

Clearly, $0 \leq t\lambda_i + (1 - t)\mu_i$; therefore it follows that $t\lambda_i + (1 - t)\mu_i \leq 1$. Hence $p = tw + (1 - t)w' \in S$, which proves that S is convex because p is a typical point on the line segment from w to w'.

Next, we show that S belongs to every convex set U containing $v_0, v_1, \cdots, v_k$. If $k = 0$ or 1, everything is obvious and there is nothing to prove. Assume that $k \geq 2$ and that the statement has been proved for smaller values than k. Let $w = \sum_{i=0}^{k} \lambda_i v_i \in S$. Since not all λ_i can equal 0, we may assume that $\lambda_0 \neq 0$. Furthermore, we may assume that $\lambda_0 < 1$, for if $\lambda_0 = 1$, then $w = v_0$, and v_0 clearly belongs to U. Thus, we can write

$$w = \lambda_0 v_0 + (1 - \lambda_0)\left(\sum_{i=1}^{k} (\lambda_i/(1 - \lambda_0))v_i\right)$$

But

$$\sum_{i=1}^{k} \lambda_i/(1 - \lambda_0) = (1/(1 - \lambda_0))\left(\sum_{i=1}^{k} \lambda_i\right) = (1/(1 - \lambda_0))(1 - \lambda_0) = 1$$

and $0 \leq \lambda_i/(1 - \lambda_0) \leq 1$. By our inductive hypothesis $\bar{w} = \sum_{i=1}^{k} (\lambda_i/(1 - \lambda_0))v_i$ belongs to every convex set containing $v_1, v_2, \cdots, v_k$. In particular, $\bar{w} \in U$. Since $v_0 \in U$, it follows that $w = \lambda_0 v_0 + (1 - \lambda_0)\bar{w} \in U$, and we are done.

This finishes the proof that $S = v_0 v_1 \cdots v_k$.

To show that every point $w \in v_0 v_1 \cdots v_k$ can be written uniquely in the form $w = \sum_{i=0}^{k} \lambda_i v_i$, suppose that $w = \sum_{i=0}^{k} \lambda_i' v_i$. Then

$$\begin{aligned} 0 = w - w &= \sum_{i=0}^{k} \lambda_i v_i - \sum_{i=0}^{k} \lambda_i' v_i \\ &= \sum_{i=0}^{k} (\lambda_i - \lambda_i')v_i \\ &= \sum_{i=1}^{k} (\lambda_i - \lambda_i')(v_i - v_0) + \left(\sum_{i=0}^{k} (\lambda_i - \lambda_i')\right)v_0 \\ &= \sum_{i=1}^{k} (\lambda_i - \lambda_i')(v_i - v_0) \end{aligned}$$

The last equality sign follows from the fact that

$$\sum_{i=0}^{k} (\lambda_i - \lambda_i') = \sum_{i=0}^{k} \lambda_i - \sum_{i=0}^{k} \lambda_i' = 1 - 1 = 0$$

Since the vectors $v_1 - v_0, v_2 - v_0, \cdots, v_k - v_0$ are linearly independent, we get that $\lambda_i = \lambda_i'$ for $i = 1,2,\cdots,k$. But we must then also have that $\lambda_0 = \lambda_0'$. This proves that the representation $w = \sum_{i=0}^{k} \lambda_i v_i$ is unique.

DEFINITION. A point $w = \sum_{i=0}^{k} \lambda_i v_i$ is called an interior point of the simplex $v_0 v_1 \cdots v_k$ if $\lambda_i > 0$ for all i. The interior of a simplex σ, int σ, is defined to be the set of interior points of σ.

CLAIM 1: Every interior point of a k-simplex, $k \geq 1$, is a midpoint of a line segment lying entirely within the simplex.

To see this, let $w = \sum_{i=0}^{k} \lambda_i v_i \in v_0 v_1 \cdots v_k$ and $\lambda_i > 0$ for all i. Choose an ϵ such that $0 < \epsilon < \min\{\lambda_1, 1 - \lambda_1, \lambda_0, 1 - \lambda_0\}$. Define $u_1 = w + \epsilon(v_1 - v_0)$ and $u_2 = w - \epsilon(v_1 - v_0)$. Consideration of the barycentric coordinates of u_1 and u_2 show that $u_1, u_2 \in v_0 v_1 \cdots v_k$. Furthermore, $w = (u_1 + u_2)/2$, and so the claim is proved.

CLAIM 2: A vertex of a simplex cannot be the midpoint of any line segment lying within the simplex.

If Claim 2 is false, then, for some i, $v_i = (x + y)/2$, where $x = \sum_{j=0}^{k} \lambda_j v_j \neq y = \sum_{j=0}^{k} \mu_j v_j$. Thus,

$$v_i = (x + y)/2 = \sum_{j=0}^{k} ((\lambda_j + \mu_j)/2) v_j$$

It follows from the uniqueness of the barycentric coordinates that $(\lambda_j + \mu_j)/2 = 0$ for $j \neq i$ and $(\lambda_i + \mu_i)/2 = 1$. Hence $\lambda_j = 0 = \mu_j$

for $j \neq i$ and $\lambda_i = \mu_i = 1$, that is, $x = y = v_i$. This contradicts the initial hypothesis that $x \neq y$ and proves Claim 2.

The proof of the remaining part of Theorem 1 now follows easily from Claim 1 and Claim 2.

Next, we shall use simplices as building blocks to build more general spaces.

DEFINITION. A simplicial complex K, is a finite collection of simplices in some $\mathbb{R}^n$ satisfying:

1. If $\sigma \in K$, then all faces of σ belong to K.
2. If $\sigma, \tau \in K$, then either $\sigma \cap \tau = \emptyset$ or $\sigma \cap \tau$ is a common face of σ and τ.

DEFINITION. Let K be a simplicial complex. The subset $|K| = \bigcup_{\sigma \in K} \sigma$ of $\mathbb{R}^n$ is called the underlying space of K. The dimension of K, dim K, is defined to be -1 if $K = \emptyset$ and the maximum of the dimensions of the simplices of K otherwise.

Examples of simplicial complexes and their corresponding underlying spaces are given in Fig. 2.6. At times we shall abbreviate "simplicial complex" to simply "complex".

REMARK. We have defined what is sometimes called a finite simplicial complex, that is, a complex which has only a finite number of simplices. It is possible to define infinite complexes, but one has to be a little careful. Otherwise, one would have examples such as $\mathbb{R}$ being the underlying space of the 0-dimensional complex consisting of all points in $\mathbb{R}$. This unpleasant situation would violate our intuition about dimension. The correct definition of an infinite complex is to substitute the word "countable" for "finite" in our definition and add the condition:

3. Every point of $\mathbb{R}^n$ has a neighborhood which meets only a finite number of simplices of K.

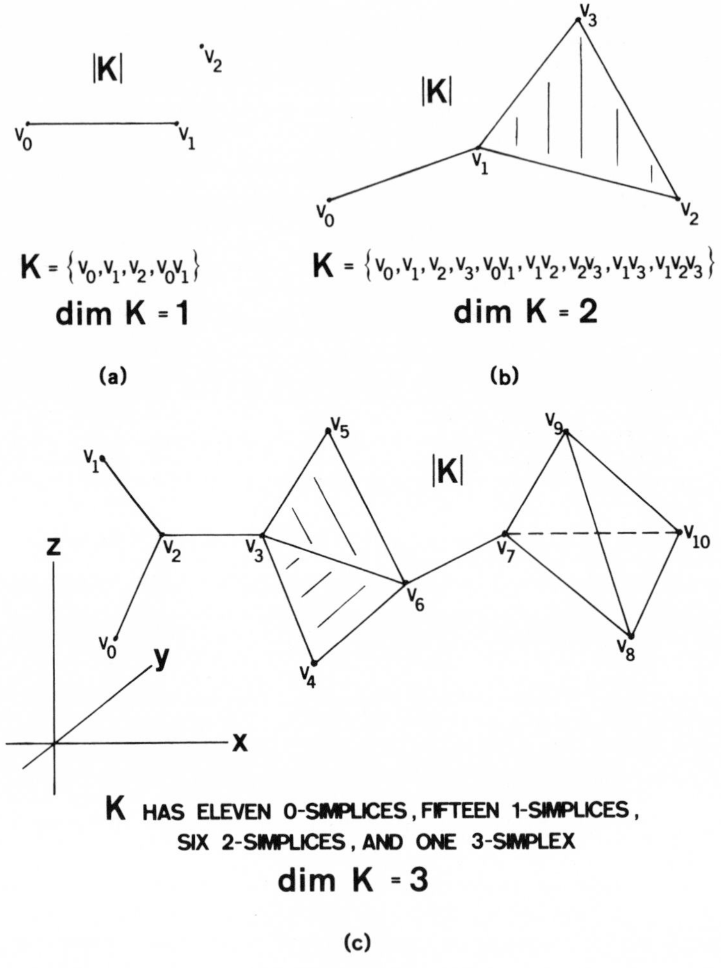

FIGURE 2.6

The reason that we restrict ourselves to finite complexes is that they suffice for our purposes and we also avoid certain complications in proofs and definitions in this way. Nevertheless, essentially everything we do in this book could, with a few slight modifications, be done in the infinite case.

Figure 2.7 shows that two distinct simplicial complexes can have the same underlying spaces because one can divide a subset of $\mathbb{R}^n$

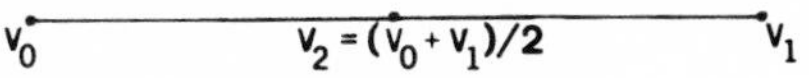

$$K = \{v_0, v_1, v_0v_1\}$$

$$L = \{v_0, v_1, v_2, v_0v_2, v_2v_1\}$$

$$|K| = |L|$$

FIGURE 2.7

into simplices in many ways. In fact, one can think of a simplicial complex K as nothing but one of many possible schemes for subdividing the space $X = |K| \subset \mathbb{R}^n$ into simplices.

A wrong way of getting a simplicial decomposition for a space is shown in Fig. 2.8a. Although the set A is a union of simplices, the two 1-simplices do not intersect in a face and therefore A is not a simplicial complex. Note that "$|A|$" is the underlying space of the simplicial complex K in Fig. 2.8b, where $v_4 = v_0v_1 \cap v_1v_3$.

The formal distinction between K and $|K|$ must be emphasized: K is a set of simplices and therefore not itself a subset of $\mathbb{R}^n$; it is $|K|$ which is a subset of $\mathbb{R}^n$. Although these two concepts are often identified in the language, one should remember the difference in the back of one's mind. In particular, sometimes we shall refer to a space such as in Fig. 2.6c and speak of "that simplicial complex K." As long as the simplices are clearly indicated in the figure, there should be no confusion.

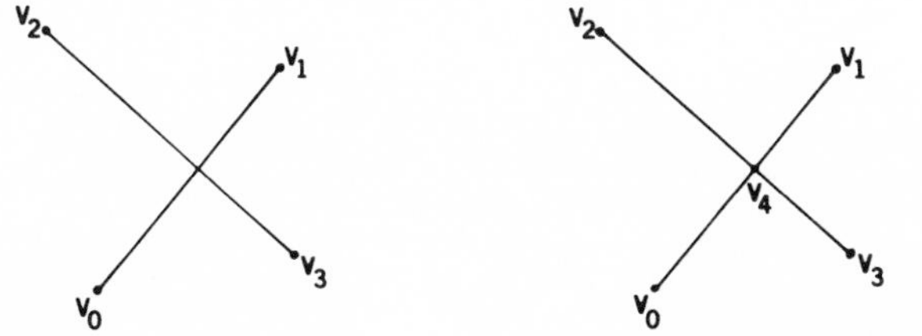

$A = \{v_0, v_1, v_2, v_3, v_0v_1, v_2v_3\}$ $\qquad$ $K = \{v_0, v_1, v_2, v_3, v_4, v_0v_4, v_4v_1, v_2v_4, v_4v_3\}$

(a) (b)

FIGURE 2.8

Next, here are a few definitions that will be useful in future discussions about complexes.

DEFINITION. Let K be a simplicial complex. A subcomplex of K is a simplicial complex L such that $L \subset K$.

For example, in Fig. 2.7 the set $M = \{v_0, v_2, v_0v_2\}$ is a subcomplex of L but not of K, even though $|M| \subset |K|$.

DEFINITION. The boundary of a simplicial complex K, denoted be ∂K, is defined by

$$\partial K = \{ \tau \mid \tau \text{ is a face of a simplex } \sigma^k \in K \text{ which belongs to a unique } (k+1)\text{-simplex of } K \}$$

The underlying space of ∂K is what one would usually call the point set boundary of $|K|$. It is easy to see that ∂K is a simplicial complex. For example, if K is the simplicial complex in Fig. 2.6b, then $\partial K = \{v_0,v_1,v_2,v_3,v_1v_2,v_2v_3,v_1v_3\}$.

DEFINITION. If σ is a simplex, then $\bar{\sigma}$, the simplicial complex determined by σ, is defined by $\bar{\sigma} = \{\tau | \tau < \sigma\}$.

For example, $\overline{v_0v_1v_2} = \{v_0,v_1,v_2,v_0v_1,v_1v_2,v_0v_2,v_0v_1v_2\}$. Clearly, $\bar{\sigma}$ is a simplicial complex and $|\bar{\sigma}| = \sigma$.

DEFINITION. We say that a simplicial complex K is connected if, for each pair of vertices v and w of K, there is a sequence $v_0,v_1,\cdots,v_n$ of vertices in K such that $v_0 = v$, $v_n = w$, and v_iv_{i+1} is a 1-simplex of K for all $i = 0,1,\cdots,n-1$.

In Fig. 2.6a, K is not connected, whereas in Fig. 2.6b and c, it is connected. There is a simple geometric interpretation to when a simplicial complex is connected.

DEFINITION. Let $X \subset \mathbb{R}^n$ and let $x,y \in X$. A path from x to y in X will mean a continuous map $\gamma : [a,b] \to X$ with $\gamma(a) = x$

and $\gamma(b) = y$. The set $\gamma([a,b])$ is called a curve from x to y in X. If the end points x and y are unimportant, then we shall call γ and $\gamma([a,b])$ simply a path and curve in X, respectively. The points $\gamma(a)$ and $\gamma(b)$ are said to be connected by the path γ. The path γ and the curve $\gamma([a,b])$ are called closed if $\gamma(a) = \gamma(b)$. Since closed paths correspond naturally to maps defined on the circle and vice versa, we shall also refer to any continuous map $f : S^1 \to X$ as a closed path in X and call the set $f(S^1)$ a closed curve.

The difference between a curve and a path is similar to the difference between a simplicial complex and its underlying space. A curve corresponds to many paths and one should think of a path as being a particular parametrization of a given curve. A path may "traverse" points on the "underlying" curve many times. In some discussions, however, we may refer to a curve in a figure and call it a path, but we do this only if it is possible to find a path which is a homeomorphism of $[a,b]$ or S^1 onto the curve and if the particular homeomorphism that is chosen is not important within the context of that discussion.

DEFINITION. Let $X \subset \mathbb{R}^n$. Define an equivalence relation $(\sim)$ among the points of X by the condition that $x \sim y$ if and only if x and y can be connected by a path in X. The equivalence classes of X under $(\sim)$ shall be called the (path) components of X. If X has only one component we shall say that X is (path) connected.

One can show that a simplicial complex K is connected if and only if its underlying space $|K|$ is connected [Problem 2.14(a)]. A space X is connected if and only if every pair of points in X can be joined by a path in X.

Invariably, when one defines certain objects in mathematics, it is also useful to define the corresponding natural maps between them which preserve the relevant structure. For example, with groups we have homomorphisms; with vector spaces, linear maps; with spaces, continuous maps; etc.

DEFINITION. Let K and L be simplicial complexes. A simplicial map $f : K \to L$ is a map f from the vertices of K to the vertices of L with the property that if $v_0, v_1, \cdots, v_q$ are vertices of a simplex in K, then $f(v_0), f(v_1), \cdots, f(v_q)$ are the vertices of a simplex in L. If f is bijective, then we call f an isomorphism and K and L are said to be isomorphic (written $K \approx L$). The identity simplicial map $1_K : K \to K$ is defined by $1_K(v) = v$ for all vertices $v \in K$. (This should not cause any confusion with our usual convention that 1_X denotes the identity map on the set X.)

Simplicial maps induce continuous maps on the underlying spaces:

DEFINITION. Let $f : K \to L$ be a simplicial map. Define a map $|f| : |K| \to |L|$ as follows: If $x \in |K|$, then x belongs to the interior of some unique simplex $v_0 v_1 \cdots v_q$ of K (Problem 2.10). Write $x = \sum_{i=0}^{q} t_i v_i$, where the t_i are the barycentric coordinates of x and define $|f|(x) = \sum_{i=0}^{q} t_i f(v_i)$. If $|f|$ is a homeomorphism, then $|f|$ will be called a linear homeomorphism.

Since barycentric coordinates are unique, $|f|$ is well defined. Furthermore, $|f|$ is continuous [Problem 2.17(a)]. The map $|f|$ is sometimes called a simplicial map from $|K|$ to $|L|$. Clearly, $|f|$ is a linear homeomorphism if and only if f is an isomorphism [Problem 2.17(b)].

Given two simplicial maps $f : K \to L$ and $g : L \to M$, there is an obvious composite $g \circ f : K \to M$ which is easily shown to be a simplicial map also and $|g \circ f| = |g| \circ |f|$ (Problem 2.16).

Finally,

DEFINITION. A triangulation of a space X is a pair (K, φ), where K is a simplicial complex and $\varphi : |K| \to X$ is a homeomorphism. The complex K is said to triangulate X. A polyhedron is any space which admits a triangulation.

Note that even simple spaces such as S^1 or S^2 are not the underlying space of any simplicial complex because they are curved and simplices are "flat." (A more precise observation is that simplices are convex whereas neither S^1 nor S^2 contain convex sets other than single points.) It is clear though that S^1 and S^2 are polyhedra. In fact, essentially all the spaces that a person would normally think of are polyhedra. The only exceptions are exceptions for obvious reasons. For example, neither $\mathbb{R}^n$ nor any open subset of $\mathbb{R}^n$ are polyhedra because they are not compact and every polyhedron is necessarily compact (Problem 2.19). We could easily have extended our definition of polyhedron to include such nice spaces by admitting "infinite triangulations," that is, by allowing the K in a triangulation (K,φ) to be an infinite complex. Figure 2.9a indicates a nice infinite triangulation of $\mathbb{R}^2$. Two examples of spaces which are not polyhedra for more fundamental reasons are the rational numbers and the space X in Fig. 2.9b. Although the latter is ac-

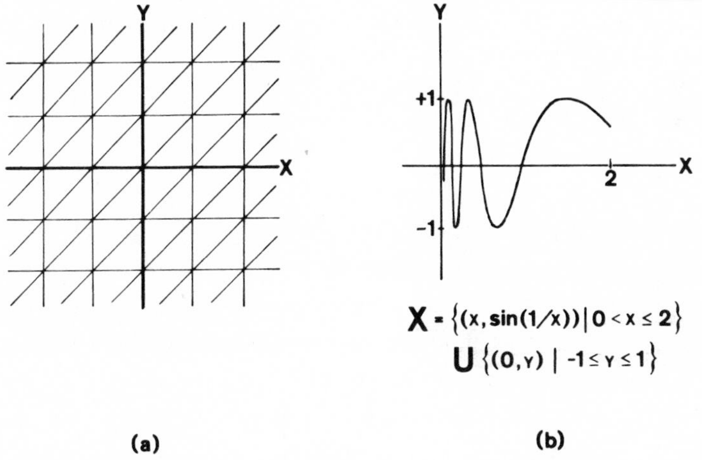

FIGURE 2.9

tually compact, neither admits even an infinite triangulation. Fortunately, such spaces do not arise in the geometric problems considered in this book and so we lose nothing by concentrating exclusively on polyhedra from now on.

DEFINITION. Let X be a polyhedron. Define the boundary of X, ∂X, and the interior of X, int X, by $\partial X = \varphi(|\partial K|)$ and $\text{int } X = X - \partial X$, where (K,φ) is a triangulation of X. Points in ∂X or int X are called boundary or interior points of X, respectively.

It will be shown later (Theorem 6 in Sec. 6.3) that the boundary of a polyhedron, and hence its interior, is a well-defined subset which does not depend on any particular triangulation. In the meantime, the definition can be treated as merely convenient notation. For example, the n-sphere S^n is the boundary of the (n + 1)-disk D^{n+1}, that is, $S^n = \partial D^{n+1}$, and $\text{int } D^n = \{x \in \mathbb{R}^n : |x| < 1\}$. Note also that, in the case of a simplex $\sigma = v_0v_1\cdots v_k$, our new definition of interior extends the earlier definition. This follows easily from the fact that (1) a point $w = \sum_{i=0}^{k} \lambda_i v_i$ in σ belongs to a j-dimensional face of σ with $j < k$ if and only if for some i the barycentric coordinate λ_i equals zero, and (2) $\partial\sigma = \bigcup \{ \tau \mid \tau < \sigma \text{ and } \dim \tau < \dim \sigma = k \}$. In general, the boundary and interior of a polyhedron are the obvious sets which are connoted by the terms "boundary" and "interior," respectively.

Problems

2.6. (a) Suppose that $v_0, v_1, \cdots$, and v_k are linearly independent points in $\mathbb{R}^n$. Prove that $v_i, v_1, v_2, \cdots, v_{i-1}, v_0, v_{i+1}, \cdots$, and v_k are linearly independent points. Use this to prove that whether or not points are linearly independent does not depend on the order in which they are listed.

(b) Prove that any subset of a set of linearly independent points in $\mathbb{R}^n$ consists of linearly independent points.

2.7. Prove that the intersection of an arbitrary number of convex sets is convex.

2.8. Show that every ℓ-dimensional face τ of a simplex σ is an ℓ-simplex contained in σ.

2.9. Let σ be the 2-simplex in $\mathbb{R}^2$ with vertices $(2,-2)$, $(4,2)$, and $(-1,4)$.

(a) Find the barycentric coordinates of the points $(1,0)$, $(2,2)$, $(0,3)$, and $(1,3) \in \mathbb{R}^2$ with respect to σ.

(b) Find the Cartesian coordinates of the points of σ whose barycentric coordinates are $(1/3,1/3,1/3)$, $(1/4,0,3/4)$, $(1/5,1/10,7/10)$, and $(1/2,1/3,1/6)$.

2.10. Let K be a simplicial complex and suppose that $x \in |K|$. Show that there is a unique simplex $\sigma \in K$ such that $x \in \operatorname{int} \sigma$.

2.11. If L and M are subcomplexes of a simplicial complex K, prove that $L \cap M$ is a subcomplex of K.

2.12. Prove that if K is a simplicial complex, then so is ∂K.

2.13. Prove that a finite set of two or more points in $\mathbb{R}^n$ is not connected.

2.14. Let K be a simplicial complex.

(a) Show that K is connected if and only if $|K|$ is connected. (This generalizes Problem 2.13.)

(b) Define a component of K to be a maximal connected subcomplex L of K (maximal in the sense that if L' is a subcomplex of K and $L \subsetneq L'$, then L' is not connected). Show that $K = L_1 \cup L_2 \cup \cdots \cup L_t$, where L_i is a component of K and $L_i \cap L_j = \emptyset$ if $i \neq j$.

(c) Show that a subcomplex L of K is a component of K if and only if $|L|$ is a component of $|K|$.

2.15. Let K be a simplicial complex and suppose that $f : |K| \to \mathbb{R}^m$. Prove that f is continuous if and only if $f|\sigma$ is continuous for every simplex $\sigma \in K$.

2.16. Let $f : K \to L$ and $g : L \to M$ be simplicial maps. Show that the composite $g \circ f : K \to M$ is a simplicial map and that $|g \circ f| = |g| \circ |f|$. Show also that $|1_K| = 1_{|K|}$.

2.17. Let $f : K \to L$ be a simplicial map.

(a) Prove that $|f| : |K| \to |L|$ is continuous. (*Hint:* Use Problem 2.15.)

(b) Prove that f is an isomorphism if and only if $|f|$ is a homeomorphism.

2.18. Prove that S^1 is a polyhedron by defining a triangulation. Find simplicial complexes K_1, K_2, and K_3 in $\mathbb{R}^3$ such that $|K_1|$, $|K_2|$, and $|K_3|$ are homeomorphic to S^2, the torus, and the Möbius strip, respectively.

2.19. Prove that every polyhedron is compact. (A definition and some properties of compactness can be found in Appendix A.)

2.3. ABSTRACT SIMPLICIAL COMPLEXES; CUTTING AND PASTING

There is a slight generalization of a simplicial complex which we shall now discuss briefly for two reasons:

1. Simplicial complexes, which are sometimes called geometric complexes, play only an intermediary role in our overall goal which is to study polyhedra. Furthermore, we shall see that it is the **abstract** part of their definition which one exploits and not the geometric fact that they happen to correspond to a subdivision into simplices of an actual space in $\mathbb{R}^n$. We want to bring out this aspect more forcefully in the future.

2. In topology one often talks about "cutting" a space in a certain place or "pasting together" (or "identifying") parts of a space. This idea of cutting and pasting sometimes makes otherwise complicated spaces and constructions much easier to understand. It is very important for the rigorous development of the material in this book that the reader understand the contents of this section thoroughly; otherwise, the cutting and pasting techniques which are used on occasion, especially in Chapter 3, will have

to be accepted on an intuitive basis. Particular attention should be paid to Theorem 2, which has a much greater mathematical significance than its trivial proof would suggest.

DEFINITION. An abstract simplicial complex $\mathcal{K}$ is a set of nonempty subsets of a given finite set V such that

1. $\{v\} \in \mathcal{K}$ for every $v \in V$, and

2. if $S \in \mathcal{K}$, then every nonempty subset of S belongs to $\mathcal{K}$.

The elements of $\mathcal{K}$ are called simplices. If $S \in \mathcal{K}$ and if S has $q + 1$ elements, then S is called an (abstract) q-simplex. The elements of V are called the vertices of $\mathcal{K}$ and we shall identify the vertex $v \in V$ with the 0-simplex $\{v\} \in \mathcal{K}$.

Clearly, every simplicial complex K determines an abstract simplicial complex $\mathcal{K}_K$, namely,

$$\mathcal{K}_K = \{\ \{v_0,v_1,\cdots,v_q\}\ |\ v_0v_1\cdots v_q \text{ is a q-simplex of } K\ \}$$

For example, if L is as in Fig. 2.7, then $\mathcal{K}_L = \{v_0,v_1,v_2,\{v_0,v_2\},\{v_1,v_2\}\}$. To show the converse, that every abstract simplicial complex corresponds to a unique geometric complex in some $\mathbb{R}^n$, we just have to "fill in" the points that are missing in the abstract simplices.

DEFINITION. Let $\mathcal{K}$ be an abstract simplicial complex. A geometric realization of $\mathcal{K}$ is a pair (φ,K), where K is a simplicial complex and φ is a bijective map from the vertices of $\mathcal{K}$ to the vertices of K such that $\{v_0,v_1,\cdots,v_q\}$ is a q-simplex of $\mathcal{K}$ if and only if $\varphi(v_0)\varphi(v_1)\cdots\varphi(v_q)$ is a q-simplex of K.

THEOREM 2. Every abstract simplicial complex $\mathcal{K}$ has a unique (up to isomorphism) geometric realization.

Proof: Let V be the set of vertices of $\mathcal{K}$ and assume that V has $n + 1$ points. Let φ be any bijective set map between the points of V and the vertices of any n-simplex σ in $\mathbb{R}^n$ and define a subcomplex K of $\overline{\sigma}$ by

$$K = \{\ \varphi(v_0)\varphi(v_1)\cdots\varphi(v_q)\ |\ \{v_0,v_1,\cdots,v_q\} \text{ is a q-simplex of } \mathcal{K}\ \}$$

It is easy to check that (φ,K) is the desired geometric realization. If (φ',K') is another geometric realization of $\mathcal{K}$, then $\varphi' \circ \varphi^{-1}$: $K \to K'$ is an isomorphism.

The quickest way to describe a space to someone is to draw a picture of it. The only problem is that most spaces are too complicated to admit a meaningful 2-dimensional projection of themselves on a piece of paper or a blackboard. However, sometimes it is possible to draw a space after it has been "cut" in suitable places and, if the resulting figure is labeled correctly to indicate where the cuts were made, then a person will be able to reconstruct the original space in his mind from such a labeled figure. For example, taking into account the labeling, are we not justified in saying that the spaces being referred to in Fig. 2.10a-c are the circle, cylinder, and the 2-sphere, respectively? How can we make this step rigorous in general?

DEFINITION. A labeled (simplicial) complex is a pair $(L,\mathcal{L})$, where L is a simplicial complex and $\mathcal{L}$ is a labeling of the vertices of L. (In mathematical terms, $\mathcal{L}$ will be considered as a surjective map from the vertices of L to a set V whose elements are the labels.)

Suppose that $(L,\mathcal{L})$ is a labeled complex.

DEFINITION. Define an abstract simplicial complex $\mathcal{K}_{(L,\mathcal{L})}$ by

$$\mathcal{K}_{(L,\mathcal{L})} = \{\ \mathcal{L}(S)\ |\ S \in \mathcal{K}_L\ \}$$

Let $(\varphi_{(L,\mathcal{L})},K_{(L,\mathcal{L})})$ denote any geometric realization of $\mathcal{K}_{(L,\mathcal{L})}$ and let $X_{(L,\mathcal{L})} = |K_{(L,\mathcal{L})}|$. Define $c_{(L,\mathcal{L})} : L \to K_{(L,\mathcal{L})}$ to be the simplicial map defined by $c_{(L,\mathcal{L})}(v) = \varphi_{(L,\mathcal{L})}(\mathcal{L}(v))$. The map $p_{(L,\mathcal{L})} = |c_{(L,\mathcal{L})}| : |L| \to X_{(L,\mathcal{L})}$ is often referred to as the natural projection.

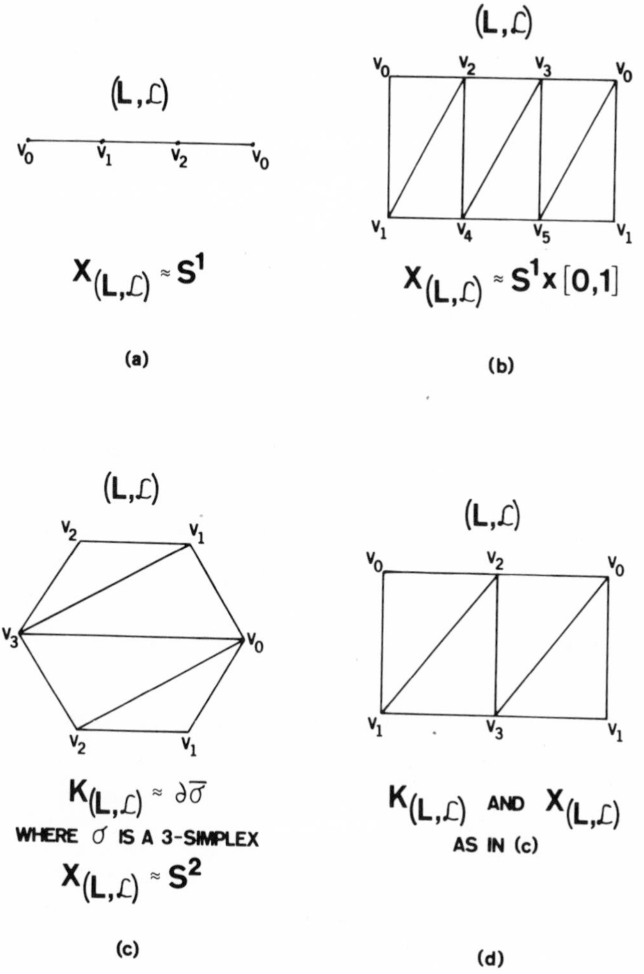

FIGURE 2.10

Topologists would call $X_{(L,\mathcal{L})}$, whose homeomorphism type is uniquely determined by $(L,\mathcal{L})$, a "quotient space" of $|L|$. Note that $c_{(L,\mathcal{L})}$ and $p_{(L,\mathcal{L})}$ are an isomorphism and homeomorphism, respectively, if and only if $\mathcal{L}$ is a bijection. The reader should again consider Fig. 2.10a-c and convince himself that the space $X_{(L,\mathcal{L})}$ is in fact the one indicated. One word of caution though: Fig. 2.10d shows that we must be a little careful with the labeling because, although the space being referred to may seem to be again a cylinder, it is actually the 2-sphere.

NOTE. The above notation will be used in the rest of the book. Furthermore, statements such as "consider the simplicial complex K in Figure A ---," where Figure A consists of a labeled complex $(L,\mathcal{L})$, will always mean the complex $K = K_{(L,\mathcal{L})}$.

Now we come to the mathematics behind "cutting" and "pasting." Consider Fig. 2.11. Intuitively, the space Z in Fig. 2.11b can be thought of as the space Y in Fig. 2.11a "cut" in the 1-simplex v_1v_4. Conversely, Y is obtained from Z by "pasting together" the two arcs from v_1 to v_4 in Z. How does one describe the relationship between Y and Z mathematically? The task is made easy by the definitions in this section. Choose a labeled complex $(L,\mathcal{L})$ with $|L| = Y$ as in Fig. 2.11c and define Z to be the space $|L'| = X_{(L',\mathcal{L}')}$, where $(L',\mathcal{L}')$ is the labeled complex in Fig. 2.11d. Reversing these steps corresponds to "pasting," that is, given $Z = |L'|$ and $(L',\mathcal{L}')$ one forms a labeled complex $(L'',\mathcal{L}'')$ as in Fig. 2.11e (that is, the vertex v_i' which is to be identified with the vertex v_i is relabeled as v_i) and defines $Y = X_{(L'',\mathcal{L}'')}$. Note that $X_{(L'',\mathcal{L}'')} \approx X_{(L,\mathcal{L})}$. Sometimes most of the labels in a figure are omitted and only the relevant ones are included. Not even the triangulation may be indicated since it is not important as far as the homeomorphism type of the resulting space is concerned. For example, Fig. 2.11f is also meant to indicate the same type of pasting as Fig. 2.11e.

It is obvious how to extend the definitions above to the situation where one wants to cut or paste along more than one simplex. One can also talk about cutting or pasting along "curved" simplices in a polyhedron. In that case one simply works with an appropriate triangulation since only the homeomorphism types of spaces are important here. We shall see more examples of cutting and pasting as we go along. In the first part of the next chapter we shall still

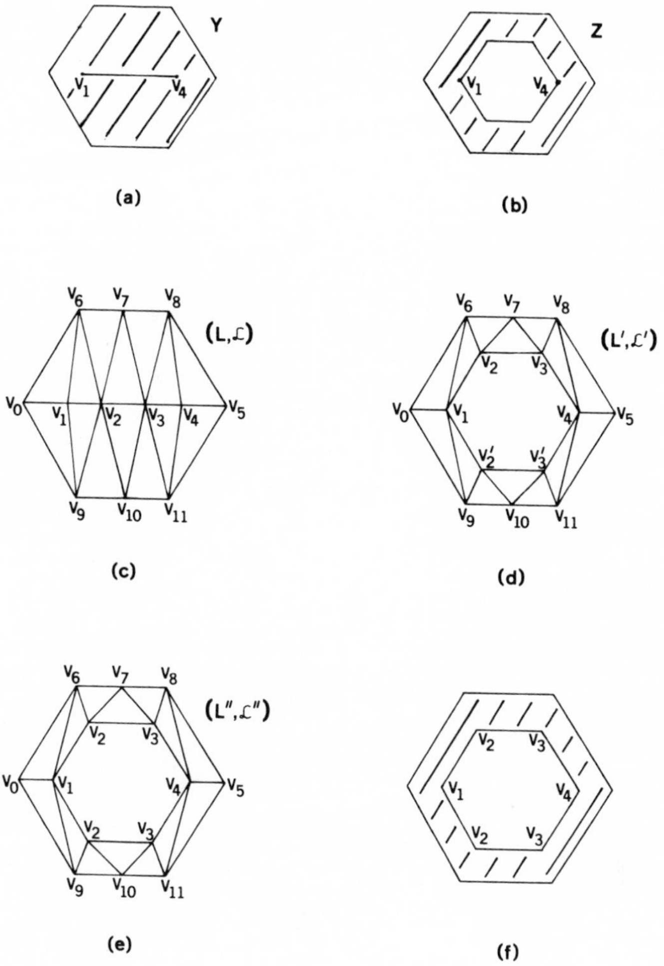

FIGURE 2.11

present all the details, but gradually we shall use the terms "cut" and "paste" freely without any further explanations.

Problem

2.20. Determine $K_{(L,\mathcal{L})}$ and $X_{(L,\mathcal{L})}$ (up to isomorphism and homeomorphism, respectively) for the labeled complexes $(L,\mathcal{L})$ in Fig. 2.12.

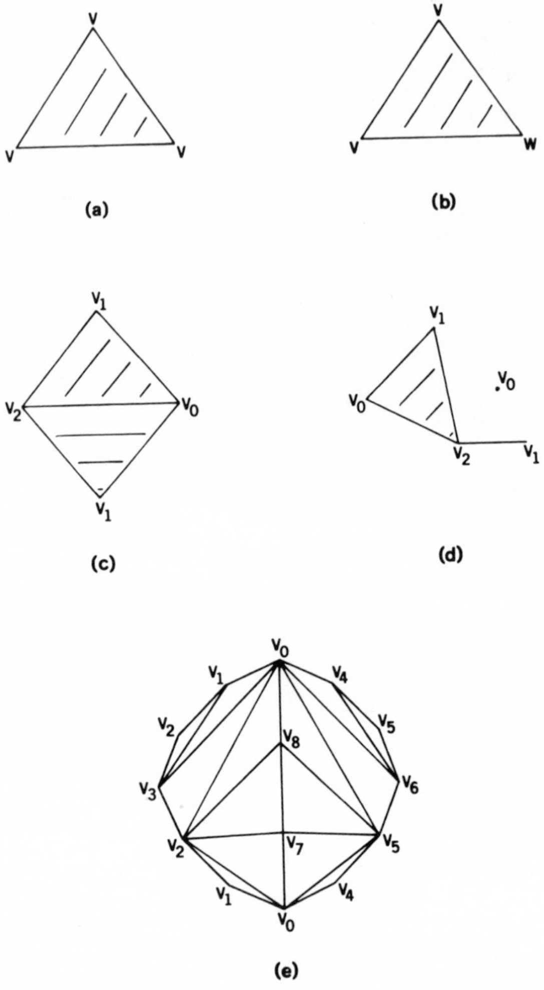

FIGURE 2.12

2.4. HISTORICAL COMMENTS

The underlying unity of the topics in Chapter 1 was not appreciated until the middle of the nineteenth century and the concept of homeomorphism is basically due to Möbius [Möb 1]. In 1863 he introduced the term "Elementarverwandtschaft" to describe the relationship between two figures when there is a bijective map from one to the other which takes "adjacent" points to "adjacent" points. This notion of preserving the relation of "adjacency" between points is, of course, what continuity of functions is all about; however, a precise definition of continuity was not formulated until the 1870s. The last half of the nineteenth century saw a great many important developments in analysis and set theory and these led to an increasing interest in the "topology" of point sets in $\mathbb{R}^n$. In the years 1879-1884 G. Cantor (1845-1918) had created set theory and considered properties of point sets such as limits and closure. The notion of limit was generalized to sets of functions by M. Fréchet (1878-1974) [Fre] in 1906 and F. Riesz (1880-1956) [Ries] in 1909. A few years later in 1914 the book, *Grundzüge der Mengenlehre*, by F. Hausdorff (1868-1942) [Hau] was published and modern point set topology was born. The primary concept becomes that of an abstract topological space which essentially means an abstract set together with a notion of when two points are "close" or "adjacent." In this way one gets rid of any superfluous surrounding space and one can concentrate completely on the intrinsic properties of the space. The definition of abstract topological spaces and their theory was already envisioned by Riemann. Some references for an elementary introduction to general topology are [Eis], [Fra 1], [Hoc-Y], and [Wil]. See [Man] and [Ti-V] for more on the early history of the subject.

Listing was the first to use the term "complex." Precise definitions of simplicial complexes grew gradually out of the study of polygons in the plane and elementary 3-dimensional polyhedra (in the sense of Sec. 1.1). Our definition is the one first used by Alexander in his important paper [Alex 6] in 1926. The term "triangulation" appears to have been introduced by H. Weyl (1885-1955) [Wey 1, p. 21]. The abstract aspect of simplicial decompositions was realized quite early. In particular, the survey article [De-H] by Dehn and P. Heegaard (1871-   ) in 1907 is written in terms of abstract complexes. See also the paper by Weyl [Wey 2] in 1923 and that by P. S. Alexandroff (1896-   ) [Alexf 1] in 1925. A more thorough analysis of abstract simplicial complexes was begun by W. Mayer [May 1] in 1929, by A. W. Tucker (1905-) [Tu] in 1933, and by S. Lefschetz (1884-1972) [Lef 5] in 1941. Simplicial maps were first used by Brouwer [Brou 3] in 1910. The definition of a connected space goes back to Jordan.

Chapter 3

THE CLASSIFICATION OF SURFACES

3.0. INTRODUCTION

In this chapter we shall study a certain nice subclass of polyhedra with which we are already somewhat familiar, namely, surfaces. However, since we want to use them only as examples for the general theory, some of the proofs will be sketchy or even omitted in order not to get bogged down in lengthy detailed discussions which would distract us from our overall goal. The interested reader can find a particularly good recent account of surfaces in [Mas]. Our motivation for studying surfaces is that they will get us used to working with simplicial complexes and will also provide us with an example of the usefulness of associating computable algebraic invariants to spaces. In fact, because surfaces can be completely classified in terms of such invariants, the solution to this particular classification problem is usually held up as a model of the kind of result we should strive for in general.

3.1. THE DEFINITION OF A SURFACE

Intuitively, the word "surface" should conjure up a space which is locally flat, that is, a space which locally looks like the plane $\mathbb{R}^2$.

For practical reasons, however, we do not choose to make this notion the basis for our definition.

DEFINITION. A surface (without boundary) is a polyhedron $S \subset \mathbb{R}^n$ which admits a triangulation (K,φ) satisfying:

1. K is a 2-dimensional connected simplicial complex
2. Each 1-simplex of K is a face of precisely two 2-simplices of K
3. For every vertex $v \in K$ the distinct 2-simplices $\sigma_1, \sigma_2, \cdots$, and σ_s of K to which v belongs can be ordered in such a way that σ_i, $1 \leq i \leq s$, meets σ_{i+1} in precisely one 1-simplex, where $\sigma_{s+1} = \sigma_1$

REMARK 1. The condition in (1) above that K be connected, or equivalently, that a surface S be a connected space [see Problem 2.14(a)], is put in only for convenience.

REMARK 2. Let us call a triangulation (K,φ) of a space "proper" if it satisfies (1-3) in the definition above. Because a space can have many different triangulations, a very natural question to ask is whether every triangulation of a surface is proper? Fortunately, this is the case (see Problem 6.9, which is an easy consequence of the results in Sec. 6.3), otherwise it would be an unsatisfactory state of affairs.

REMARK 3. It is easy to show that every surface satisfies property (*) below (Problem 3.1):

(*): Every point has a neighborhood which is homeomorphic to $\mathbb{R}^2$

The converse is also true. In 1925 T. Radó (1895-) [Ra 2] (see also [Ra 1]) proved the following theorem:

THEOREM A (Radó). Let S be a compact subset of $\mathbb{R}^n$. If S satisfies property (*), then S admits a triangulation.

The triangulation in Radó's theorem is actually proper whenever S is connected (see Problem 6.9). In other words, we see that a set in $\mathbb{R}^n$ is a surface if and only if it is compact, connected, and satisfies property (*). In light of this fact, the reader may wonder why we have chosen the rather technical definition of a surface above over the more natural definition in terms of property (*) which corresponds to one's usual intuitive notion about the kind of space that a surface really is. The reason is that our goal is to give a complete classification of surfaces and for this one needs proper triangulations. Thus, we would not have been able to avoid the technical aspects anyway. Furthermore, with our choice of definition we do not have to appeal to a theorem which we have no intention of proving since its proof would involve a lengthy digression.

Incidentally, the requirement in Theorem A that S be compact is not essential and could be dropped if we allowed infinite triangulations. It is necessitated by the existence of "noncompact surfaces" such as $\mathbb{R}^2$ (see Remark 5 and Sec. 3.5). Since every polyhedron is compact by Problem 2.19, $\mathbb{R}^2$ cannot be triangulated by a finite complex and is therefore not a surface using our definitions.

REMARK 4. Surfaces are examples of what are called (topological) manifolds. An n-dimensional manifold is a space with the property that every point has a neighborhood which is homeomorphic to $\mathbb{R}^n$. Allowing infinite triangulations, Radó had proved that every 2-dimensional manifold can be triangulated. The proof used methods from complex analysis. The obvious generalization, namely that every n-dimensional manifold can be triangulated, remained a famous unsolved problem. In 1952 E. E. Moise (1918-) [Moi] proved that all 3-dimensional manifolds could be triangulated. One of the major achievements of the 1960s was the solution to the triangulation problem by R. C. Kirby (1938-) and L. C. Siebenmann [Ki-S]. The thrust of their main result is that not all n-dimensional manifolds can be triangulated. (It should be cautioned however that they only consider triangulations which satisfy a weak regularity condition on the "star" of each vertex.)

REMARK 5. A surface as we have defined it is sometimes also called a closed surface because the term "surface" (and also "manifold") is often used in a broader sense. The adjective "closed" in the context of surfaces or manifolds then means compact without boundary. In Sec. 3.5 we briefly discuss the definition and classification of surfaces with boundary. Examples of such would be the unit disk, D^2, and the torus with two open disks removed (see Fig. 3.1). (The term "n-disk" will always be used to denote a space homeomorphic to D^n. If the dimension n is obvious from the context, we shall simply say "disk.") Another topic mentioned in Sec. 3.5 is that of noncompact surfaces with or without boundary such as the open unit disk or the open unit disk with an open disk removed from its interior.

After these preliminary remarks about surfaces we are confronted by the rather natural question: Just exactly "how many" distinct (up to homeomorphism) surfaces are there and what are they? We already know many examples of "orientable" surfaces such as the sphere, torus, etc. We shall now describe the surface P^2 called the projective plane, which is "nonorientable." Orientability was discussed in Sec. 1.2; however, a precise definition will be postponed until Sec. 6.5.

Let us consider the Möbius strip in $\mathbb{R}^3$ (see Fig. 1.9b), which is a surface with boundary. How can we form a surface without boundary from it? Observe that if we cut out an open disk from the torus

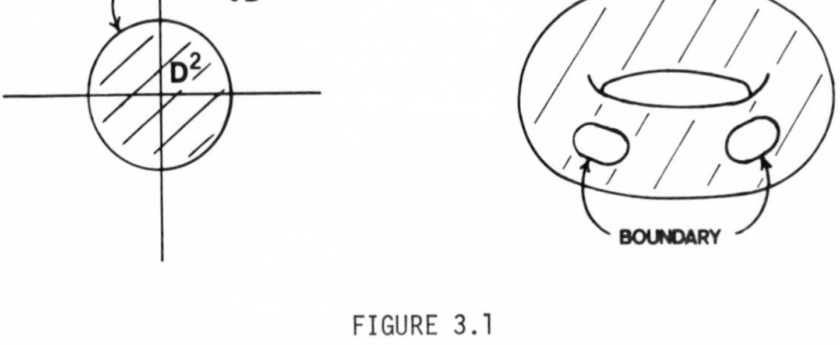

FIGURE 3.1

then we are left with a surface X whose boundary would be a circle. On the other hand, suppose that we had started with X and asked ourselves how we could obtain a surface without boundary from X. Clearly, all we have to do is to paste back the disk, whose boundary is also a circle. In fact, we can obtain a closed surface from X in other ways. For example, take another torus from which an open disk has been removed and call it X'. Since the boundary of both X and X' is a circle, we could paste the two spaces together along their common boundary. This would also produce a closed surface (see Fig. 3.2).

Now let us see what the boundary of the Möbius strip M really is. We have drawn it in Fig. 3.3, and it is obvious that it is nothing but a twisted circle. Therefore, why not proceed as above and form a closed surface P^2 from M by pasting M and a disk together along their common boundary? If this procedure seems more difficult than before, it is only because $\mathbb{R}^3$ is too small and any attempt to construct P^2 in $\mathbb{R}^3$ would fail--there would always be some self-

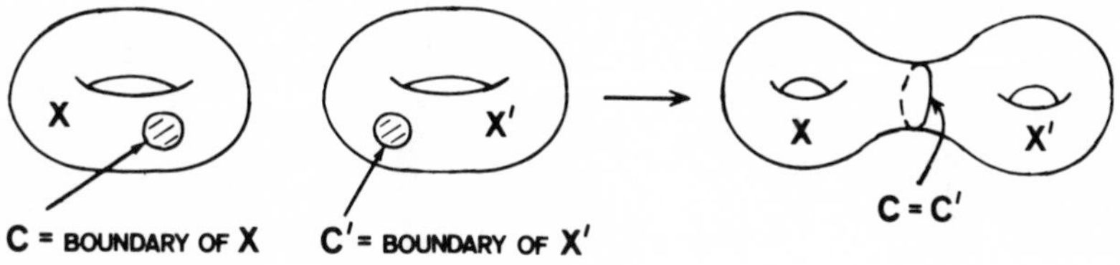

FIGURE 3.2

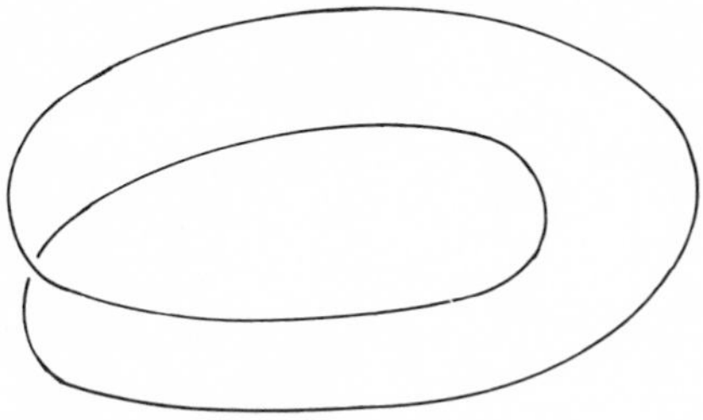

FIGURE 3.3

intersections somewhere. Consider the problem faced by a 2-dimensional person living in $\mathbb{R}^2$ who wants to form a closed surface from D^2 by pasting another disk D to D^2 along their common boundary. He would be unable to do this in $\mathbb{R}^2$ but a 3-dimensional person in $\mathbb{R}^3$ could do it by making D into a "cap" (see Fig. 3.4). The point of this story has been made before: We should think of spaces more intrinsically and not as being imbedded in some special way in a particular Euclidean space. The surface P^2 is called the projective plane. It is "nonorientable," since the Möbius strip was, and can be realized in $\mathbb{R}^4$. We can think of it in the same way that a 2-dimensional man in $\mathbb{R}^2$ thinks of S^2. A particularly nice "immersion" of P^2 in $\mathbb{R}^3$ was found by W. Boy [Boy] in 1901. (See also [Ph] for related aspects of "immersing" surfaces in $\mathbb{R}^3$.) We shall have more to say about P^2 later, but first we shall describe an extremely fruitful way to represent an arbitrary surface.

Problems

3.1. Prove that every surface is a 2-dimensional manifold.

3.2. Find a simplicial complex in $\mathbb{R}^4$ whose underlying space fits the description of P^2 that was given in this section.

3.2. REPRESENTING SURFACES BY SYMBOLS

The basic procedure in our classification of surfaces will be to cut the surface into pieces and then to reassemble the pieces in some

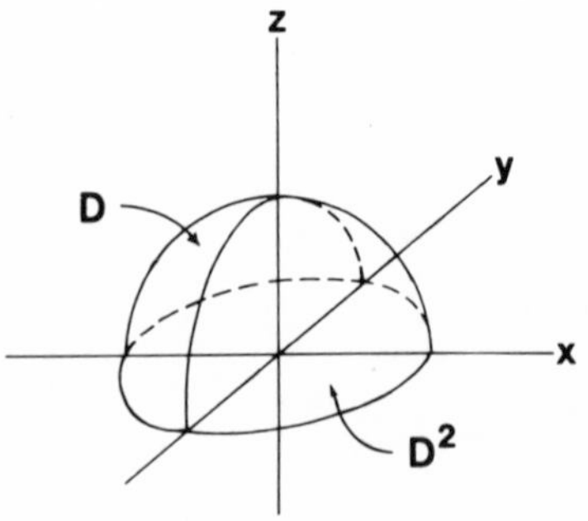

FIGURE 3.4

recognizable form. Of course, each time we make a cut we will have to remember that the two edges which are created are actually identified in the surface. We can do this with the use of labels; however, the labeling will be an abbreviated version of that in Sec. 2.3. For example, consider the torus in Fig. 3.5a. If we cut along the circle A_1, then we obtain Fig. 3.5b. Next, if we cut along the edge A_2 in Fig. 3.5b, or along the circles A_1 and A_2 in Fig. 3.5a simultaneously, then we obtain Fig. 3.5c. Clearly, Fig. 3.5b and c are a good representation of the torus if we understand the labeling and arrows properly. We want to show now that an arbitrary surface S can be represented in a similar manner by a labeled polygon in the plane, that is, S is merely a polygon with its sides identified (or pasted together) in an appropriate way. Corresponding to this geometric presentation there will be an "algebraic" presentation by means of a formal symbol.

Let $k \geq 3$ and define points $w_j(k) \in S^1$ by $w_j(k) = (\cos 2\pi j/k, \sin 2\pi j/k)$. Note that the $w_j(k)$ divide S^1 into k parts of equal length. Let Q_k be the solid regular k-gon which is the convex hull of the points $w_j(k)$, and let $e_j(k) = w_{j-1}(k)w_j(k)$ denote the j-th edge of Q_k (see Fig. 3.6a). Later we shall also find it convenient to have a two-sided "polygon." Therefore, let $Q_2 = D^2$ be the "polygon" with vertices $w_0(2)$ and $w_1(2)$ whose two edges $e_1(2)$ and $e_2(2)$ are the upper and lower half S^1_+ and S^1_- of the circle S^1, respectively (see Fig. 3.6b). In general, we shall call

$$S^n_+ = \{ (x_1,x_2,\cdots,x_{n+1}) \in S^n \mid x_{n+1} \geq 0 \}$$

and

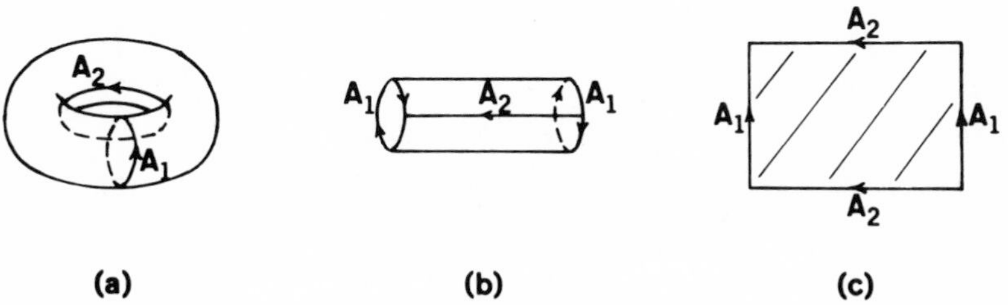

FIGURE 3.5

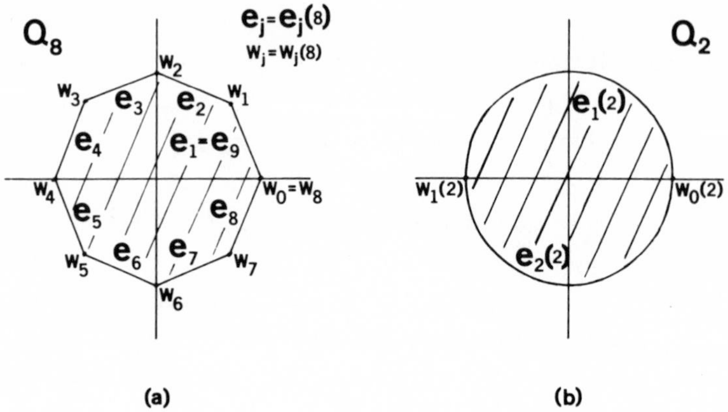

FIGURE 3.6

$$S^n_- = \{ (x_1, x_2, \cdots, x_{n+1}) \in S^n \mid x_{n+1} \le 0 \} = c\ell(S^n - S^n_+)$$

the upper and lower half (or hemisphere) of S^n, respectively.

LEMMA 1. Given a surface S we can always find a labeled complex $(L_S, \mathcal{L}_S)$ satisfying:

(1) $|L_S| = Q_k$ for some k

(2) The vertices of ∂L_S are precisely the points $w_0(k)$, $w_1(k)$, $\cdots$, and $w_{k-1}(k)$

(3) There is a homeomorphism $h : X_{(L_S, \mathcal{L}_S)} \to S$

(4) If $\psi_S : Q_k \to S$ is given by $\psi_S = h \circ p_{(L_S, \mathcal{L}_S)}$, then $\psi_S|\text{int } Q_k$ and $\psi_S|e_i(k)$ are injections

Proof: Let (K, φ) be a proper triangulation of S and let $k - 2$ be the number of 2-simplices in K. List the 2-simplices of K in some order, say σ_1, σ_2, $\cdots$, σ_{k-2}, so that σ_i meets $\sigma_1 \cup \sigma_2 \cup \cdots \cup \sigma_{i-1}$, $2 \le i \le k - 2$, in at least one edge, say e_i (see Problem 3.3). Let v_0, v_1, and v_2 be the vertices of σ_1 and let v_{i+1}, $2 \le i \le k - 2$, denote the vertex of σ_i which is not in e_i.

(Not all of the v_i may be distinct.) Now choose any 2-simplex $\sigma_1' = v_0'v_1'v_2'$ in $\mathbb{R}^2$ and let $\alpha_1 : \overline{\sigma_1'} \to \overline{\sigma_1}$ be the isomorphism defined by $\alpha_1(v_j') = v_j$, for $j = 0,1,2$. Let $e_2' = |\alpha_1|^{-1}(e_2)$. Denote the vertices of e_2 by u and w and let $u' = \alpha_1^{-1}(u)$, $w' = \alpha_1^{-1}(w)$. Next, choose any point $v_3' \in \mathbb{R}^2$ so that $\sigma_2' = u'w'v_3'$ is a 2-simplex and $\sigma_2' \cap \sigma_1' = e_2'$. Let $\alpha_2 : \overline{\sigma_2'} \to \overline{\sigma_2}$ be the isomorphism given by $\alpha_2(u') = u$, $\alpha_2(w') = w$, and $\alpha_2(v_3') = v_3$. Set $e_3' = |\alpha_2|^{-1}(e_3)$. If we continue in this way, we shall get, using induction, a connected 2-dimensional simplicial complex L in $\mathbb{R}^2$ and a simplicial map $\alpha : L \to K$ with the following properties:

1. Each 1-simplex of L belongs to at least one but no more than two 2-simplices of L.

2. The 2-simplices of L can be listed in a sequence $\sigma_1', \sigma_2', \cdots, \sigma_{k-2}'$, such that σ_i' meets $\sigma_1' \cup \sigma_2' \cup \cdots \cup \sigma_{i-1}'$ in a 1-dimensional face e_i' for $i = 2,3,\cdots,k-2$.

3. If v_0', v_1', v_2' are the vertices of σ_1' and v_i' is the vertex of σ_{i-1}' not in e_{i-1}' for $i = 3,4,\cdots,k-1$, then $\alpha(v_i') = v_i$.

4. For $1 \le i \le k - 2$, $\alpha_i = \alpha|\overline{\sigma_i'} : \overline{\sigma_i'} \to \overline{\sigma_i}$ is an isomorphism.

Figure 3.7 gives an example to show how this construction might work out in the case $S = S^2$ and $K = \partial(\overline{v_0v_1v_2v_3})$.

The next step is to define the simplicial complex L_S in the conclusion of Lemma 1 and an isomorphism $\beta : L \to L_S$. (Since L has

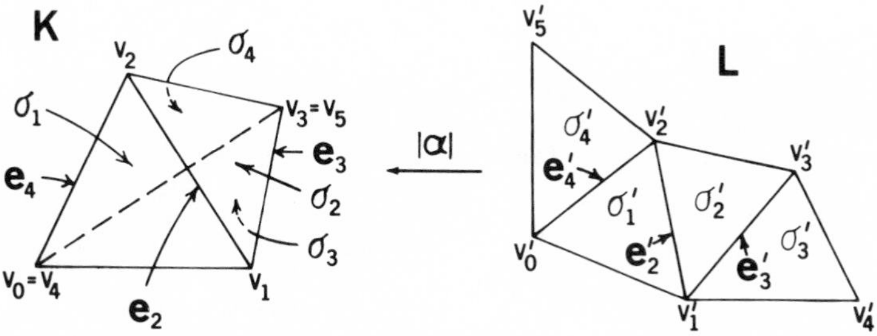

FIGURE 3.7

$k - 2$ 2-simplices, an easy induction using property (2) of L shows that ∂L has k 1-simplices.) We shall use induction on k with respect to all connected simplicial complexes L satisfying (1) and (2) above. If $k = 3$, then we can let $L_S = \overline{w_0(3)w_1(3)w_2(3)}$ and define β to be the isomorphism between $\overline{\sigma_1'}$ and L_S which sends v_i' to $w_i(3)$ for $i = 0,1$, and 2. If $k > 3$, then consider the subcomplex L' of L determined by the first $k - 3$ 2-simplices of L, that is, $L' = \overline{\sigma_1'} \cup \overline{\sigma_2'} \cup \cdots \cup \overline{\sigma_{k-3}'}$. Assume inductively that $Q_{k-1} = |L_S'|$, where L_S' is a simplicial complex with vertices $w_0(k - 1)$, $w_1(k - 1)$, $\cdots$, and $w_{k-2}(k - 1)$, and that there is an isomorphism $\beta' : L' \to L_S'$. If $\beta'(e_{k-2}') = e_{i+1}(k - 1)$, then let $v_{k-1}'' = w_i(k - 1) + w_{i+1}(k - 1)$, $\sigma_{k-2}'' = w_i(k - 1)w_{i+1}(k - 1)v_{k-1}''$, and $L_S'' = L_S' \cup \overline{\sigma_{k-2}''}$. Let $\beta'' : L \to L_S''$ be the isomorphism determined by the condition that $\beta''|L' = \beta'$ and $\beta''(v_{k-1}') = v_{k-1}''$. Next, define a map γ from the vertices of L_S'' to the vertices of Q_k by $\gamma(w_j(k - 1)) = w_j(k)$ if $0 \leq j \leq i$, $\gamma(w_j(k - 1)) = w_{j+1}(k)$ if $i + 1 \leq j \leq k - 2$, and $\gamma(v_{k-1}'') = w_{i+1}(k)$. The desired simplicial complex L_S with $|L_S| = Q_k$ is defined uniquely by the condition that $\gamma : L_S'' \to L_S$ be an isomorphism. See Fig. 3.8 to follow the definition of β'' and γ in the special case where $k = 4$ and $i = 0$. Finally, let $\beta = \gamma \circ \beta'' : L \to L_S$. This finishes the inductive construction of L_S. The labeling $\mathcal{L}_S$ of L_S can be arbitrary as long as it satisfies the condition that two vertices x and y of L_S are given the same label if and only if $(\alpha \circ \beta^{-1})(x) = (\alpha \circ \beta^{-1})(y)$. This condition clearly implies that $K_{(L_S,\mathcal{L}_S)} \approx K$. In other words, there is a homeomorphism $h : X_{(L_S,\mathcal{L}_S)} \to S$. The fact that $\psi_S = h \circ p_{(L_S,\mathcal{L}_S)}$ satisfies the required properties is obvious from the construction and Lemma 1 is proved.

The labeled complex $(L_S,\mathcal{L}_S)$ in Lemma 1 should be thought of intuitively as a presentation of the surface S after it has been cut along $\psi_S(\partial Q_k)$. Conversely, we can obtain S from Q_k by past-

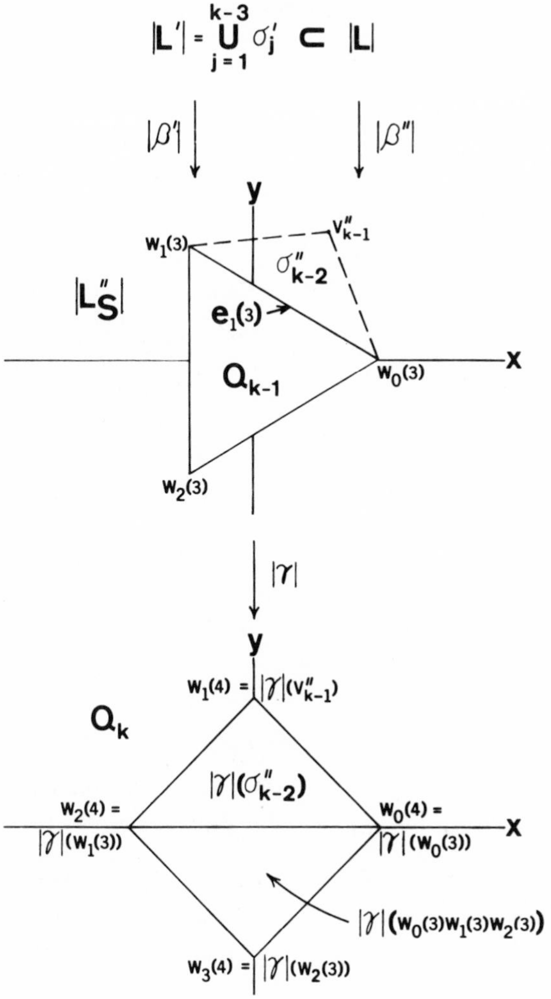

THE CASE $i = 0, k = 4$

FIGURE 3.8

ing together those edges of ∂Q_k which are mapped onto the same set in S by ψ_S. In other words, the study of surfaces could be reduced to the study of certain labeled complexes since each surface S gives rise to a labeled complex $(L_S, \mathcal{L}_S)$ which in turn determines the surface (namely, $S \approx X_{(L_S,\mathcal{L}_S)}$). The next step is to show how to get an even simpler presentation of S by replacing $(L_S,\mathcal{L}_S)$ with a certain symbol.

Let $\mathcal{A}^+ = \{A_1,A_2,\cdots\}$ be the infinite set of distinct symbols A_i. Define $\mathcal{A} = \{A_1,A_2,\cdots\} \cup \{A_1^{-1},A_2^{-1},\cdots\}$, where each A_i^{-1} is considered as a purely formal symbol and no algebraic significance is attached to the superscript "-1." If we identify the symbol $(A_i^{-1})^{-1}$ with A_i, then a^{-1} will belong to $\mathcal{A}$ whenever $a \in \mathcal{A}$. Let $\mathcal{W}$ be the set of all nonempty finite strings $a_1a_2\cdots a_q$, where $a_i \in \mathcal{A}$. For example, the strings $A_1A_1 = A_1(A_1^{-1})^{-1}$, $A_1A_2A_1^{-1}A_2^{-1}$, and $A_1A_1A_1^{-1}A_1A_3$ belong to $\mathcal{W}$. Given $(L_S,\mathcal{L}_S)$ and $\psi_S : Q_k \to S$ as above we define an element $w_S = a_1a_2\cdots a_k \in \mathcal{W}$ as follows: a_1 is an arbitrary element of $\mathcal{A}$. Assume that $2 \le i \le k$ and that $a_1,a_2,\cdots$, and a_{i-1} have already been defined. If $\psi_S(e_i(k)) \ne \psi_S(e_j(k))$ for $1 \le j < i$, then a_i in an arbitrary element of $\mathcal{A} - \{a_1,a_1^{-1},a_2,a_2^{-1},\cdots, a_{i-1}, a_{i-1}^{-1}\}$. On the other hand, suppose that $\psi_S(e_i(k)) = \psi_S(e_j(k))$ for some j, $1 \le j < i$. Define $a_i = a_j$ if $\psi_S(w_{i-1}(k)) = \psi_S(w_{j-1}(k))$ and $a_i = a_j^{-1}$, otherwise.

DEFINITION. The string $w_S \in \mathcal{W}$ is called the symbol associated to $(L_S,\mathcal{L}_S)$ or simply a symbol for the surface S.

It is possible to define the symbol w_S in another way. We start with the first edge $e_1(k)$ of Q_k and label it with an arbitrary element from $\mathcal{A}$. Then we continue around to the other edges of Q_k in a counterclockwise fashion associating a different label from $\mathcal{A}$ to each of them (if a has been used, a^{-1} does not count as different) unless that edge, say $e_i = e_i(k)$, is mapped by ψ_S onto the same set in S as a previously labeled edge, say $e_j = e_j(k)$. In that case, since we want to think of e_i and e_j as being the two edges of a cut in S, the labels for e_i and e_j should reflect this identification. Assume that e_j has been labeled a. We cannot simply label e_i also a because, although e_i and e_j are to be identified when constructing S, there are two distinct possibilities for the identification depending on how the vertices of e_i and e_j are mapped by ψ_S. By using the labels a or a^{-1} for e_i we can distinguish between these two possibilities. Equivalently, let us consider the edges of Q_k as having been oriented in a counterclockwise fashion and label e_i either a or a^{-1} depending on whether the orientations on $\psi_S(e_i) = \psi_S(e_j) \subset S$ induced by $\psi_S|e_i$ and $\psi_S|e_j$ agree or not. We now obtain the symbol associated to $(L_S, \mathcal{L}_S)$ from this labeled polygon Q_k by writing down the labels from the edges of Q_k in sequence, starting with the label for e_1 and continuing around Q_k in a counterclockwise fashion. We also can see that there is a natural correspondence between symbols and labeled polygons Q_k, so that these two notions will be used interchangeably in the future.

For example, consider the case $S = S^2$ and let $(L_0, \mathcal{L}_0)$ be the labeled complex in Lemma 1 with respect to this surface. If we suppose in the proof of Lemma 1 that K, L, and $\alpha : L \to K$ are as in Fig. 3.7 and that $\beta : L \to L_0$ satisfies $\beta(v_i') = w_i(6)$, $i = 0,1,3,5$, $\beta(v_2') = w_4(6)$, and $\beta(v_4') = w_2(6)$ (see Fig. 3.9a), then one possibility for $(L_0, \mathcal{L}_0)$ is shown in Fig. 3.9b. A symbol associated to this particular $(L_0, \mathcal{L}_0)$ would be $A_1A_1^{-1}A_2A_3A_3^{-1}A_2^{-1}$ and the corresponding labeled polygon is shown in Fig. 3.9c. However, there was a certain amount of freedom in the labeling of Q_6 and we might equally well have ended up with $A_4^{-1}A_4A_1A_6A_6^{-1}A_1^{-1}$. Nevertheless, all symbols for $(L_0, \mathcal{L}_0)$ will have the form $a_1a_1^{-1}a_2a_3a_3^{-1}a_2^{-1}$ for $a_i \in \mathcal{A}$. In general, although the labeled complex $(L_S, \mathcal{L}_S)$ from Lemma 1 does not really determine a unique symbol, the basic structure of the symbol is invariant. In other words, if $a_1a_2\cdots a_k$ and $b_1b_2\cdots b_k$ are two symbols associated to $(L_S, \mathcal{L}_S)$, then there is a permutation σ of $\mathcal{A}$ satisfying $\sigma(a^{-1}) = \sigma(a)^{-1}$ for all $a \in \mathcal{A}$ such that $b_i = \sigma(a_i)$. This justifies our talking about "the" symbol w_S associated to $(L_S, \mathcal{L}_S)$.

Now consider the example $(L_0, \mathcal{L}_0)$ again. Even though Fig. 3.9c is a good pictorial representation for its symbol, it is not the one

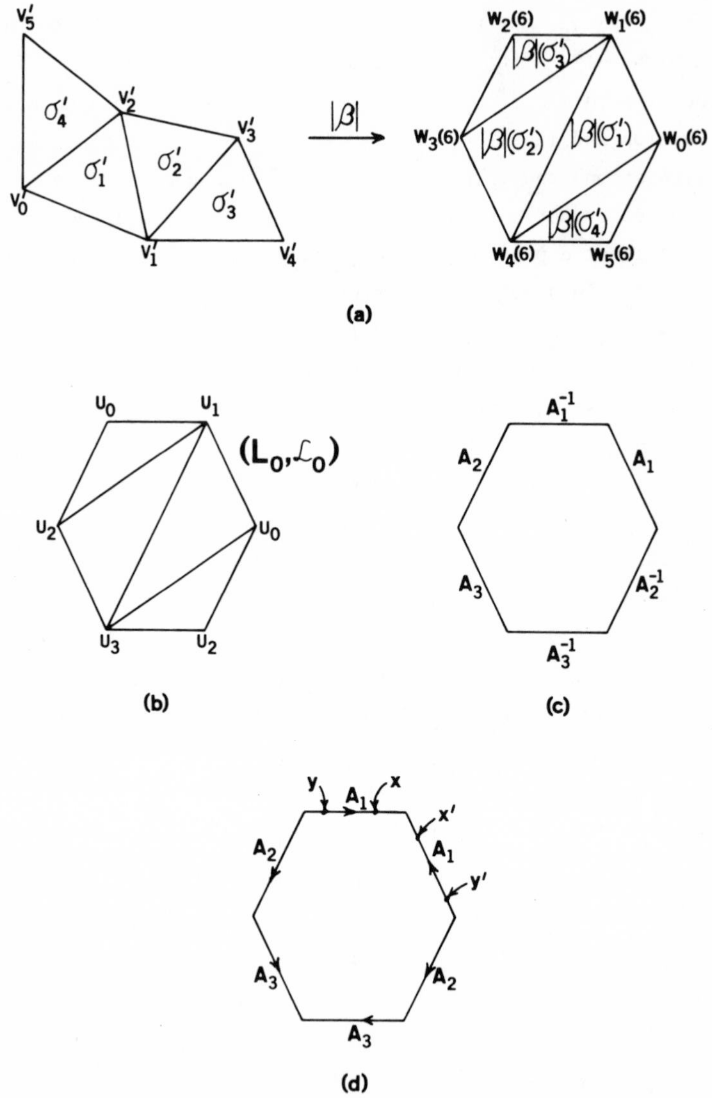
v_5'
v_2'
v_3'
v_0'
v_1'
v_4'
σ_4'
σ_1'
σ_2'
σ_3'
$|\beta|$
$w_2(6)$
$w_1(6)$
$w_3(6)$
$w_0(6)$
$w_4(6)$
$w_5(6)$
$|\beta|(\sigma_3')$
$|\beta|(\sigma_2')$
$|\beta|(\sigma_1')$
$|\beta|(\sigma_4')$
(a)
u_0
u_1
$(L_0, \mathcal{L}_0)$
u_2
u_0
u_3
u_2
(b)
A_1^{-1}
A_2
A_1
A_3
A_2^{-1}
A_3^{-1}
(c)
y
A_1
x
x'
A_2
A_1
y'
A_3
A_2
A_3
(d)

FIGURE 3.9

which is usually adopted. By drawing little arrows in the edges of Q_6 as indicated in Fig. 3.9d one can incorporate, without superscripts on the symbols, the same information that was contained in Fig. 3.9c. The two possible ways of identifying edges (via linear maps) are specified by the direction of the arrows. For example, the arrows tell us that the points x and y in Fig. 3.9d are to be identified with x' and y', respectively. Observe that the direction of the arrows is not uniquely specified by a symbol. Simultaneously reversing their direction on two edges which are to be identified changes nothing. The only important property which is an invariant is whether these arrows are both in the same or opposite direction. Most of the labeled polygons we shall refer to from now on will have the arrows in their sides rather than superscripts on the labels, but we should remember that either or both methods simultaneously is permissible.

Although we have shown how certain symbols can be associated to each surface, the symbols will not be of much use to us unless there is a procedure whereby one can pass from a symbol directly to its associated surface without making use of the labeled complex from which it was defined. Of course, on an intuitive level it is obvious what one has to do, because an element $w = a_1a_2\cdots a_k \in \mathcal{W}$ tells us how to construct a space S_w as follows: We start with Q_k and paste together the edges of ∂Q_k according to w, that is, $e_i(k)$ and $e_j(k)$ are pasted together appropriately whenever $a_i = a_j$ or a_j^{-1}. What is a precise definition of S_w and when is it a surface?

DEFINITION. If $a \in \mathcal{A}$, then define $n_w(a)$ to be the number of times that the symbol a or a^{-1} appear in the string w. The length of w, $\ell(w)$, is defined by

$$\ell(w) = \sum_{a \in \mathcal{A}^+} n_w(a)$$

For example, if $w = A_1A_2A_1^{-1}A_1A_2A_3^{-1}$, then $n_w(A_1) = 3$, $n_w(A_2) = 2$, $n_w(A_3) = 1$, and $n_w(A_i) = 0$ for $i > 3$. Also, $\ell(w) = 6$.

DEFINITION. Define a subset $\mathcal{W}^*$ of $\mathcal{W}$ by

$$\mathcal{W}^* = \{ w \in \mathcal{W} \mid n_w(a) = 0 \text{ or } 2 \quad \text{for all} \quad a \in A \}$$

It is easy to see that if w_S is a symbol for a surface S, then $w_S \in \mathcal{W}^*$. In fact, one can show that S_w is a surface if and only if $w \in \mathcal{W}^*$. We shall only prove half of this result, namely, if $w \in \mathcal{W}^*$, then S_w is a surface. The other part will not be needed in subsequent discussions and will therefore be left to the reader. In both cases, the main problem is to give a precise definition of S_w.

Let $w \in \mathcal{W}^*$ and let $k = \ell(w)$.

CASE 1. $k > 2$: We shall use the special case $w_0 = A_1A_2A_1^{-1}A_2^{-1}A_3A_3A_4A_4^{-1}$ and Fig. 3.10 to illustrate the steps in the definition of S_w in general.

First of all, w_0 corresponds to the labeled polygon shown in Fig. 3.10a. Triangulate Q_8 by means of a simplicial complex L_{w_0} and label the vertices of L_{w_0} as shown in Fig. 3.10b to obtain a labeled complex $(L_{w_0}, \mathcal{L}_{w_0})$. In the general case we introduce two new vertices, $w_{i,1}(k)$ and $w_{i,2}(k)$, in each edge $e_i(k)$, where $w_{i,j}(k) = w_{i-1}(k) + j(w_i(k) - w_{i-1}(k))/3$ for $j = 1$ or 2, and L_w is a simplicial complex which is the obvious generalization of L_{w_0} such that $|L_w| = Q_k$ and the vertices of ∂L_w are the points $w_i(k)$ and $w_{i,j}(k)$ for $0 \leq i \leq k - 1$ and $j = 1$ or 2. Each $w_i(k)$ is labeled v_0 and all other vertices of L_w are given distinct labels with the following exceptions: If $w = a_1a_2 \cdots a_k$ and $s = 1$ or 2, then (i) $w_{i,s}(k)$ and $w_{j,s}(k)$ are given the same label whenever $a_i = a_j$, and (ii) $w_{i,s}(k)$ and $w_{j,s+1}(k)$ are given the same label whenever $a_i = a_j^{-1}$ [$w_{j,s+1}(k)$ is defined to equal $w_{j,1}(k)$ if $s = 2$]. Let $(L_w, \mathcal{L}'_w)$ denote the labeled complex which we obtain in this way.

Next, let $K'_w = K_{(L_w, \mathcal{L}'_w)}$ and let $p'_w = p_{(L_w, \mathcal{L}'_w)} : Q_k = |L_w| \to |K'_w|$ be the natural projection. We would like $|K'_w|$ to be a surface. Unfortunately, condition (3) in the definition of a surface might not

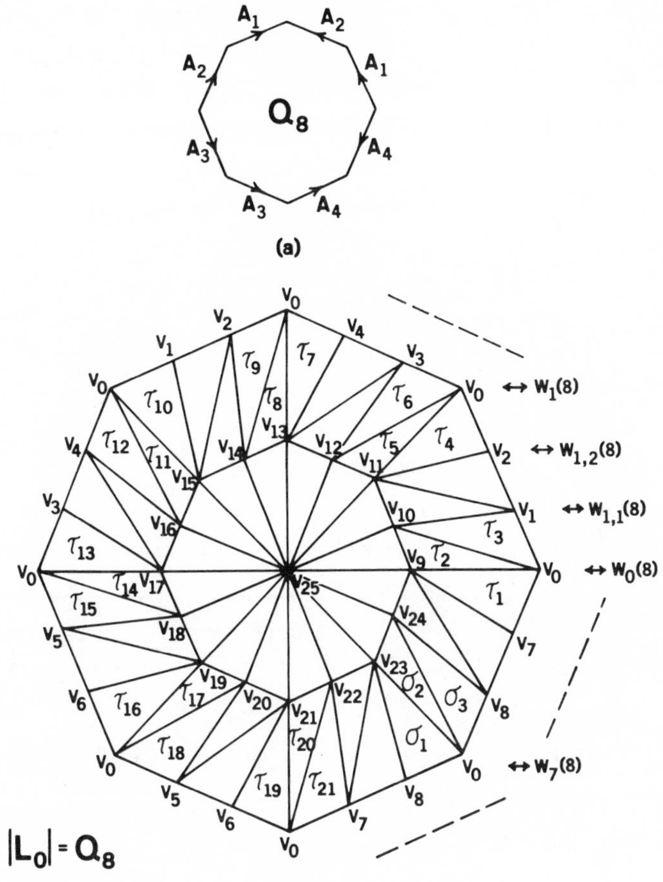

(b)

THE CASE $w = A_1A_2A_1^{-1}A_2^{-1}A_3A_3A_4A_4^{-1}$

FIGURE 3.10

be satisfied at the point $p'_w(w_i(k))$ with respect to the obvious triangulation $(K'_w, 1_{|K'_w|})$. [Conditions (1) and (2) are satisfied because $w \in \mathcal{W}^*$.] The problem is that, although $|K'_w|$ is certainly the correct space as far as pasting together the appropriate edges of Q_k according to w is concerned, we were careless about identifying the vertices of ∂Q_k correctly. One can show, however, that the 2-simplices of K'_w which have $p'_w(w_i(k))$ as a vertex can be partitioned into sets $T_1, T_2, \cdots, T_n$ such that the 2-simplices of each T_i separately can be ordered to satisfy (3). In our example in Fig. 3.10, $n = 2$ and the two sets are $\{p'_{w_0}(\sigma_1), p'_{w_0}(\sigma_2), p'_{w_0}(\sigma_3)\}$ and $\{p'_{w_0}(\tau_1), p'_{w_0}(\tau_2), \cdots, p'_{w_0}(\tau_{21})\}$. Let

$$T_i^{-1} = \{ \sigma \mid \sigma \text{ is a 2-simplex of } L_w \text{ and } p'_w(\sigma) \in T_i \}$$

The sets T_i partition the vertices of Q_k into sets $V_1, V_2, \cdots, V_n$, where

$$V_i = \{ w_j(k) \mid w_j(k) \in \sigma \text{ for some } \sigma \in T_i^{-1} \}$$

We now relabel the vertices $w_0(k)$, $w_1(k)$, $\cdots$, and $w_{k-1}(k)$ of L_w as follows: Two vertices $w_s(k)$ and $w_t(k)$ are given the same label if and only if they belong to the same V_i for some i. Returning to our example in Fig. 3.10, this can be accomplished by relabeling $w_7(8)$ with v_{26}. Let $(L_w, \mathcal{L}_w)$ denote the new labeled complex and let $K_w = K_{(L_w, \mathcal{L}_w)}$. It is easy to check that $S_w = |K_w|$ is a surface with proper triangulation $(K_w, 1_{|K_w|})$ and that the homeomorphism type of this surface is uniquely determined by this construction. This finishes our definition of S_w in Case 1.

CASE 2. $k = 2$: The only complication in this case is that Q_2 is not the underlying space of a simplicial complex, but the steps in Case 1 can easily be modified to apply here. There are two subcases.

1. Suppose that $w = aa^{-1}$ for some $a \in A$: Let $(L_w, \mathcal{L}_w)$ be the labeled complex in Fig. 3.11a, so that $|L_w| = Q_6$. Let $S_w = X_{(L_w, \mathcal{L}_w)}$. Since the natural symbol associated to $(L_w, \mathcal{L}_w)$ is $w_1 = abcc^{-1}b^{-1}a^{-1}$, we are, in essence, defining S_w to equal S_{w_1}, where the latter is defined as in Case 1. This is intuitively correct because w suggests that we are to paste together the upper and lower

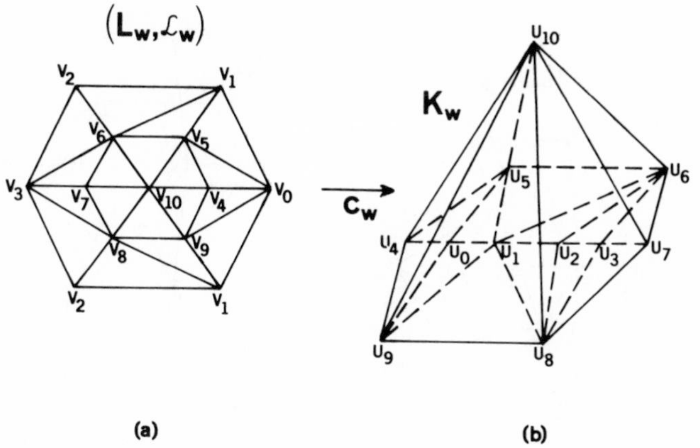

FIGURE 3.11

hemisphere of ∂Q_2 and w_1 suggests the same in the case of ∂Q_6, thereby leading to basically the same space in both instances. In order to identify S_w, consider the 2-dimensional complex K_w in Fig. 3.11b. Clearly, $K_w = K_{(L_w, \mathcal{L}_w)}$ and the natural map $c_w = c_{(L_w, \mathcal{L}_w)} : L_w \to K_w$ is given by $c_w(v_i) = u_i$. Therefore, $S_w = |K_w| \approx S^2$.

2. Suppose that $w = aa$ for some $a \in A$: Let $(L_w, \mathcal{L}_w)$ be the labeled complex in Fig. 3.12a and let $S_w = X_{(L_w, \mathcal{L}_w)}$. (This time we are essentially defining S_w to equal S_{abcabc}.) To see which surface S_w actually is, we shall split $(L_w, \mathcal{L}_w)$ into two pieces, namely, the labeled complexes $(L_w^{(1)}, \mathcal{L}_w^{(1)})$ and $(L_w^{(2)}, \mathcal{L}_w^{(2)})$ as shown in Fig. 3.12b. This splits S_w into two pieces homeomorphic to $X_{(L_w^{(1)}, \mathcal{L}_w^{(1)})}$ and $X_{(L_w^{(2)}, \mathcal{L}_w^{(2)})}$, respectively, with the latter clearly being homeomorphic to the disk D^2. Now consider the labeled complex $(L, \mathcal{L})$ in Fig. 3.12c and the natural simplicial map from $L_w^{(1)}$ to L which sends v_i to u_i. It follows easily that $X_{(L_w^{(1)}, \mathcal{L}_w^{(1)})}$ is

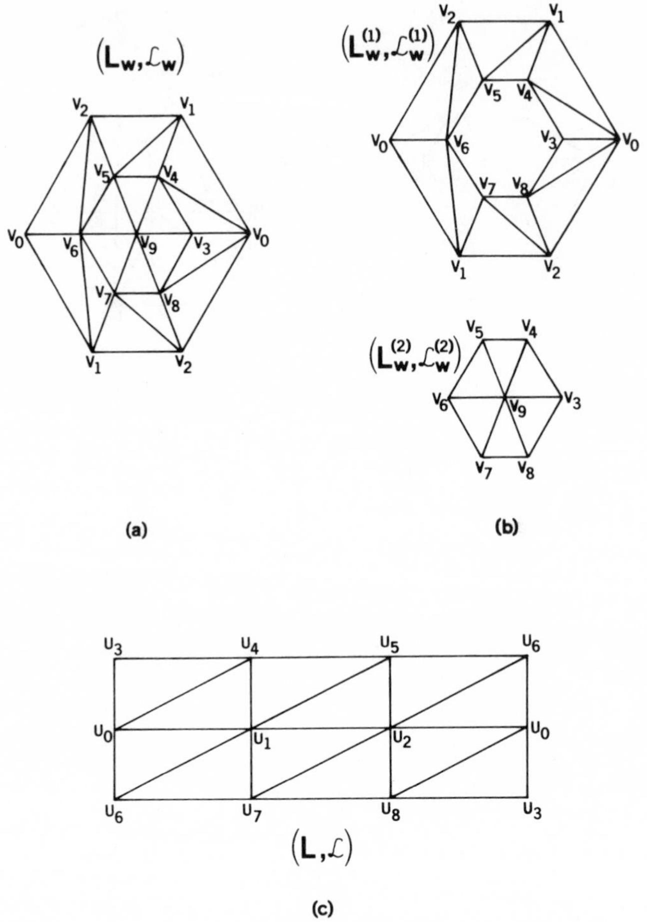

FIGURE 3.12

homeomorphic to $X_{(L,\mathcal{L})}$. Since $X_{(L,\mathcal{L})}$ is obviously a Möbius strip, we have shown that S_W is the union of a Möbius strip and a disk which intersect along their common boundary. In other words, S_W fits the intuitive description of the projective plane P^2 given at the end of Sec. 3.1. Not only does this identify S_W, but, turning the observation around, we can now give a rigorous definition of P^2:

DEFINITION. The projective plane P^2 is defined to be any space homeomorphic to S_w.

Let us summarize this discussion in a lemma.

LEMMA 2. There is a construction which associates to each $w \in \mathcal{W}^*$ a well-defined labeled complex $(L_w, \mathcal{L}_w)$ such that

1. $|L_w| = Q_{\ell(w)}$ if $\ell(w) > 2$ and $|L_w| = Q_6$ if $\ell(w) = 2$.
2. The space $S_w = X_{(L_w, \mathcal{L}_w)}$ is a surface.
3. If S is a surface and if w_S is any symbol for S, then $S \approx S_{w_S}$.
4. Let $a \in \mathcal{A}$. If $w = aa^{-1}$, then $S_w \approx S^2$; and if $w = aa$, then $S_w \approx P^2$.

The only part of Lemma 2 which still needs justification is (4), but it is easy and will be left to the reader.

DEFINITION. If S is a surface, then any $w \in \mathcal{W}^*$ such that $S_w \approx S$ will be called a symbol for S.

By (3) in Lemma 2 this definition is compatible with our previous notion of a symbol for a surface.

The task that we set for ourselves in this section is now completed. The following diagram gives an overview of the various cor-

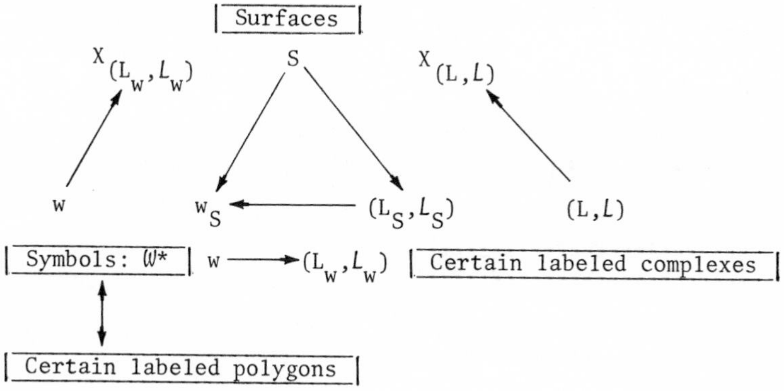

respondences that we have established. In particular, we have shown that to every surface there is associated a collection of symbols, each of which determines the surface uniquely. If two surfaces have a symbol in common, then they are homeomorphic. From now on our study of surfaces will proceed by analyzing their symbols. Actually, we shall not work so much with the symbols as with their corresponding labeled polygons. In the future we shall use the language of cutting and pasting polygons quite freely with little further explanations. It is hoped that the precise mathematical definition of these operations (in terms of labeled complexes) have been indicated in enough situations so that the reader will be able to translate this intuitive terminology into mathematical language by himself.

As a final remark, we would like to point out that there is nothing special about the polygons Q_k. They were convenient, but any other k-sided polygon would serve our purposes equally well when interpreting a symbol as a polygon whose sides are to be pasted together in a certain way. (The term "polygon" has not been defined, but it will always be clear what is meant from the context in which it appears.) In fact, future discussions will be simplified if other polygons are admitted and we shall do so on such occasions. It will be left to the reader to translate everything back in terms of the Q_k or, alternatively, to develop the theory in the context of arbitrary polygons from the start.

Problems

3.3. Prove the claim made in this section that if (K,φ) is a proper triangulation of a surface S, then the 2-simplices of K, $\sigma_1, \sigma_2, \cdots, \sigma_t$, can be ordered so that σ_i meets $\sigma_1 \cup \sigma_2 \cup \cdots \cup \sigma_{i-1}$, $2 \leq i \leq t$, in at least one edge. (*Hint:* Choose a maximal collection of 2-simplices of K with that property and use the connectedness of K and condition (3) in the definition of a surface to show that this collection must already contain all of the 2-simplices of K. Compare this problem with Problem 1.1.)

3.4. (a) Show that $A_1A_2A_3A_4A_5A_6A_3^{-1}A_2^{-1}A_1^{-1}A_6^{-1}A_5^{-1}A_4^{-1}$ is a symbol for the torus $S^1 \times S^1$ which can be obtained via Lemma 1. In other words, choose a proper triangulation (K,φ) of $S^1 \times S^1$, construct the complex L and simplicial map $\alpha : L \to K$ (cf. Fig. 3.7), and find a homeomorphism $|\beta| : |L| \to Q_{12}$, so that an appropriate labeling of the edges of Q_{12} leads to this symbol. (*Hint:* Figure 3.5 will help in finding the appropriate triangulation.)

(b) Show that $A_1A_2A_1^{-1}A_2^{-1}$ is another (and simpler) symbol for $S^1 \times S^1$ but that this symbol does not come from a construction as in (a). (*Hint:* Use Fig. 3.5.)

3.5. Find a symbol for the double torus (see Fig. 1.8b).

3.6. Fill in the details left out in the proof of Lemma 2. Also, prove the statement that if two surfaces have a symbol in common, then they are homeomorphic.

3.7. Give a precise definition of S_w for an arbitrary $w \in \mathcal{W}$ which extends the definition given in this section for the case $w \in \mathcal{W}^*$. Prove that S_w is a surface if and only if $w \in \mathcal{W}^*$.

3.8. There is an alternate definition for P^2 which is very common: Define P^2 to be the space obtained from S^2 by identifying antipodal points on the sphere, where $-x$ is said to be the antipodal point of $x \in S^2$. Although we have not gone into the question of how one can form spaces from other spaces by identifying certain points, which would force us to discuss the notion of an abstract topological space, it is a good exercise for one's geometric insight to convince oneself, on intuitive grounds at least, that the two definitions for P^2 are equivalent by describing a homeomorphism between the two spaces. (*Hint:* P^2 in this new definition is nothing but the upper hemisphere of S^2 with antipodal points in its boundary S^1 identified. But S^1 with its antipodal points identified is just another circle and a neighborhood of this circle in P^2 is the Möbius strip.)

3.3. THE NORMAL FORM FOR SOME SURFACES

If we are going to use symbols as invariants for surfaces, then it will be useful to know at least one symbol for some of the so-called standard surfaces. We already know symbols for S^2, P^2, and $S^1 \times S^1$ (Problem 3.4). Other surfaces can be formed from these by means of the "connected sum" operation. Intuitively, the connected sum of two surfaces S_1 and S_2 is defined to be the surface which is obtained by cutting out a disk from both S_1 and S_2 and pasting the remainders together along the boundaries of the holes (see Fig. 3.13). More precisely, choose a proper triangulation (K_i, φ_i) for S_i and let σ_i be a 2-simplex of K_i. Position K_2 so that $\sigma_1 = \sigma_2 = |K_1| \cap |K_2|$.

DEFINITION. Any space which is homeomorphic to the underlying space of the simplicial complex $L = (K_1 \cup K_2) - \{\sigma_1\}$ is defined to be the connected sum of S_1 and S_2 and is denoted by $S_1 \# S_2$.

The operation # is well defined, but proving this is not at all trivial. What must be shown is that if (K_i', φ_i') is another proper triangulation for S_i and if σ_i' is a 2-simplex of K_i' such that $\sigma_1' = \sigma_2' = |K_1'| \cap |K_2'|$, then $|(K_1' \cup K_2') - \{\sigma_1'\}| \approx |(K_1 \cup K_2) - \{\sigma_1\}|$. Basically, this problem reduces to one of showing that deleting a disk from a surface produces a space which (up to homeomorphism)

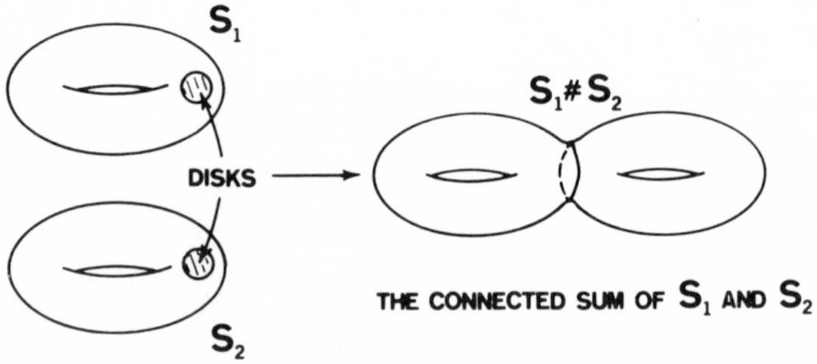

FIGURE 3.13

does not depend on the particular disk that is removed. One needs the following theorem which we state without proof. We shall have more to say about it at the end of Sec. 3.6.

DEFINITION. Let $X, Y \subset \mathbb{R}^n$. A map $f : X \to Y$ is called an imbedding if f is a homeomorphism of X onto its image $f(X)$.

Every homeomorphism is an imbedding. Although every imbedding is injective, it is not true that every continuous injective map is an imbedding [see Problem 2.1(b)].

THEOREM B. Let S be a surface. If $\nu_1, \nu_2 : D^2 \to S$ are imbeddings, then there is a homeomorphism $h : S \to S$ such that $\nu_2 = h \circ \nu_1$.

The fact that # is well defined is now an easy consequence of Theorem B. The main properties of # are listed below:

1. $S_1 \# S_2$ is a surface.
2. (Commutativity) $S_1 \# S_2 \approx S_2 \# S_1$.
3. (Associativity) $S_1 \# (S_2 \# S_3) \approx (S_1 \# S_2) \# S_3$.
4. (Identity) $S_1 \# S^2 \approx S_1$.

Properties 1 and 4 are proved directly from the definition of #. The latter is obvious from the fact that S^2 is the union of two disks, so that in forming $S_1 \# S^2$ one is in effect removing a disk from S_1 and then pasting another in its place. Properties 2 and 3 are obvious once one knows that # is well defined.

NOTE. Although the operation # will be used in what follows, we must point out that it will not be essential to know that it is well defined. The only properties of # that we really need are ones which are easily deducible directly from the definition. The reason for this disclaimer is that Theorem B is proved using the classification theorem for surfaces (see the end of Sec. 3.6) so that to use it now would involve us in a circular argument.

LEMMA 3. (a) Let S_1 and S_2 be surfaces with symbols $u = a_1a_2\cdots a_s$ and $v = b_1b_2\cdots b_t$, respectively, where $a_i \neq b_j$ or b_j^{-1} for all i and j. Let $1 \leq k \leq s + 1$. Then $a_1\cdots a_{k-1}b_1\cdots b_t a_k\cdots a_s$ is a symbol for $S_1 \# S_2$.

(b) Conversely, let $a_1\cdots a_{k-1}b_1\cdots b_t a_k\cdots a_s$ be a symbol for a surface S, where $s,t \geq 1$ and $1 \leq k \leq s + 1$. Let $u = a_1a_2\cdots a_s$ and $v = b_1b_2\cdots b_t$. If u and v belong to $\mathcal{U}^*$, then $S \approx S_1 \# S_2$, where S_1 and S_2 are surfaces with symbols u and v, respectively.

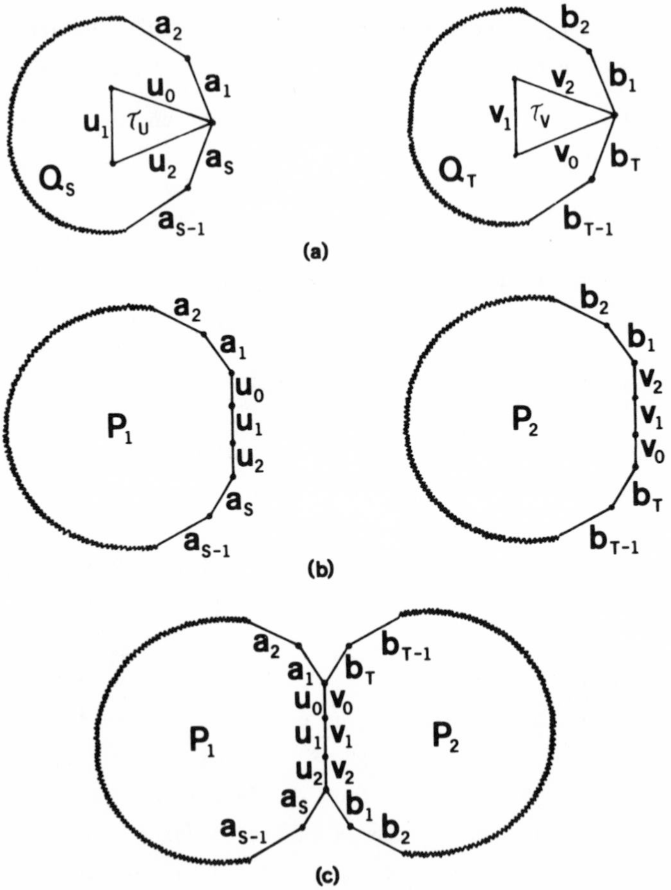

FIGURE 3.14

Proof: (a) We may assume for simplicity that $k = s$, because a similar argument works when $k \neq s$. First, consider the case where $s,t > 2$. The surfaces S_1 and S_2 correspond, via their symbols u and v, to labeled polygons Q_s and Q_t, respectively. Choose any 2-simplex $\tau_u \subset Q_s$ which has $w_0(s)$ as one vertex and whose other two vertices lie in int Q_s. Let τ_v be a similar 2-simplex in Q_t and label the edges of τ_u and τ_v as shown in Fig. 3.14a. Now delete int τ_u from Q_s, cut Q_s - int τ_u in $w_0(s)$, and straighten out the edges of τ_u to obtain the labeled polygon P_1 in Fig. 3.14b. The labeled polygon P_2 is gotten in a similar way from Q_t. It is clear that the labeled polygon P_i corresponds to (S_i - open disk). Next, paste P_1 and P_2 together along the edges u_i and v_i as shown in Fig. 3.14c. This new labeled polygon corresponds to $S_1 \# S_2$ and, looking at the labels on the edges, one sees immediately from the picture that a symbol for $S_1 \# S_2$ is $a_1 \cdots a_s b_1 \cdots b_t$. If either s or t is equal to 2, then a slight modification is necessary in the above construction which is left to the reader. This finishes the proof of (a).

(b) The fact that there are surfaces S_1 and S_2 with symbols u and v, respectively, follows from Lemma 2. By part (a), $a_1 \cdots a_{k-1} b_1 \cdots b_t a_k \cdots a_s$ is also a symbol for $S_1 \# S_2$. Therefore, $S \approx S_1 \# S_2$ (Problem 3.6).

Lemmas 2 and 3 allow us to make the following table of surfaces and symbols, where $a, a_i \in A$ and $a_i \neq a_j$ or a_j^{-1} if $i \neq j$:

TABLE 1

Surface	Symbol
S^2	aa^{-1}
$\underbrace{(S^1 \times S^1) \# \cdots \# (S^1 \times S^1)}_{n \text{ times}}$	$a_1 a_2 a_1^{-1} a_2^{-1} \cdots a_{2n-1} a_{2n} a_{2n-1}^{-1} a_{2n}^{-1}$
$\overbrace{P^2 \# \cdots \# P^2}$	$a_1 a_1 \cdots a_n a_n$

DEFINITION. The symbols of Table 1 will be called the normal forms for the corresponding surfaces.

Problems

3.9. Assuming the validity of Theorem B, prove that the connected sum operation # is well defined.

3.10. Given that the operation # is well defined, prove that it satisfies the four properties we listed.

3.11. Fill in the details in the proof of Lemma 3(a).

3.4. THE CLASSIFICATION THEOREMS

We are finally ready to state and prove the principal theorems of this chapter. The amazing fact is that the surfaces listed in Sec. 3.3, Table 1, are all the surfaces there are.

THEOREM 1. Every surface is homeomorphic either to the sphere, or to a connected sum of tori, or to a connected sum of projective planes.

This theorem, together with the next one, provides us with a complete classification of surfaces.

THEOREM 2. The sphere, the connected sum of n tori, $n \geq 1$, and the connected sum of m projective planes, $m \geq 1$, are nonhomeomorphic spaces.

Proof of Theorem 1: It suffices to show that every surface S admits a symbol of one of the three types given in Table 1. What is needed therefore is a finite algorithm whereby a given symbol $v = a_1a_2\cdots a_{2k}$, $a_i \in A$, for a surface S can be reduced to a normal form without changing the homeomorphism class of S. If $k = 1$, then $v = aa$ or aa^{-1} and S is homeomorphic to either S^2 or P^2. Assume inductively that the theorem is true for $k - 1$ and $k \geq 2$. Let

$(L_v, \mathcal{L}_v)$ be the labeled complex associated to v as in Lemma 2 and let $p_v = p_{(L_v,\mathcal{L}_v)} : Q_{2k} = |L_v| \to X_{(L_v,\mathcal{L}_v)}$ $(\approx S)$ be the natural projection.

DEFINITION. Interpreting the symbol as a labeled polygon, let e be an edge of Q_{2k} which has been labeled a. If $v = \cdots a \cdots a^{-1} \cdots$ or $\cdots a^{-1} \cdots a \cdots$, then we shall call e an edge of the first kind for v. Otherwise, $v = \cdots a \cdots a \cdots$ and e is called an edge of the second kind.

Step 1: Elimination of adjacent edges of the first kind.

Suppose that $e_i(2k)$ and $e_{i+1}(2k)$ are adjacent edges of the first kind and $p_v(e_i(2k)) = p_v(e_{i+1}(2k))$. By Lemma 2 there is a surface S' with symbol $w = a_1 \cdots a_{i-1} a_{i+2} \cdots a_{2k}$ or $a_2 a_3 \cdots a_{2k-1}$, depending on whether $1 \le i \le 2k - 1$ or $i = 2k$, respectively. It follows from Lemma 3 that $S \approx S' \# S^2$. This implies that $S \approx S'$. But by our inductive hypothesis S' is homeomorphic to a sphere or a connected sum of tori or projective planes and Theorem 1 is proved in this case. Therefore, we can suppose that there are no adjacent edges of the first kind.

Step 2: We may assume that $p_v(w_i(2k)) = p_v(w_j(2k))$ for all i and j.

If not all the vertices of Q_{2k} are mapped to the same point by p_v, then let i, $0 < i \le 2k$, be the smallest integer such that $p_v(w_i(2k)) \ne p_v(w_0(2k))$. Then $a_{i+1} \ne a_i^{-1}$ because Step 1 has already been carried out. Also, $a_{i+1} \ne a_i$ because $p_v(w_i(2k)) \ne p_v(w_{i-1}(2k))$. Hence $p_v(e_j(2k)) = p_v(e_{i+1}(2k))$ for some j, where $i + 1 < j \le 2k$. It follows that $a_j = a_{i+1}$ (which is the case shown in Fig. 3.15) or $a_j = a_{i+1}^{-1}$. Now we cut Q_{2k} in the 1-simplex $w_{i+1}(2k)w_{i-1}(2k)$ [labeled c in Fig. 3.15a] and paste the 2-simplex $\Delta = w_{i-1}(2k)w_i(2k)w_{i+1}(2k)$ to the edge $e_j(2k)$ along the edge $e_{i+1}(2k)$, where we send the vertex $w_i(2k)$ into the vertex $u = w_{j-1}(2k) + w_j(2k)$ [see Fig. 3.15b].

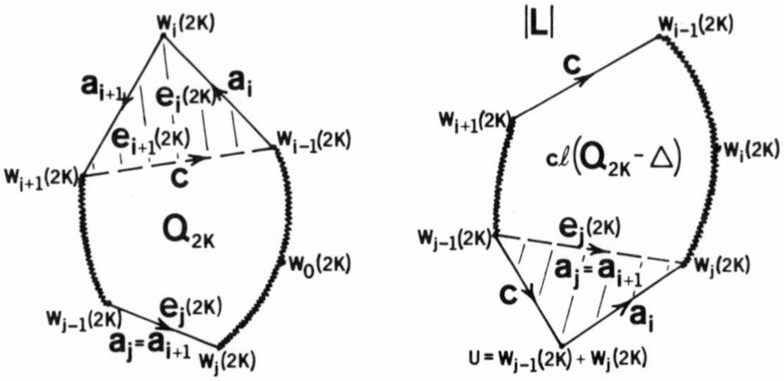

(a) (b)

FIGURE 3.15

This gives rise to a new labeled complex (L,L) with $|L| = c\ell(Q_{2k} - \Delta) \cup uw_{j-1}(2k)w_j(2k)$ and $X_{(L,\mathit{L})} \approx S$. If Step 1 is applicable to the symbol for (L,L), then Theorem 1 will have been proved by our inductive hypothesis. Otherwise, we see that the number of vertices of the polygon $|L|$ which get mapped by $p_{(L,\mathit{L})}$ to $p_{(L,\mathit{L})}(w_0(2k))$ is larger than the corresponding number for (L_v,L_v). Repeating this process a finite number of times will therefore prove Step 2.

Step 3: Making pairs of edges of the second kind adjacent.

Assume that $e_i(2k)$ and $e_j(2k)$ are two edges of the second kind for v which are not adjacent and such that $p_v(e_i(2k)) = p_v(e_j(2k))$. Let $a = a_i$. Figure 3.16 shows how we can make the edges adjacent. Cut Q_{2k} in the 1-simplex $w_i(2k)w_j(2k)$ [labeled c in Fig. 3.16a] and paste the shaded region X in Fig. 3.16a to $c\ell(Q_{2k} - X)$ along $e_j(2k)$ and $e_i(2k)$ as shown in Fig. 3.16b. The number of pairs of edges of the second kind which are not adjacent for the symbol of the resulting labeled polygon is now less than the corresponding number for v. After repeating this process at most a finite number of times we will have finished Step 3.

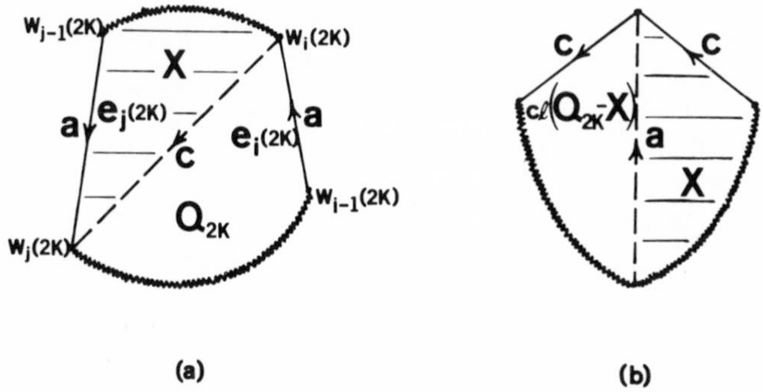

FIGURE 3.16

If v has no edges of the first kind, then the symbol v is of the form $a_1a_1a_2a_2\cdots a_ka_k$, $a_i \in A$, and Theorem 1 is proved. Assume therefore that $e_i(2k)$ and $e_j(2k)$, $0 \leq i < j \leq 2k$, are a pair of edges of the first kind for v with $a_j = a_i^{-1}$.

LEMMA 4. There is another pair of edges, $e_s(2k)$ and $e_t(2k)$, of the first kind for v with $a_t = a_s^{-1}$ and $0 \leq i < s < j < t \leq 2k$.

The proof of Lemma 4 is postponed temporarily, but assuming its validity we go on to the next step. Let $a = a_i$ and $b = a_s$.

Step 4: Collecting pairs of edges of the first kind.

Suppose that v has edges of the first kind. Then we have a picture as in Fig. 3.17a. Cut Q_{2k} in the edge labeled c and paste the two pieces together along the edges labeled b to get Fig. 3.17b. That same figure is now cut in the edge labeled d (see Fig. 3.17c) and the new two pieces are pasted together along the edges labeled a. The final result is shown in Fig. 3.17d. We can now keep repeating the process and also collect all the edges of the first kind into adjacent groups of four by Lemma 3. Thus, we finally end up with a labeled polygon associated to S whose symbol is of the form $w =$

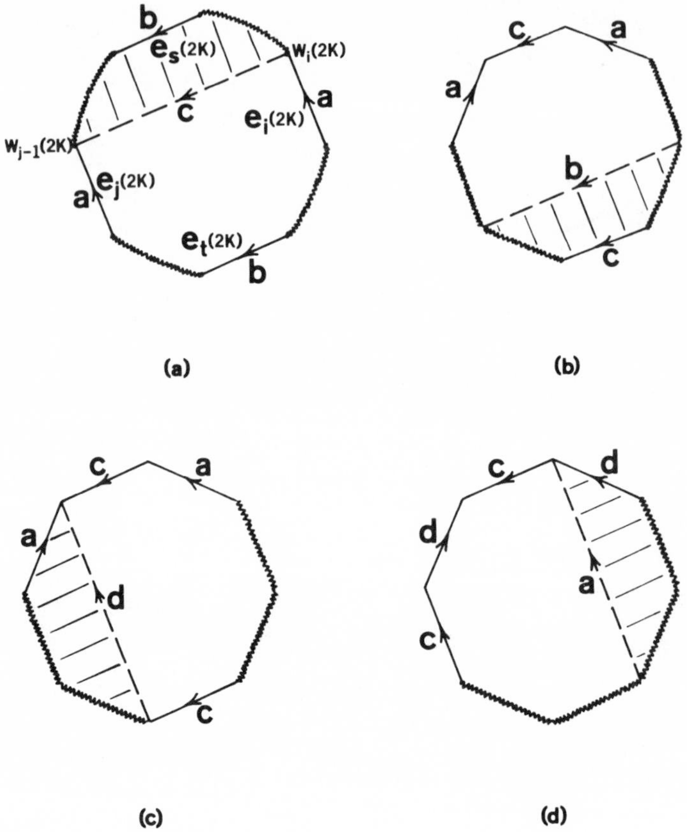

FIGURE 3.17

$a_1a_1\cdots a_ma_mb_1c_1b_1^{-1}c_1^{-1}\cdots b_nc_nb_n^{-1}c_n^{-1}$, where $m + 2n = k$ and $a_i, b_j, c_q \in A$. This concludes Step 4.

If either m or n equals 0, then we are of course finished. Assume that $m,n \geq 1$. Then we have the following, perhaps somewhat surprising, lemma.

LEMMA 5. $P^2 \# (S^1 \times S^1) \approx P^2 \# P^2 \# P^2$.

The proof of this lemma is also postponed in order to first finish the proof of Theorem 1. Lemma 5 clearly implies that we may replace the $a_m a_m b_1 c_1 b_1^{-1} c_1^{-1}$ in the symbol w by $a_m a_m a_{m+1} a_{m+1} a_{m+2} a_{m+2}$. Repeating this procedure n times will show that S admits a symbol of the type $a_1 a_1 a_2 a_2 \cdots a_{m+2n} a_{m+2n}$. Theorem 1 is proved.

Proof of Lemma 4: Let $a = a_i$ and express v in the form $AaBa^{-1}C$, where A, B, and C are strings of symbols (see Fig. 3.18a). Cut Q_{2k} in the edge labeled d separating it into two pieces X and Y (see Fig. 3.18b). If $e_i(2k)$ and $e_j(2k)$ are not separated by a pair of edges of the first kind, then B and $Aaa^{-1}C$ belong to $\mathcal{W}^*$. (Here we need the fact that Step 3 has been carried out.) This means that if we paste together the edges of the polygon Y as indicated by the labeling, then we will get a "surface with boundary." The boundary would be a circle and would come from the edge d in Y.

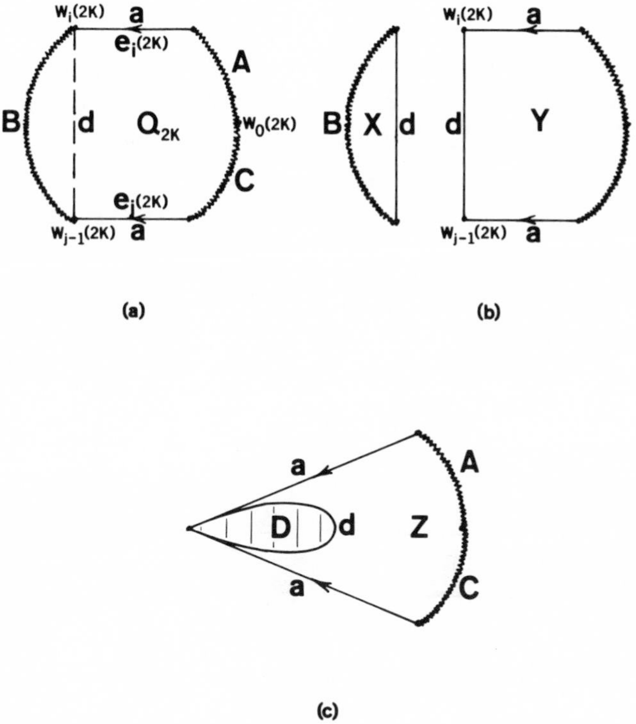

FIGURE 3.18

(Recall that Step 2 has been carried out so that the vertices $w_i(2k)$ and $w_{j-1}(2k)$ get pasted together.) In order to eliminate this boundary, let us paste together the vertices $w_i(2k)$ and $w_{j-1}(2k)$ of Y and paste a disk D inside the loop generated by the edge d. The resulting labeled polygon Z (see Fig. 3.18c) then has the property that pasting together the edges according to the labeling would produce a surface S_1 (without boundary). The symbol associated to Z is $Aaa^{-1}C$. However, since Step 2 had been carried out for v, it would also follow from our construction that all of the vertices of Z are identified to the same point x in S_1. This fact and the existence in Z of a pair of edges of the first kind which are adjacent (namely, the pair aa^{-1}) would imply that condition (3) in the definition of a surface is not satisfied at the point $x \in S_1$. Thus, we have arrived at a contradiction and Lemma 4 is proved.

Proof of Lemma 5: We shall sketch two proofs of this lemma. One is geometric and the other involves shuffling around symbols.

First, let us show that the connected sum of two projective planes is the surface called the Klein bottle one of whose symbols is $A_1A_2A_1A_2^{-1}$. From Fig. 3.19a we see that the Klein bottle K is formed from the square Q_4 by pasting together the edges $e_1(4)$ and $e_3(4)$, forming a cylinder, and then pasting together the ends of the cylinder after giving one end a 180° twist. (This is different from the torus where the ends are identified without a twist.) Unfortunately, the Klein bottle, like the projective plane, cannot be realized in $\mathbb{R}^3$. In Fig. 3.19b-d we have tried to suggest what it looks like. The right end of a cylinder is brought around, made to intersect the cylinder in X so that now the end is in the interior of the tube, and then connected to the left end. The actual Klein bottle K of course has no intersection at X, but we cannot get around that if we want to draw it in $\mathbb{R}^3$. The space K is another example of a nonorientable surface because it has only one side. It was introduced by Klein [Kle 3] in 1882.

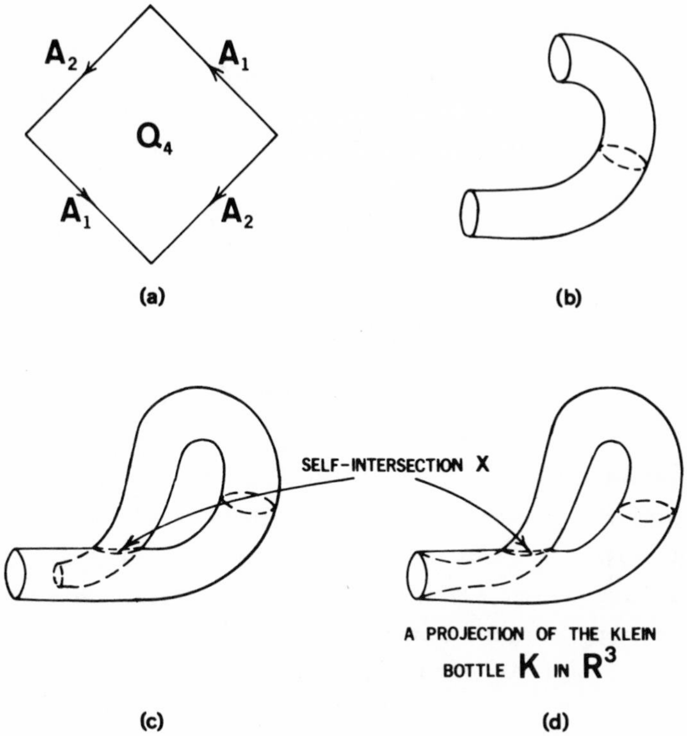

FIGURE 3.19

If we cut the Klein bottle along the lines labeled c and d in Fig. 3.20 which intersect the edges $e_1(4)$ and $e_3(4)$, we will get two Möbius strips. The shaded region in Fig. 3.20 corresponds to one of them. In other words, the Klein bottle is formed by taking two Möbius strips and pasting them together along the common edge. But this is precisely what we get with $P^2 \# P^2$ since the projective plane minus an open disk is just the Möbius strip. This proves that $K \approx P^2 \# P^2$.

Therefore, to show that $P^2 \# (S^1 \times S^1) \approx P^2 \# P^2 \# P^2$ it suffices to show that $P^2 \# (S^1 \times S^1) \approx P^2 \# K$. Let us write $P^2 = M \cup D$,

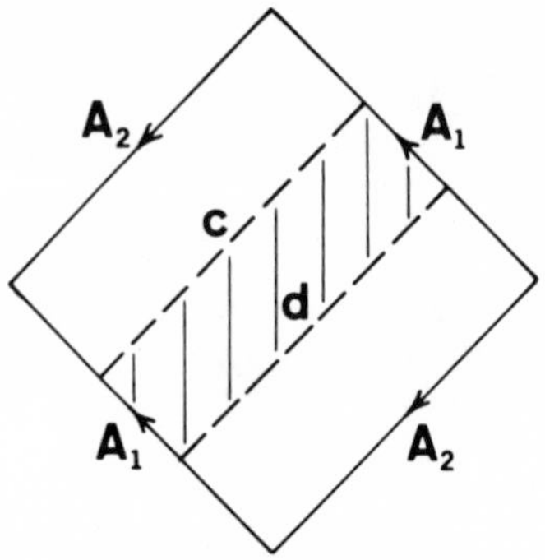

FIGURE 3.20

where M is the Möbius strip and $D = c\ell(P^2 - M)$ is a closed disk. In forming the connected sum of P^2 with $S^1 \times S^1$ or K we must cut out a disk D_1 from P^2 and we assume that $D_1 \subset M$. Figure 3.21a and b represent $S^1 \times S^1$ and K, respectively, as Q_4 with the edges appropriately identified. The D_i, $i = 2,3$, are disks. Cut $S^1 \times S^1$ and K in the lines labeled c, A_1 and d, A_1, respectively, and let X, Y, X', and Y' be the regions into which the surfaces are divided. Then $P^2 \# (S^1 \times S^1) = D \cup (M \# Y) \cup X$ and $P^2 \# K = D \cup (M \# Y') \cup X'$. But Y and Y' are just cylinders, so that $M \# Y \approx M \# Y'$ and both spaces are essentially M with two holes cut out. Now X and X' are also cylinders, and they are attached to $M \# Y$ and $M \# Y'$, respectively, along the boundaries of the holes. The only difference is that one attaching map has a twist in it (see Fig. 3.21c and d). However, since the Möbius strip is one-sided one can slide the right side of the handle in Fig. 3.21d around in M so that the resulting figure looks just like Fig. 3.21c. In other words, $(M \# Y) \cup X \approx (M \# Y') \cup X'$ and $P^2 \# (S^1 \times S^1) \approx P^2 \# K$. This finishes the first and geometric proof of Lemma 5.

The second proof, which is quicker but less geometric, is an exercise in symbol manipulation and is followed best by glancing at Fig. 3.22. We know that one symbol for $P^2 \# (S^1 \times S^1)$ is $A_1A_1A_2A_3A_2^{-1}A_3^{-1}$. First cut Q_6 in the edge labeled A_4 in Fig. 3.22a

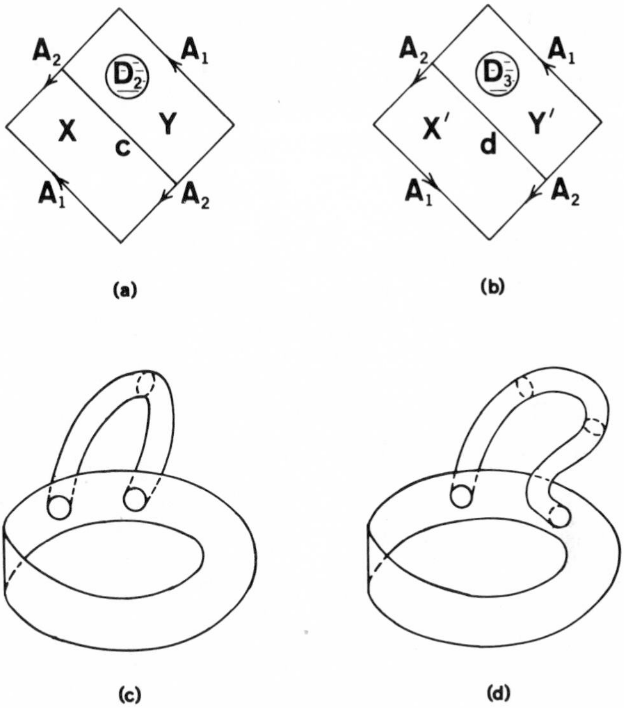

FIGURE 3.21

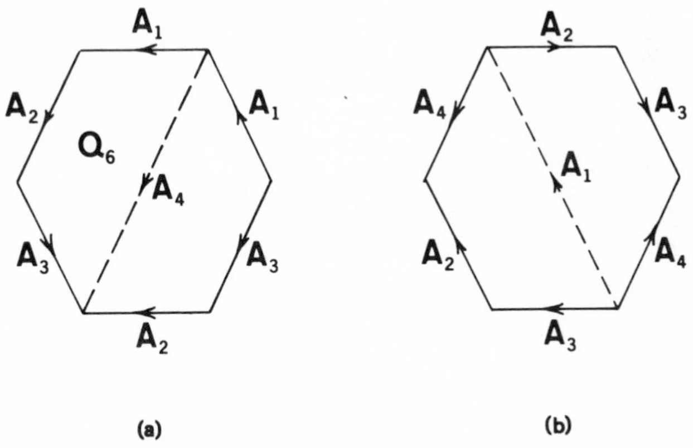

FIGURE 3.22

and paste the two pieces together along the edges labeled A_1 to get Fig. 3.22b. We obtain the symbol $A_3A_2A_4A_2A_3A_4$. Now all we have to do is apply Step 3 in the proof of Theorem 1 three times to get that $P^2 \# (S^1 \times S^1)$ admits a symbol of the type $A_1A_1A_2A_2A_3A_3$. Since $P^2 \# P^2 \# P^2$ also has this symbol we are done. This completes the proof of Lemma 5.

Next, let S be a surface and (K,φ) a proper triangulation of S. Let $n_2(K)$, $n_1(K)$, and $n_0(K)$ denote the number of 2-simplices, 1-simplices, and 0-simplices of K, respectively.

DEFINITION. The Euler characteristic of S, $\chi(S)$, is defined by $\chi(S) = n_0(K) - n_1(K) + n_2(K)$.

The Euler characteristic has been discussed already in Chapter 1 when we considered map coloring problems. In Sec. 6.6 we shall show that $\chi(S)$ is a well-defined integer which depends only on the homeomorphism class of S and not on the particular triangulation (K,φ). This is a fact that we shall assume for now. From Sec. 1.1 and Problem 1.5, we know that $\chi(S^2) = 2$ and $\chi(S^1 \times S^1) = 0$. It follows from Problem 3.14 that $\chi(P^2) = 1$.

LEMMA 6. Let S_1 and S_2 be surfaces. Then $\chi(S_1 \# S_2) = \chi(S_1) + \chi(S_2) - 2$.

Proof: We shall use the notation in the definition of # in Sec. 3.3 and let K_1, K_2, L, σ_1, and σ_2 denote the same spaces and simplices that they did there. Now L, the space which triangulates $S_1 \# S_2$, has all the simplices of K_1 and K_2 except for σ_1 and σ_2. Hence, $n_2(L) = n_2(K_1) + n_2(K_2) - 2$. But the boundary of σ_1 has been identified to the boundary of σ_2 in L. Therefore, $n_1(L) = n_1(K_1) + n_1(K_2) - 3$ and $n_0(L) = n_0(K_1) + n_0(K_2) - 3$, because we do not want to count the 0- and 1-simplices in the boundary of σ_1 and σ_2 twice. The result follows easily from these three equations.

Using Lemma 6 and Theorem 1 we can now compute the Euler characteristic of any surface. But first recall the discussion of orientability in Sec. 1.2. It is easy to see that S^2 and the connected sum of tori are orientable surfaces, whereas any connected sum of projective planes is nonorientable. (We could turn this observation around and use it to define orientability. Namely, we could say that a surface is orientable if it is homeomorphic to S^2 or a connected sum of tori and that it is nonorientable otherwise. This is not a good definition, however, because it is not intrinsic. Given an arbitrary surface, we would not be able to determine its orientability until we knew just exactly which surface it was, and this should not be necessary.) Next, let us introduce an additional bit of standard terminology and define the genus g of a surface S to equal 0 if $S \approx S^2$ and to equal n if S is homeomorphic to a connected sum of n tori or n projective planes. Table 2 summarizes what we have established about the Euler characteristic, orientability, and genus of a surface.

TABLE 2

Surface	Euler characteristic	Orientability	Genus
S^2	2	Orientable	0
$\underbrace{(S^1 \times S^1) \# \cdots \# (S^1 \times S^1)}_{n \text{ times}}$	$2 - 2n$	Orientable	n
$\overbrace{P^2 \# \cdots \# P^2}$	$2 - n$	Nonorientable	n

Proof of Theorem 2: Theorem 2 is now a direct consequence of Table 2 given the topological invariance of the Euler characteristic and orientability. The invariance of the latter is proved in Sec. 6.5.

We conclude our study of surfaces with a summary of the main results. Suppose that we are given a surface S. We can determine

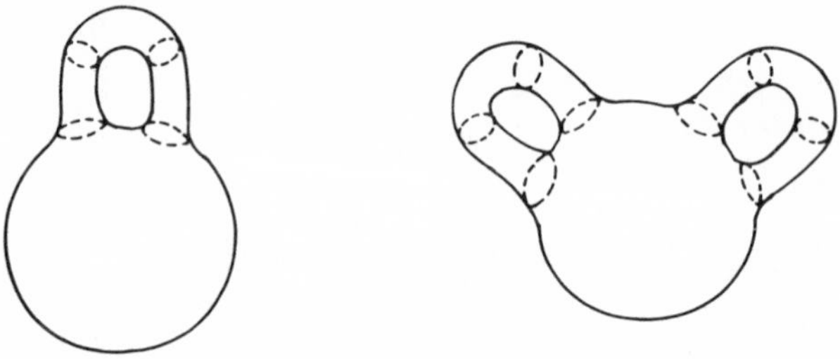

$S^1 \times S^1$ — A SURFACE OF GENUS 1

$(S^1 \times S^1) \# (S^1 \times S^1)$ — A SURFACE OF GENUS 2

FIGURE 3.23

from Table 2 just exactly which unique standard surface it is in two steps: First, decide whether or not S is orientable; and second, compute the Euler characteristic from a triangulation. Step 2 is a simple counting procedure but Step 1 is a little more subtle. We have not yet given a formal definition of orientability and shall return to that point later. Note that a surface is also completely determined by its genus and orientability and that the genus is a more geometric notion than the Euler characteristic because an orientable surface of genus g can then be thought of as a sphere with g handles (see Fig. 3.23). There is a simple formula relating the genus of a surface S to its Euler characteristic: $g = (2 - \chi)/2$ if S is orientable and $g = 2 - \chi$ otherwise.

Problems

3.12. Give a rigorous proof of Lemma 4.

3.13. By transforming the following symbols into their normal forms use Table 1 to determine which surfaces they represent:

(a) $A_1 A_2 \cdots A_n A_1^{-1} A_2^{-1} \cdots A_{n-1}^{-1} A_n$

(b) $A_1 A_2 \cdots A_n A_1^{-1} A_2^{-1} \cdots A_{n-1}^{-1} A_n^{-1}$

(*REMARK*. (a) and (b) could be used as alternative normal forms for surfaces.)

3.14. Show that $\chi(P^2) = 1$. Compute also the Euler characteristic of the Klein bottle from a triangulation.

3.15. Sketch a proof of the topological invariance of $\chi(P^2)$ similar to the way it was done for $\chi(S^2)$ and $\chi(S^1 \times S^1)$ in Sec. 1.1 and Problem 1.6. Given that the Euler characteristic of S^2, P^2, and $S^1 \times S^1$ are topological invariants, sketch a proof of the topological invariance of the Euler characteristic of an arbitrary surface using induction on the genus.

3.16. (a) Sketch a proof of the fact that the genus of an orientable surface is equal to the maximum number of pairwise disjoint simple closed curves that can be drawn on the surface without dividing it into two or more parts. (*Hint:* Cut the surface along the circles, add disks, show that the new surface must be S^2, and determine how the Euler characteristic was changed in the process.)

(b) Define the connection number of a surface to equal the maximum number of distinct simple closed curves (not necessarily disjoint) that can be drawn on the surface without dividing it into two or more parts. Find formulas relating the connection number to the Euler characteristic and genus.

3.17. Let S be a surface and C an imbedded circle in S. If we consider the strip T which consists of all points of S within a sufficiently small distance from C, then T is either a cylinder or a Möbius strip. If S is nonorientable (or one-sided), sketch a proof of the fact that we can always choose C so that $S - C$ is orientable (or two-sided) and that for such a C the strip T will be the cylinder or Möbius strip depending on whether the genus of S is even or odd, respectively. In other words, one can always obtain an orientable surface (with boundary) from a nonorientable one by cutting along a suitable circle. (*Hint:* Use Problem 3.13.)

3.18. Let (K,φ) be a proper triangulation of a surface S.

(a) Prove that

$$3n_2(K) = 2n_1(K),$$

$$n_1(K) = 3(n_0(K) - \chi(S)), \text{ and}$$

$$n_0(K) \geq (7 + \sqrt{49 - 24\chi(S)})/2.$$

(b) Apply (a) to show that

(i) if $S = S^2$, then $n_0(K) \geq 4$, $n_1(K) \geq 6$, and $n_2(K) \geq 4$;

(ii) if $S = P^2$, then $n_0(K) \geq 6$, $n_1(K) \geq 15$, and $n_2(K) \geq 10$; and

(iii) if $S = S^1 \times S^1$, then $n_0(K) \geq 7$, $n_1(K) \geq 21$, and $n_2(K) \geq 14$.

(c) Use (b.iii) to find a minimal (proper) triangulation of the torus. ("Minimal" here means the smallest number of simplices.)

3.5. BORDERED AND NONCOMPACT SURFACES

Having completed the study of surfaces, an obvious next step is to generalize this theory to the case where we allow our surfaces to have boundaries and/or where we drop the hypothesis of compactness. A good reference for the topics in this section is [Mas]. We shall only give a brief outline of the main results.

DEFINITION. A surface with boundary is a polyhedron $S \subset \mathbb{R}^n$ which admits a triangulation (K,φ) satisfying:

1. K is a 2-dimensional connected simplicial complex

2. Each 1-simplex of K is a face of at least one but not more than two 2-simplices of K

3. For every vertex $v \in K$ the distinct 2-simplices $\sigma_1, \sigma_2, \cdots$, and σ_s of K to which v belongs can be ordered in such a way that σ_i meets σ_{i+1} in precisely one 1-simplex for $1 \leq i \leq s-1$

If S is a surface with boundary and if $\partial S = \emptyset$, then S is just a surface as defined in Sec. 3.1. (Recall that the boundary of S, ∂S, is just the set $\varphi(|\partial K|)$.) We shall follow Massey [Mas] and call a surface with boundary whose boundary is nonempty a bordered surface. The classification of bordered surfaces proceeds in an analogous fashion to what we did in the previous sections for surfaces. One shows that each bordered surface can be represented as a polygon some (but not necessarily all) of whose sides have been identified in pairs and that such representatives can be put into a "normal form."

Suppose that S is a bordered surface. Observe that ∂S comes from those 1-simplices in a triangulation of S which belong to only one simplex. From this it is easy to see that each component of ∂S is homeomorphic to the circle S^1. By pasting one 2-disk along its boundary to each of these components we obtain a surface S^*, that is, $S^* = S \cup D_1 \cup D_2 \cup \cdots \cup D_k$, where k is the number of components of ∂S, D_i is a 2-disk, and $D_i \cap S = \partial D_i$ is the i-th component of ∂S. In other words, a bordered surface is nothing but a surface (without boundary) from which one has cut out a certain number of disks. Therefore, let us look at some special bordered surfaces:

(I) A sphere with k holes: This bordered surface can be represented by the labeled "polygon" in Fig. 3.24a. In order to get a representation with a connected boundary we cut along the line segments $c_1, c_2, \cdots$, and c_k shown in Fig. 3.24b. Figure 3.24c shows the result after we have straightened out the edges. The symbol

$$aa^{-1}c_1B_1c_1^{-1}c_2B_2c_2^{-1}\cdots c_kB_kc_k^{-1}$$

will be called the normal form for the sphere with k holes.

(II) The connected sum of n tori with k holes: The normal form which represents this bordered surface is

$$a_1b_1a_1^{-1}b_1^{-1}\cdots a_nb_na_n^{-1}b_n^{-1}c_1B_1c_1^{-1}\cdots c_kB_kc_k^{-1}$$

See Fig. 3.25.

(III) The connected sum of n projective planes with k holes: The normal form is

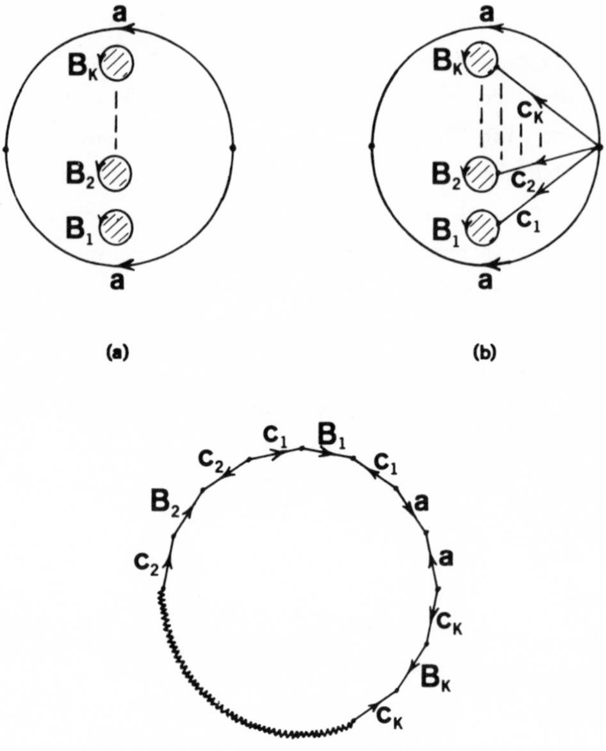

(a) (b)

(c)

FIGURE 3.24

$$a_1 a_1 \cdots a_n a_n c_1 B_1 c_1^{-1} \cdots c_k B_k c_k^{-1}$$

One arrives at this symbol by a procedure as in (I) and (II).

Having shown how certain standard bordered surfaces can be represented by symbols, one now tries to do the same in the general case. Given a bordered surface S whose boundary has k components one uses the triangulation to show that it can be represented as a

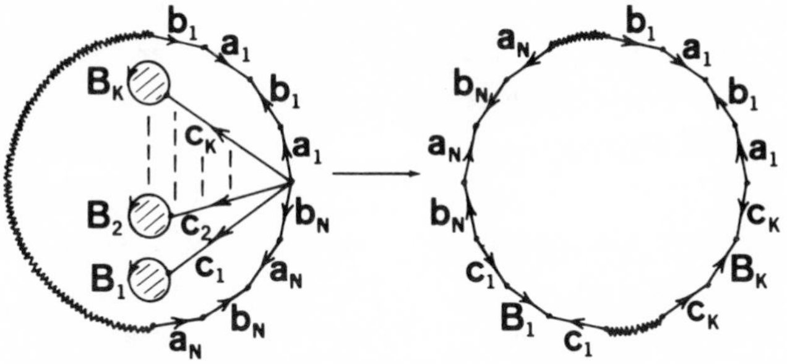

FIGURE 3.25

polygon P all of whose edges have been identified pairwise and from whose interior k disks have been removed. Using the same steps as in the proof of Theorem 1 (the only proviso being that all the cutting and pasting be done away from the holes) and then making cuts from the boundary of the resulting polygon to the boundary of the various holes it is possible to reduce P to one of the three types associated to the bordered surfaces in (I)-(III). This will prove that every bordered surface is homeomorphic to one of the three types (I)-(III). To prove that these three types are actually topologically distinct is not hard. Let S be a bordered surface with triangulation (K,φ) and define the Euler characteristic as in the case of surfaces to be the alternating sum of the number of simplices in each dimension. Let S' be the bordered surface obtained from S by removing the interior of a 2-simplex σ from $|K|$, where σ does not meet $|\partial K|$. Clearly, $\chi(S') = \chi(S) - 1$ and S' has one more boundary component that S. Therefore, a simple inductive argument shows that $\chi(S) = \chi(S^*) - k$, where k is the number of components of ∂S. Furthermore, we also have that S is orientable if and only if S* is orientable. (Orientability for bordered surfaces is defined as in the case of surfaces.) These two facts together with Theorem 2 now easily imply that the bordered surfaces in (I)-(III) are distinct. Summarizing, we get

THEOREM 3 (The Classification Theorem for Bordered Surfaces). Two bordered surfaces are homeomorphic if and only if they have the same number of boundary components, they are both orientable or nonorientable, and they have the same Euler characteristic.

Proof: See [Mas, p. 37-43] for details in the proof we have sketched above.

We can use Theorem 3 to give a nice geometric representation of bordered surfaces. First, let us describe a simple way to construct examples of bordered surfaces. We start with a disk D and then successively paste the two ends of rectangular strips to D along its boundary (see Fig. 3.26). One can paste each strip in two ways, depending on whether or not they are given a half-twist (see Fig. 3.26a and b). Furthermore, each time we attach a new strip we can attach it in a manner so that it interlocks with previous strips if we so desire (compare Fig. 3.26c and d). By computing the number of boundary components, the orientability, and the Euler characteristic of these bordered surfaces it can be shown, using Theorem 3, that all possible bordered surfaces can be obtained via the construction we have just described (see [Mas, p. 43-45]). It follows from this that all bordered surfaces can be realized in $\mathbb{R}^3$. Recall that not all surfaces (without boundary) had this property.

Finally, we come to the topic of noncompact surfaces with or without boundary. This topic is extremely complicated because the number of possibilities is so much greater than in the compact case. Some examples are:

1. Any open subset of a compact surface, or, to put it another way, the complement of an arbitrary closed set
2. The surface of a ladder with an infinite number of rungs

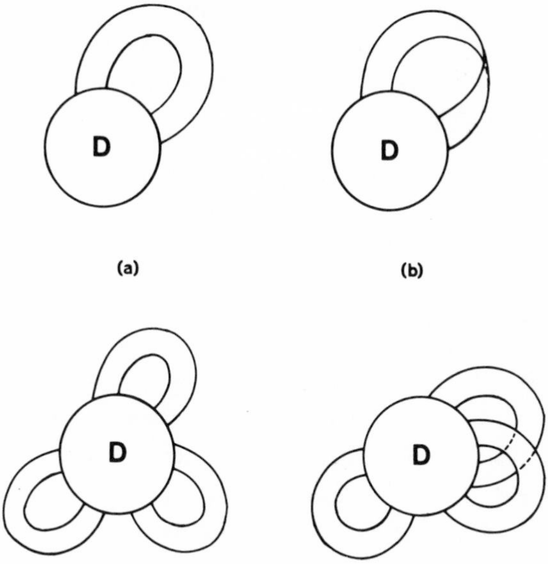

(a) (b)

(c) (d)

FIGURE 3.26

3. The surface of an infinite wire grid which is infinite in both directions

4. Infinite connected sums of surfaces

The interested reader is referred to [Mas, p. 47-51]. The classification results here are a far cry from the neat results in the compact case. Of course, any triangulation here will necessarily be infinite.

Problems

3.19. Fill in the details left out in our discussion of bordered surfaces.

3.20. (a) Which bordered surfaces can be imbedded in the plane $\mathbb{R}^2$?

(b) Which bordered surfaces can be constructed as follows: Cut holes in a disk D by removing the interior of a finite number of pairwise disjoint disks in the interior of D. Join the boundaries of certain pairs of holes by tubes, each of which may be attached in two different ways (see Fig. 3.27).

(The answer to (a) and (b) should be put in terms of the number of boundary components, the orientability, and the Euler characteristic.)

3.6. HISTORICAL COMMENTS

As we pointed out in Sec. 1.4 Riemann was the first to essentially solve the homeomorphism problem for orientable surfaces, or Riemann surfaces as they are usually called in complex function theory. Riemann discovered that one could make multiple-valued analytic functions single-valued by extending their domain from a region in $\mathbb{C}$ to a suitable surface. This was more than just a new way of looking at these functions. By showing that surfaces were their natural domains of definition, Riemann brought to the foreground the fact that analytic functions possessed certain geometric properties which could be studied using topology. His work on the theory of algebraic functions and integrals is one example which showed the importance of topological considerations. Riemann's solution to the classification problem of surfaces, including those with boundary, involved studying the possible cuts that could be made on them. In 1851 in [Riem 1] he used the method of "Querschnitte" (cuts that connect two points on a boundary) and in 1857 in [Riem 3], the method of "Rückkehr-

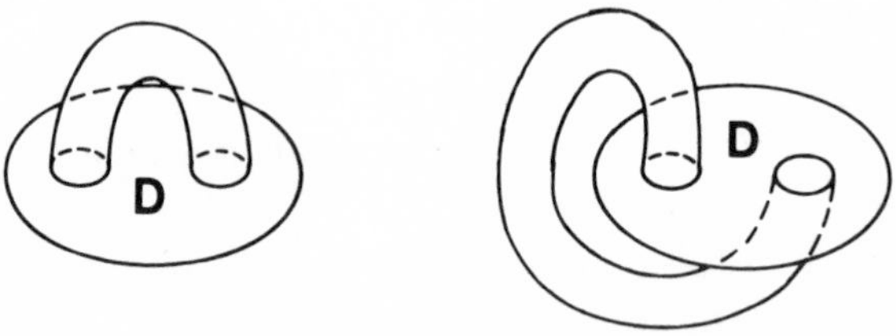

FIGURE 3.27

schnitte" (cuts along an imbedded circle). The latter led Riemann to define the genus of a surface as the maximal number of nonintersecting circles which do not separate it [see Problem 3.16(a)].

The definition of a surface which we have given can already be found in work of Möbius from 1865 (see [Möb 3]). The idea of representing a surface by its normal form is due to Listing [Lis 2] in 1861 (see also [Möb 2, Sec. 9]). Möbius [Möb 1] used this approach to solve the classification problem in 1863; however, the idea probably goes back to Gauss [G 2] who had studied the torus in this way. Other representations of surfaces as spheres with holes or handles were given by W. K. Clifford (1845-1879) [Cli] and Klein [Kle 3] in 1877 and 1882, respectively. The classification problem for surfaces basically ended with the investigations of Jordan [J 1] in 1866, L. Schläfli (1814-1895) [Schl 2] in 1872, and Dyck [Dyc 1] in 1888. However, the proofs in these early papers were geometric and usually made certain implicit assumptions (see [De-H, p. 189]). The first complete proof is attributed by some to Dehn and Heegaard [De-H] and by others to H. R. Brahana (1895-   ) [Bra].

The projective plane first appeared in the context of projective geometry. The roots of this subject can be found in work of Pappus of Alexandria (third century A.D.) and in the theory of perspectives as applied to architecture, pictorial art, and technical projections, which dates to the Renaissance and Leonardo da Vinci (1452-1516). Two other important figures in the early development of projective geometry are G. Desargues (1593-1662) and B. Pascal (1623-1662). The projective plane P^2 was invented historically in order to eliminate the distinction between lines that were parallel and lines that intersected. J. V. Poncelet (1788-1867) was the first one to define P^2, and it was his book [Pon] in 1822 that really made projective geometry into a definite branch of mathematics. In modern times the space P^2 and its higher dimensional analogues have also become important to topology and algebraic geometry.

The intrinsic definition of an n-dimensional manifold as an abstract space which locally looks like $\mathbb{R}^n$ originates with Riemann

[Riem 2] in 1854 although geometers had begun to talk about n-dimensional space even before that. In particular, higher dimensions were discussed by Cayley [Cay 1] in 1843 and H. Grassmann (1809-1877) [Gra] in 1844 (see also [Cay 4]). In 1852 Schläfli [Schl 1] had extended Euler's theorem to n-dimensional convex polyhedra.

Finally, there are some subtle points about Theorem B which should be pointed out. The obvious extension of Theorem B to arbitrary dimensions is false unless one takes orientability into account. The correct extension is the following:

THEOREM B'. Let M be a connected "oriented" n-dimensional manifold. If $\nu_1, \nu_2 : D^n \to M$ are "orientation-preserving" imbeddings, then there is a homeomorphism $h : M \to M$ such that $\nu_2 = h \circ \nu_1$.

The term "oriented" in Theorem B' is defined in Sec. 6.5. "Orientation-preserving" in this context means that D^n has been given some orientation and that the orientation induced by ν_i at the points of $\nu_i(D^n)$ agrees with the given orientation of M (see Problem 6.18). For example, an oriented surface is an orientable surface where we have made a particular choice of a rotational direction (clockwise or counterclockwise) at each point. It is assumed that these orientations are induced from the orientation at one point by walking along the surface and carrying along that given orientation. An imbedding ν of D^2 is orientation preserving if the standard counterclockwise orientation of D^2 gets carried over into the given orientation of the surface. Because every oriented surface happens to admit an "orientation-reversing" homeomorphism, Theorem B does not need any extra hypotheses; but there are higher dimensional manifolds which do not have this property. One such in three dimensions was discovered by H. Kneser (1898-1973) [Kn 2] in 1929. The best way to prove the existence of "orientation-reversing" homeomorphisms h for surfaces is to use their classification and actually construct a specific h for each connected sum of tori. Fortunately, this aspect is not really needed to prove Theorems 1 and 2. If one is a little careful in the constructions all that is necessary is a weakened version of Theorem B which is not hard to prove. On the other hand, one does need the full strength of Theorem B to prove that the connected sum operation is well defined; but this fact was also not necessary for the main steps in our proofs, so that they are not based on circular reasoning.

REMARK. One can show that the imbeddings ν_1 and ν_2 in Theorem B' are actually "isotopic." In fact, we may assume that h is "isotopic" to 1_M. (See Sec. 5.2 for a definition of "isotopic.")

Chapter 4

THE HOMOLOGY GROUPS

4.1. SOME MOTIVATION

We have stated that the central problem of algebraic topology is to classify spaces up to homeomorphism by means of computable algebraic invariants. As an example of what was meant by this we showed in the last chapter that a complete classification of surfaces can be given in terms of two such invariants, namely, the Euler characteristic and the orientability. We now begin our study of arbitrary polyhedra and ideally we would like to classify them as completely as we did surfaces. (One-dimensional (triangulable) manifolds had been classified by Möbius [Möb 1]. The only possibilities are the circle or a line segment. Möbius had also briefly considered the homeomorphism problem for compact 3-dimensional manifolds, and Heegaard later attempted to find a normal form for these in his Dissertation, Kopenhagen, 1898. However, to this day many problems remain unsolved in three dimensions.)

Consider the torus T in Fig. 4.1. The circles α and β intuitively correspond to distinct "holes" in T and it is now our object to find out how one might detect such circles. However, we are not simply looking for a list of all imbedded circles of a space. For one thing, although γ and α are distinct circles, they correspond to the same "hole" and so we would not want them listed sep-

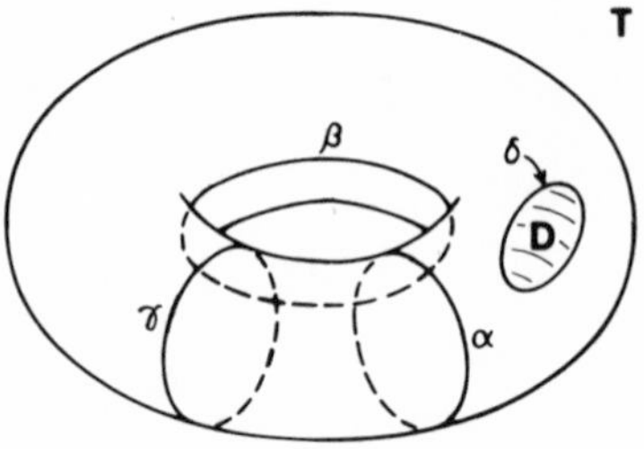

FIGURE 4.1

arately. For quite a different reason circles such as δ should not be listed at all because they do not even correspond to a "hole." Roughly speaking, therefore, what we really want is a list of equivalence classes of circles or closed paths under some appropriate equivalence relation, that is, we want a kind of factor group of all closed paths modulo the uninteresting ones. With this in mind, let us analyze in greater detail just why δ is uninteresting. One obvious property of δ is that δ bounds a disk D in T, which is not the case for α and β, and so we ought to take a closer look at what it means to bound something. For this, it is convenient to look at the situation from a simplicial point of view.

Suppose that a simplicial complex K contains two 2-simplices $\sigma_1 = v_0v_1v_2$ and $\sigma_2 = v_1v_2v_3$ as in Fig. 4.2. Let $\tau_1 = v_0v_1$, $\tau_2 = v_0v_2$, $\tau_3 = v_1v_2$, $\tau_4 = v_1v_3$, and $\tau_5 = v_2v_3$. The boundaries of σ_1, σ_2, and $\sigma_1 \cup \sigma_2$ are $\tau_1 \cup \tau_2 \cup \tau_3$, $\tau_3 \cup \tau_4 \cup \tau_5$, and $\tau_1 \cup \tau_2 \cup \tau_4 \cup \tau_5$, respectively. Is there any relationship between the boundary of $\sigma_1 \cup \sigma_2$ and the boundaries of σ_1 and σ_2 separately? Let us make some simple-minded observations in an attempt to find some algebraic relationships. We can see that in some sense τ_3, considered as part of the boundary of σ_1, and τ_3, considered as part of the boundary of σ_2, have cancelled each other. To put this idea of cancellation more in focus, let us use more suggestive notation and write "+" instead of "∪." If we want "boundary" to be an additive function, then we want the equation

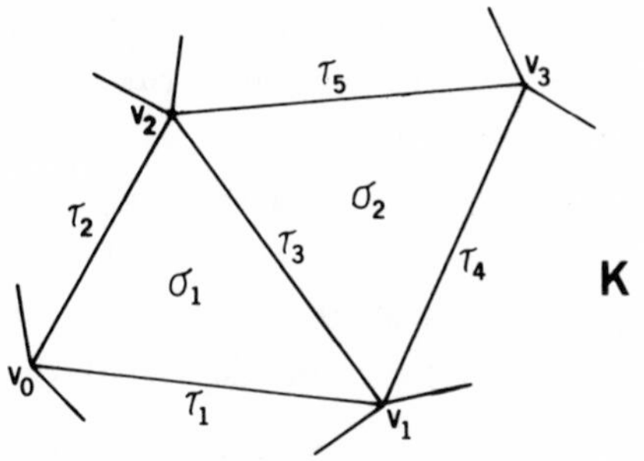

FIGURE 4.2

$$\tau_1 + \tau_2 + \tau_4 + \tau_5 = (\tau_1 + \tau_2 + \tau_3) + (\tau_3 + \tau_4 + \tau_5)$$

One way to satisfy this equation is to treat simplices as formal symbols and to identify $\tau_3 + \tau_3 = 2\tau_3$ with 0. This approach to studying the "holes" of $|K|$ will be taken up in Sec. 6.4. Right now we want to describe a second approach which initially may seem slightly more complicated but which will lead to better invariants.

To begin with, let us denote simplices by means of their vertices and think of the boundaries above as more than just a set of points but also as being a path along which one can walk in one of two possible directions. Observe that by writing v_0v_1 for the 1-simplex τ_1 we specify not only the simplex τ_1 but also an ordering of its vertices v_0 and v_1 so that we can interpret v_0v_1 as a path which is to be traversed in the direction from v_0 to v_1 rather than the other way around. Carrying this one step further, we see that to write $v_0v_1v_2$ for the 2-simplex σ_1 induces an ordering of the vertices of the 1-simplices in its boundary. Thus, we can think of the boundary of $v_0v_1v_2$ as the directed closed path $v_0v_1 + v_1v_2 + v_2v_0$ which, in Figure 4.2, corresponds to walking in a counterclockwise direction. We can interpret the boundary of $v_1v_3v_2$ and $v_0v_1v_2 + v_1v_3v_2$ in a similar fashion. Continuing this line of thought we see that the sum of the two directed closed paths which are the boundary of $v_0v_1v_2$ and $v_1v_3v_2$ corresponds to someone walking along the closed path from v_1 to v_3, from v_3 to v_2, from v_2 to v_1, from v_1 to v_2, from v_2 to v_0, and from v_0 to v_1. The only difference between this path and $v_0v_1 + v_1v_3 + v_3v_2 + v_2v_0$ is that at one point in the former we walked from v_2 to v_1 and then back again from v_1 to v_2. If we let the symbol ∂^* denote "boundary of" in the directed sense in which we are using the word "boundary" at the moment, then we shall obtain the formal equation

$$\begin{aligned} v_0v_1 + v_1v_3 + v_3v_2 + v_2v_0 &= \partial^*(v_0v_1v_2 + v_1v_3v_2) \\ &= \partial^*(v_0v_1v_2) + \partial^*(v_1v_3v_2) \\ &= (v_0v_1 + v_1v_2 + v_2v_0) + (v_1v_3 + v_3v_2 + \\ &\quad v_2v_1) \end{aligned}$$

provided that we identify $v_1v_2 + v_2v_1$ with 0. (Note that since v_1v_2 and v_2v_1 now mean different directed paths, the identification that is involved here is of a different sort than the one involved in our first approach where we referred simply to the underlying simplex, in which case v_1v_2 was equal to v_2v_1.) That we ought to do this becomes even more plausible if we formally define $v_2v_1 = -v_1v_2$, that is, $-v_1v_2$ is the path from v_1 to v_2 traversed in the opposite direction.

In the next two sections we shall show how the preceding discussion can be made rigorous, but all that we were trying to accomplish here was to bring about the realization that if one wants to study the "holes" of a space, then it is natural to want to define the idea of an orientation of a simplex and the notion of formal linear sums of simplices. These considerations will eventually lead us to important invariants of a space, called the homology groups, which are the "group" of "cycles" of the space modulo the "group" of boundaries.

4.2. THE ORIENTATION OF A SIMPLEX

Let σ be a q-simplex in $\mathbb{R}^n$. There are many ways to order the set V of vertices of σ; however, any two orderings of V differ by a permutation of V. (See Appendix B for a brief discussion of permutations.)

DEFINITION. An orientation of σ is an equivalence class of orderings of the vertices of σ, where two orderings are said to be equivalent if they differ by an even permutation of the vertices.

Note that since the inverse of an even permutation is an even permutation, we are indeed dealing with an equivalence relation. A q-simplex $\sigma = v_0v_1\cdots v_q$, $q \geq 1$, has two possible orientations: the one determined by the ordering $(v_0,v_1,\cdots,v_q)$ and the one determined by the ordering $(v_1,v_0,v_2,v_3,\cdots,v_q)$. If ν is one orientation of σ, then it will be convenient to let $-\nu$ denote the other, so that $-(-\nu) = \nu$. A 0-simplex has only one orientation.

In order to make the definition of orientation above more meaningful, we shall show how it relates to the way the term is used in everyday language. Let us begin by considering the plane $\mathbb{R}^2$. An orientation of $\mathbb{R}^2$ intuitively means that we have made a choice on what to call left and what to call right. Equivalently, it means that we have picked out one of the two possible rotations--the counterclockwise or the clockwise indicated by a curved arrow in Fig. 4.3a and b, respectively. The problem is how to do this mathematically. Well, for a start, we could write down the ordered pair (v_1,v_2), where v_1 and v_2 form a basis for $\mathbb{R}^2$ (see Fig. 4.3a), and let this suggest the idea of counterclockwise motion. The only trouble is that there are many ordered pairs of basis vectors in $\mathbb{R}^2$. For

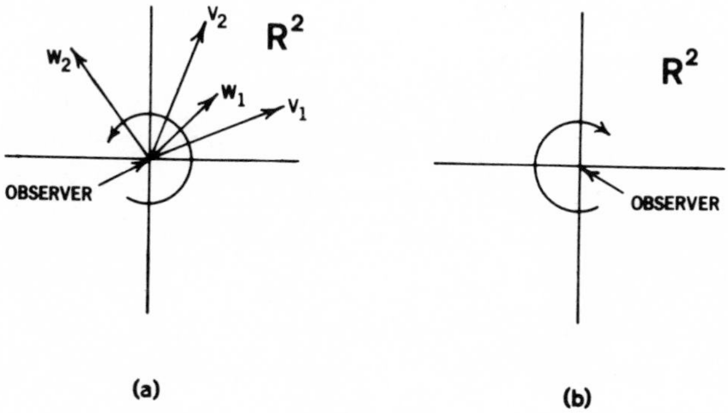

FIGURE 4.3

example, (w_1,w_2) also corresponds to counterclockwise motion in Fig. 4.3a. Therefore, we must define an appropriate equivalence relation. Given two ordered bases (v_1,v_2) and (w_1,w_2) of $\mathbb{R}^2$, express the v's in terms of the w's, say $v_i = a_{i1}w_1 + a_{i2}w_2$, for $a_{ij} \in \mathbb{R}$, and define (v_1,v_2) to be equivalent to (w_1,w_2) if the determinant of the matrix (a_{ij}) is positive. Since the v's and w's form bases we know that the matrix (a_{ij}) exists and is nonsingular. In particular, the determinant is nonzero and hence either positive or negative. This shows that we will get precisely two equivalence classes of ordered bases. Note that if $w_1 = v_2$ and $w_2 = v_1$, then

$$(a_{ij}) = \begin{pmatrix} 0 & 1 \\ 1 & 0 \end{pmatrix}$$

and the determinant of this matrix is -1. Thus, (v_1,v_2) and (v_2,v_1) are in different equivalence classes, which is exactly what we want. An orientation of $\mathbb{R}^2$ could therefore be defined to be an equivalence class of ordered bases.

In general, we could define an orientation of an n-dimensional vector space to be a similar equivalence class of ordered basis vectors. Two ordered bases $(v_1,v_2,\cdots,v_n)$ and $(w_1,w_2,\cdots,w_n)$ are said to be equivalent $[(v_1,v_2,\cdots,v_n) \sim (w_1,w_2,\cdots,w_n)]$ if the determinant of the matrix (a_{ij}) is positive, where $v_i = \sum_{j=1}^{n} a_{ij}w_j$, $a_{ij} \in \mathbb{R}$. This definition is a direct generalization of what happens in the two dimensional case, although it is no longer so easy to picture. The reader ought to take a few bases (say in $\mathbb{R}^3$) and compute the required determinants to convince himself that everything really does work out the way it should! The connection between this definition of orientation and that for simplices is that every ordered set of vertices $(v_0,v_1,\cdots,v_n)$ of an n-simplex in $\mathbb{R}^n$ gives rise to an ordered basis $(v_1 - v_0,v_2 - v_0,\cdots,v_n - v_0)$ of $\mathbb{R}^n$ and the equivalence relations are compatible (Problem 4.2). Thus, an orientation of an n-simplex in $\mathbb{R}^n$ determines a unique orientation of $\mathbb{R}^n$ and vice versa.

REMARK. What we called an orientation of $\mathbb{R}^2$ should really have been called an orientation at the origin in $\mathbb{R}^2$, because we are thinking of an observer who is standing there and deciding on what to call clockwise and counterclockwise. (A similar remark holds for $\mathbb{R}^n$.) However, the observer could equally well have been standing at another point $p \in \mathbb{R}^2$ and then an orientation at p would naturally have been an equivalence class of ordered pairs (u_1,u_2) of basis vectors at p (see Fig. 4.4). A given orientation at the origin (or at any other point of $\mathbb{R}^2$ for that matter) of course induces an orientation at every point of $\mathbb{R}^2$: The orientation determined from (v_1,v_2) is the class containing (u_1,u_2), where u_1 and u_2 are vectors at p with endpoints $p + v_1$ and $p + v_2$, respectively. It is this induced collection of orientations at points of $\mathbb{R}^2$ which ought to be called an orientation of $\mathbb{R}^2$. We make these remarks because there is a close connection between this and the discussion of orientability of surfaces in Sec. 1.2 and Chapter 3. First note that every point on a surface has a neighborhood U which is homeomorphic to the disk D^2. Identifying U with D^2 under this homeomorphism allows us to talk about an orientation of U. Then we can give the following alternate intuitive definition of an orientable surface: A surface S is ori-

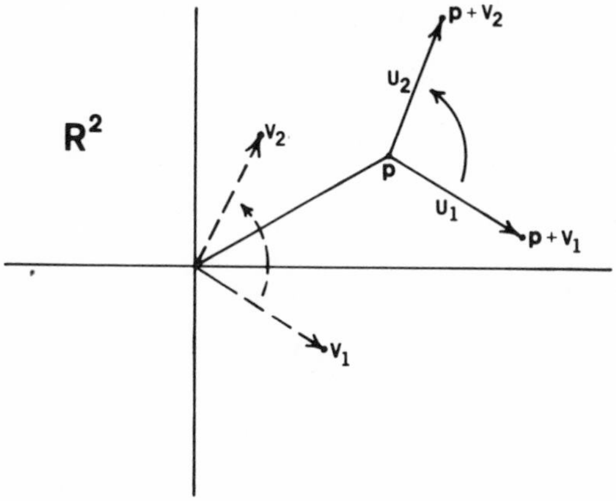

FIGURE 4.4

entable if and only if we can cover it with a collection of oriented disks $D_1, D_2, \cdots, D_k$ so that the orientations of D_i and D_j coincide on $D_i \cap D_j$ for all i and j. This definition could very easily be made completely rigorous, so that it is sound on a theoretical basis. On the other hand, it is not a good practical criterion for whether or not any given surface S is orientable because finding a covering of S by oriented disks would be no easy task in general. We shall still have to wait until Sec. 6.5 before we will get a simple computable criterion for orientability.

Let us finish this section by combining two important concepts into one.

DEFINITION. An oriented q-simplex $[\sigma]$ is a q-simplex σ together with an orientation of σ. (More precisely, $[\sigma]$ is a pair (σ,orientation of σ).) The symbol $[v_0 v_1 \cdots v_q]$ shall denote the oriented q-simplex which consists of the q-simplex with vertices $v_0, v_1, \cdots, v_q$ together with the orientation determined by the ordering $(v_0, v_1, \cdots, v_q)$. If $q = 0$, then there is only one orientation, and we shall usually write simply v_0 instead of $[v_0]$. If $q \geq 1$ and if $[\sigma]$ is an oriented q-simplex, then $-[\sigma]$ is defined to be the oriented q-simplex composed of σ together with the opposite orientation of σ, that is, if $[\sigma] = [v_0 v_1 \cdots v_q]$, then $-[\sigma] = [v_1 v_0 v_2 v_3 \cdots v_q]$.

Problems

4.1. (a) Prove that the relation $\sim$ between ordered bases of $\mathbb{R}^n$, which we defined in this section, is an equivalence relation.

(b) Which of the following ordered bases B_i of $\mathbb{R}^3$ determine the same orientation of $\mathbb{R}^3$: $B_1 = [(2,1,3), (2,2,3), (-1,1,2)]$; $B_2 = [(-2,-1,-3), (-2,-2,-3), (1,-1,-2)]$; $B_3 = [(2,2,3), (2,1,3), (-1,1,2)]$; and $B_4 = [(1,0,0), (0,1,0), (0,0,1)]$.

4.2. Let $\sigma = v_0 v_1 \cdots v_n$ be an n-simplex in $\mathbb{R}^n$ and let α be a permutation of $\{0,1,\cdots,n\}$. Show that α is even if and only if the ordered bases $(v_1 - v_0, v_2 - v_0, \cdots, v_n - v_0)$ and $(v_{\alpha(1)} - v_{\alpha(0)}, v_{\alpha(2)} - v_{\alpha(0)}, \cdots, v_{\alpha(n)} - v_{\alpha(0)})$ determine the same orientation of $\mathbb{R}^n$. Use this to show that the natural correspondence which associates $(v_0, v_1, \cdots, v_n)$ with $(v_1 - v_0, v_2 - v_0, \cdots, v_n - v_0)$ induces a well-defined bijective map between the orientations of σ and those of $\mathbb{R}^n$.

4.3. Which of the following pairs of oriented simplices are the same: (a) $[v_0v_1v_2]$ and $[v_2v_1v_0]$; (b) $[v_0v_1v_2v_3]$ and $[v_2v_0v_3v_1]$.

4.3. THE DEFINITION OF THE HOMOLOGY GROUPS

Let K be a simplicial complex and let S_q denote the set of oriented q-simplices of K.

DEFINITION. Let $0 \leq q \leq \dim K$. A q-chain of K is a function $f : S_q \to \mathbb{Z}$ with the property that if $q \geq 1$, then $f(-[\sigma^q]) = -f([\sigma^q])$ for every $[\sigma^q] \in S_q$. The set of all q-chains of K is denoted by $C_q(K)$.

For $f, g \in C_q(K)$ define a map $f + g : S_q \to \mathbb{Z}$ by $(f + g)([\sigma^q]) = f([\sigma^q]) + g([\sigma^q])$. It is easy to see that $f + g \in C_q(K)$ and that the operation $+$ makes $C_q(K)$ into an abelian group. The additive identity for $+$ is the zero function $0 : S_q \to \mathbb{Z}$, where $0([\sigma^q]) = 0$ for all $[\sigma^q] \in S_q$. The additive inverse of any $f \in C_q(K)$, denoted by $-f$, is defined by the formula $(-f)([\sigma^q]) = -(f([\sigma^q]))$ for all $[\sigma^q] \in S_q$. Finally, for notational convenience we extend the definition of $C_q(K)$ to all values of q.

DEFINITION. If $q < 0$ or $q > \dim K$, define $C_q(K) = 0$.

THEOREM 1. For all q, $(C_q(K), +)$ is an abelian group called the group of q-chains of K. The "vector" $C_\#(K) = (\cdots, C_{-1}(K), C_0(K), C_1(K), \cdots)$ is called the (oriented) chain complex of K.

NOTE. The only groups $C_q(K)$ which are interesting to us of course are those for which $0 \leq q \leq \dim K$. In particular, those with negative indices will be completely ignored in computations.

Next, let us describe the elements of $C_q(K)$ in more manageable and descriptive terms. For each oriented q-simplex $\alpha = [\sigma^q] \in S_q$ define a q-chain $\alpha_F \in C_q(K)$ by $\alpha_F(\beta) = 0$ for each $\beta \in S_q$ with $\beta \neq \pm\alpha$ and $\alpha_F(\epsilon\alpha) = \epsilon$ for $\epsilon = \pm 1$. We are going to show that $C_q(K)$ is generated by these "elementary" q-chains α_F. In fact, suppose we pick one particular orientation for each q-simplex of K and let S_q^+ be the collection of these oriented q-simplices. (The best systematic way to pick an orientation for each simplex of K is to order the vertices of K once and for all and then to take the induced orientations.) More precisely, S_q^+ is any subset of S_q with the following property: if $q = 0$, then $S_q^+ = S_q$, and if $q \geq 1$, then, for any $\beta \in S_q$, either β or $-\beta$ belongs to S_q^+ but not both.

LEMMA 1. $C_q(K) = \oplus \sum_{\alpha \in S_q^+} \mathbb{Z}\alpha_F$.

NOTE. Appendix B summarizes some of the basic notation and results about abelian groups. In particular, $\mathbb{Z}\alpha_F$ means the subgroup of $C_q(K)$ of all integer multiples of the element α_F and $\oplus$ denotes the direct sum operation. Observe that $\mathbb{Z}\alpha_F \cap \mathbb{Z}\beta_F = 0$ if $\alpha \neq \pm\beta$.

Proof: Let $f \in C_q(K)$. We must show that f can be written uniquely in the form $\sum_{\alpha \in S_q^+} n_\alpha \alpha_F$, where $n_\alpha \in \mathbb{Z}$.

(a) Uniqueness: It suffices to show that if $\sum_{\alpha \in S_q^+} n_\alpha \alpha_F = 0$ for some $n_\alpha \in \mathbb{Z}$, then $n_\alpha = 0$ for all $\alpha \in S_q^+$. This follows from the fact that if $\beta \in S_q^+$, then

$$\begin{aligned} 0 &= \Big(\sum_{\alpha \in S_q^+} n_\alpha \alpha_F \Big)(\beta) \\ &= \sum_{\alpha \in S_q^+} n_\alpha \alpha_F(\beta) \\ &= n_\beta \end{aligned}$$

(b) Representability: Let $n_\alpha = f(\alpha)$ and set $x = \sum_{\alpha \in S_q^+} n_\alpha \alpha_F$.

We shall show that $f = x$. Let $\beta \in S_q$.

Case (i): $\beta \in S_q^+$. Then

$$x(\beta) = \sum_{\alpha \in S_q^+} n_\alpha \alpha_F(\beta) = n_\beta = f(\beta)$$

Case (ii): $\beta \in S_q - S_q^+$. Then

$$x(\beta) = x(-(-\beta)) = -x(-\beta) = -f(-\beta) = f(\beta)$$

where the third equality sign follows from Case (i) and the fact that $-\beta \in S_q^+$. Incidentally, our argument here shows that any $g \in C_q(K)$ is completely determined by what it does on S_q^+.

We have now shown that $x(\beta) = f(\beta)$ for all $\beta \in S_q$, which finishes the proof of (b) and also that of Lemma 1.

Because the map $\mathbb{Z} \to \mathbb{Z}\alpha_F$ which sends n to $n\alpha_F$ is clearly an isomorphism, it follows from Lemma 1 that $C_q(K)$ is isomorphic to a free group which is a direct sum of as many copies of $\mathbb{Z}$ as there are q-simplices in K. (By convention, a sum of 0 copies of a group is 0.) Thus, if we identify an oriented q-simplex $[\sigma^q]$ with the map $[\sigma^q]_F$, which we shall do from now on, then it is obvious that $C_q(K)$ merely makes mathematically precise the notion, referred to earlier, of "formal linear combinations" of oriented q-simplices of K, where one orientation has been picked for each q-simplex σ^q. In making computations and in using our intuition with $C_q(K)$ we shall invariably treat the elements of $C_q(K)$ as such formal sums; however, note that the actual definition of $C_q(K)$ depends only on K, not on any particular choice of orientations.

Next, we formalize the notion of the boundary of a q-chain.

DEFINITION. The boundary map

$$\partial_q : C_q(K) \to C_{q-1}(K)$$

is defined as follows:

(i) If $1 \leq q \leq \dim K$, then ∂_q is the unique homomorphism with the property that

$$\partial_q([v_0v_1\cdots v_q]) = \sum_{i=0}^{q} (-1)^i[v_0\cdots\hat{v}_i\cdots v_q]$$

for each oriented q-simplex $[v_0v_1\cdots v_q]$ of K, where "$\hat{v}_i$" denotes the fact that the i-th vertex v_i has been deleted; and

(ii) if $q \leq 0$ or $q > \dim K$, then ∂_q is defined to be the zero homomorphism.

LEMMA 2. The maps ∂_q are well-defined homomorphisms and $\partial_{q-1} \circ \partial_q = 0$, for all q.

Proof: Let ν be an orientation of a q-simplex σ and let τ be a (q - 1)-dimensional face of σ. Suppose that $\sigma = v_0v_1\cdots v_q$, $\tau = v_0\cdots\hat{v}_i\cdots v_q$, and ν is the equivalence class of the ordering $\xi = (v_0,v_1,\cdots,v_q)$. Let μ_ξ denote the orientation of τ which is the equivalence class of the ordering $(v_0,\cdots,\hat{v}_i,\cdots,v_q)$.

CLAIM. The orientation $\hat{\mu}_\xi = (-1)^i\mu_\xi$ of τ depends only on ν and not on the particular ordering ξ of the vertices of σ which was chosen.

First, let us consider the effect on μ_ξ when we pass from the ordering ξ to the ordering $\xi' = (v_0,\cdots,v_t,\cdots,v_s,\cdots,v_q)$, $s < t$, which corresponds to interchanging two vertices v_s and v_t. If $s \neq i \neq t$, then all that has happened is that the same two vertices of τ have been interchanged, so that $\mu_{\xi'} = -\mu_\xi$ and $\hat{\mu}_{\xi'} = (-1)^i\mu_{\xi'} = -(-1)^i\mu_\xi = -\hat{\mu}_\xi$. If $s = i$, then the new ordering of the vertices of τ is $(v_0,\cdots,v_{i-1},v_t,v_{i+1},\cdots,v_{t-1}v_{t+1},\cdots,v_q)$ and $\hat{\mu}_{\xi'} = (-1)^t\mu_{\xi'} = (-1)^t(-1)^{t-i-1}\mu_\xi = -(-1)^i\mu_\xi = -\hat{\mu}_\xi$. A similar equation holds if $t = i$. Therefore, in all cases, interchanging two vertices of ξ results in a change of sign of $\hat{\mu}_\xi$. Since ξ is well-defined up to an even permutation of the vertices and every such permutation is the composition of an even number of transpositions, the claim is proved.

DEFINITION. The orientation $\hat{\mu}_\xi$ of τ is called the orientation of τ induced by the orientation ν of σ.

It follows from the claim that ∂_q associates a well-defined "boundary" (q - 1)-chain $\sum_{i=0}^{q} (-1)^i [v_0 \cdots \hat{v}_i \cdots v_q]$ to each oriented q-simplex $[v_0 v_1 \cdots v_q]$. This (q - 1)-chain is exactly what we would like the (oriented) boundary of the oriented q-simplex to be on intuitive grounds and justifies our calling the maps ∂_q "boundary maps." For example, the oriented 2-simplex $[v_0 v_1 v_2]$ gives rise to the oriented boundary $\partial_2([v_0 v_1 v_2]) = [v_1 v_2] - [v_0 v_2] + [v_0 v_1] = [v_1 v_2] + [v_2 v_0] + [v_0 v_1]$. Finally, recall that in order to define a homomorphism on a free group it suffices to specify it on a basis (see Theorem 5 in Appendix B), so that the first part of Lemma 2 is proved.

In order to show that the homomorphism $\partial_{q-1} \circ \partial_q : C_q(K) \to C_{q-2}(K)$ is the 0-map, it suffices to show that $\partial_{q-1} \circ \partial_q([v_0 v_1 \cdots v_q]) = 0$ for every oriented q-simplex $[v_0 v_1 \cdots v_q]$ of K, since these generate $C_q(K)$. Now

$$\begin{aligned}
\partial_{q-1} \circ \partial_q([v_0 v_1 \cdots v_q]) &= \partial_{q-1}\Big(\sum_{i=0}^{q} (-1)^i [v_0 \cdots \hat{v}_i \cdots v_q] \Big) \\
&= \sum_{i=0}^{q} (-1)^i \partial_{q-1}([v_0 \cdots \hat{v}_i \cdots v_q]) \\
&= \sum_{i=0}^{q} (-1)^i \Big(\sum_{j=0}^{i-1} (-1)^j [v_0 \cdots \hat{v}_j \cdots \hat{v}_i \cdots v_q] \\
&\qquad + \sum_{j=i+1}^{q} (-1)^{j-1} [v_0 \cdots \hat{v}_i \cdots \hat{v}_j \cdots v_q] \Big)
\end{aligned}$$

However, each term $[v_0 \cdots \hat{v}_s \cdots \hat{v}_t \cdots v_q]$, where $s < t$, appears twice in the sum above, namely, once with coefficient $(-1)^{i+j-1}$ when $i = s$ and $j = t$, and the second time with coefficient $(-1)^{i+j}$ when $j = s$ and $i = t$. Thus the terms in the sum cancel pairwise and the lemma is proved.

Next, we introduce some basic notation and terminology.

DEFINITION. Let $c \in C_q(K)$. If $\partial_q(c) = 0$, then we shall call c a q-cycle of K. If $c = \partial_{q+1}(c')$ for some $c' \in C_{q+1}(K)$, then we shall call c a q-boundary of K. The set of q-cycles and q-boundaries of K will be denoted by $Z_q(K)$ and $B_q(K)$, respectively.

Both $Z_q(K)$ and $B_q(K)$ are subgroups of $C_q(K)$ because $Z_q(K) = \operatorname{Ker} \partial_q$ and $B_q(K) = \operatorname{Im} \partial_{q+1}$. The importance of Lemma 2 lies in the fact that it shows $B_q(K)$ to be a subgroup of $Z_q(K)$. Thus, $B_q(K) \subset Z_q(K) \subset C_q(K)$ and it makes sense to talk about the quotient group of q-cycles modulo q-boundaries.

DEFINITION. The q-th homology group of K, $H_q(K)$, is defined by

$$H_q(K) = Z_q(K)/B_q(K)$$

NOTE. As in the case of $C_q(K)$, it is notationally convenient to have the groups $B_q(K)$, $Z_q(K)$, and $H_q(K)$ defined for all values of q, even though they are interesting only when $0 \leq q \leq \dim K$. Negative values of q will be ignored in computations.

With the definition of the homology groups, we have really obtained important algebraic invariants for polyhedra. However, it will take some time to show that this is the case because we have to prove some basic properties of the homology groups first. Although we could have used other regular figures as building blocks for spaces, we can now give two reasons for the choice of simplices, namely, with other figures both the orientation and the important maps ∂_q would be more complicated to define.

EXAMPLES

1. $K = \{v_0\}$: In this case, $C_q(K) = 0$ if $q > 0$, $C_0(K) = \mathbb{Z}v_0$, and $B_0(K) = 0$. Therefore,

$$H_q(K) = Z_q(K) = B_q(K) = 0 \quad \text{if} \quad q > 0$$

$$H_0(K) = Z_0(K) = C_0(K) \approx \mathbb{Z}$$

2. $K = \partial(\overline{v_0v_1v_2})$ (see Fig. 4.5a): We have that $C_q(K) = 0$ for $q > 1$, $C_1(K) = \mathbb{Z}[v_0v_1] \oplus \mathbb{Z}[v_1v_2] \oplus \mathbb{Z}[v_2v_0]$, and $C_0(K) = \mathbb{Z}v_0 \oplus \mathbb{Z}v_1 \oplus \mathbb{Z}v_2$. Right away we can say that

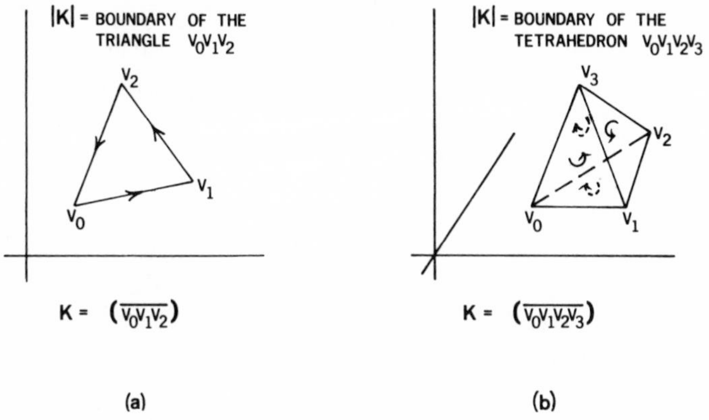

FIGURE 4.5

$$H_q(K) = Z_q(K) = B_q(K) = 0 \quad \text{if} \quad q > 1$$

Suppose that $x = a[v_0v_1] + b[v_1v_2] + c[v_2v_0] \in C_1(K)$ and $\partial_1(x) = 0$. Then $0 = \partial_1(x) = (av_1 - av_0) + (bv_2 - bv_1) + (cv_0 - cv_2) = (c - a)v_0 + (a - b)v_1 + (b - c)v_2$. It follows that $c - a = a - b = b - c = 0$, that is, $a = b = c$. Hence, $Z_1(K) = \mathbb{Z}([v_0v_1] + [v_1v_2] + [v_2v_0]) \approx \mathbb{Z}$. Since $B_1(K) = 0$, we get that

$$H_1(K) = Z_1(K) \approx \mathbb{Z}$$

DEFINITION. Let L be a simplicial complex. If $x,y \in C_q(L)$, then we shall say that x is homologous to y, or that x and y are homologous, and write $x \sim y$, provided that $x - y = \partial_{q+1}(w)$ for some $w \in C_{q+1}(L)$. Also, if $z \in Z_q(L)$, let $[z] = z + B_q(L) \in H_q(L)$. The coset $[z]$ is called the homology class in $H_q(L)$ determined by z.

Clearly, $\sim$ is an equivalence relation in $C_q(L)$ and two q-cycles $x,y \in Z_q(L) \subset C_q(L)$ determine the same homology class in $H_q(L)$ if and only if $x \sim y$.

Returning to Example 2, we still have to compute $H_0(K)$. Now $Z_0(K) = C_0(K)$ and $v_0 \sim v_1 \sim v_2$ because $\partial_1([v_0v_1]) = v_1 - v_0$ and $\partial_1([v_1v_2]) = v_2 - v_1$. This shows that $H_0(K)$ is generated by the homology class $[v_0]$. If $n[v_0] = 0$, then $nv_0 = \partial_1(x)$ for some $x = a[v_0v_1] + b[v_1v_2] + c[v_2v_0] \in C_1(K)$. Next, compute the coefficients of the vertices v_0, v_1, and v_2 in $\partial_1(x)$ and equate them to the same coefficients in $nv_0 + 0v_1 + 0v_2$. It follows easily that $n = 0$. This shows that the map from $\mathbb{Z}$ to $H_0(K)$ which sends k to $k[v_0]$ is an isomorphism, and hence

$$H_0(K) \approx \mathbb{Z}$$

3. $K = \partial(\overline{v_0v_1v_2v_3})$ (see Fig. 4.5b): Again we have that $C_q(K) = 0$ for $q > 2$, and so

$$H_q(K) = 0 \quad \text{for} \quad q > 2$$

On the other hand, $C_2(K) = \mathbb{Z}[v_0v_1v_2] \oplus \mathbb{Z}[v_0v_1v_3] \oplus \mathbb{Z}[v_1v_2v_3] \oplus \mathbb{Z}[v_0 v_2v_3]$, $C_1(K) = \mathbb{Z}[v_0v_1] \oplus \mathbb{Z}[v_1v_2] \oplus \mathbb{Z}[v_0v_2] \oplus \mathbb{Z}[v_0v_3] \oplus \mathbb{Z}[v_3v_2] \oplus \mathbb{Z}[v_1v_3]$, and $C_0(K) = \mathbb{Z}v_0 \oplus \mathbb{Z}v_1 \oplus \mathbb{Z}v_2 \oplus \mathbb{Z}v_3$. Since $B_2(K) = 0$ and $\partial_0 = 0$, it follows that $H_2(K) = Z_2(K)$ and $Z_0(K) = C_0(K)$. To compute $Z_2(K)$ we proceed as in Example 2. Let $z = a[v_0v_1v_2] + b[v_0 v_1v_3] + c[v_1v_2v_3] + d[v_0v_2v_3] \in Z_2(K)$. By definition $\partial_2(z) = 0$. Computing the coefficients of the 1-simplices in $\partial_2(z)$ and setting them equal to 0 implies that $a = -b = -c = d$. In other words, $Z_2(K) = \mathbb{Z}([v_0v_2v_1] + [v_0v_1v_3] + [v_1v_2v_3] + [v_0v_3v_2])$, and so

$$H_2(K) \approx \mathbb{Z}$$

In principle $H_1(K)$ could also be computed by explicitly calculating $Z_1(K)$ and $B_1(K)$ as above, but it is obvious that such calculations are becoming progressively more complicated and tedious. A simpler procedure is called for. Let $z \in C_1(K)$. If the oriented 1-simplex $[v_2v_3]$ appears in z with some nonzero coefficient a, then replace $a[v_2v_3]$ by $a([v_2v_1] + [v_1v_3])$ to get a 1-chain z_1 in which $[v_2v_3]$ does not appear. Note that $z \sim z_1$ because $[v_2v_3] = [v_2v_1] + [v_1v_3] + \partial_2([v_1v_2v_3])$. Similarly, we may replace any occurrence of $[v_1v_2]$ in z_1 by $[v_1v_0] + [v_0v_2]$, so that we finally get a $z_2 \in C_1(K)$ such that $z \sim z_2$ and z_2 does not contain either $[v_2v_3]$ or $[v_1v_2]$. Now assume that z actually was a 1-cycle. Then z_2 will be a 1-cycle and $[v_0v_2]$ cannot appear in z_2 either, otherwise the coefficient of v_2 in $\partial_1(z_2)$ would not vanish. These facts together with the work we did in Example 2 show that z_2 must be a multiple of $x = [v_0v_1] + [v_1v_3] + [v_3v_0]$. Thus $H_1(K)$ is generated by $[x]$. But $x = \partial_2([v_0v_1v_3])$ and so $[x] = 0$. This proves that

$$H_1(K) = 0$$

Finally, to compute $H_0(K)$ one also proceeds as in Example 2: One shows that $H_0(K)$ is generated by $[v_0]$ and that $n[v_0] = 0$ implies that $n = 0$. In other words,

$$H_0(K) \approx \mathbb{Z}$$

in this example too.

4. K is the 2-dimensional complex triangulating the torus indicated in Fig. 4.6: First of all, since there are no q-simplices for $q > 2$,

$$H_q(K) = 0 \quad \text{for} \quad q > 2$$

Orient the 2-simplices in K as indicated by the circular arrows in Fig. 4.6 and let Σ denote the sum of these oriented simplices in

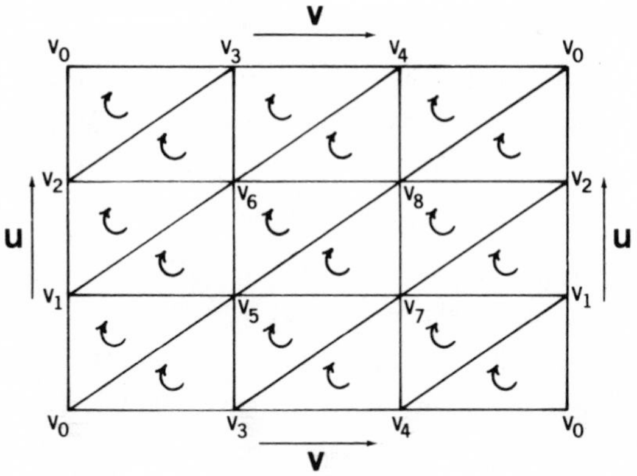

$|K| \approx S^1 \times S^1$

FIGURE 4.6

$C_2(K)$. For example, $[v_0v_2v_8]$ appears in the sum Σ. It is easy to see that $\partial_2(\Sigma) = 0$ because if τ^1 is a 1-simplex and τ^1 is a face of the two 2-simplices σ_1^2 and σ_2^2, then the coefficient of $[\tau^1]$ in $\partial_2([\sigma_1^2])$ is the negative of the coefficient of $[\tau^1]$ in $\partial_2([\sigma_2^2])$ so that the total coefficient of $[\tau^1]$ in $\partial_2(\Sigma)$ is zero. On the other hand, let $z \in Z_2(K)$. Then the only way that the coefficient of $[\tau^1]$ vanishes is if $[\sigma_1^2]$ and $[\sigma_2^2]$ appear in z with the same multiplicity. Since this is true for an arbitrary pair of adjacent 2-simplices in K, it follows that $z = a\Sigma$ for some integer a. This proves that

$$H_2(K) = Z_2(K) = \mathbb{Z}[\Sigma] \approx \mathbb{Z}$$

Next, let $z \in C_1(K)$.

CLAIM. $z \sim z' = a_1[v_0v_3] + a_2[v_3v_4] + a_3[v_4v_0] + a_4[v_0v_1] + a_5[v_1v_2] + a_6[v_2v_0] + a_7[v_3v_6] + a_8[v_6v_5] + a_9[v_4v_8] + a_{10}[v_8v_7]$ for some $a_i \in \mathbb{Z}$.

The claim follows from the fact that we may successively replace $[v_2v_3]$ by $[v_2v_6] + [v_6v_3]$, $[v_2v_6]$ by $[v_2v_1] + [v_1v_6]$, and so on. This is similar to what was done in Example 3 and we do not change the homology class of the element with each replacement. If we now also assume that $z \in Z_1(K)$, then $z' \in Z_1(K)$ and $a_i = 0$ for $i = 7,8,9$, and 10; otherwise one or more of the vertices v_5, v_6, v_7, and v_8 would appear in $\partial_1(z')$ with a nonzero coefficient. Furthermore, if we actually compute $\partial_1(z')$ and set it equal to zero, it is easy to check that we also must have $a_1 = a_2 = a_3$ and $a_4 = a_5 = a_6$. Set $u = [v_0v_1] + [v_1v_2] + [v_2v_0]$ and $v = [v_0v_3] + [v_3v_4] + [v_4v_0]$. What we have shown is that any cycle $z \in Z_1(K)$ is homologous to $au + bv \in Z_1(K)$ for some $a,b \in \mathbb{Z}$; in other words, $H_1(K)$ is generated by $[u]$ and $[v]$. Suppose that $a[u] + b[v] = 0 \in H_1(K)$. Then $au + bv = \partial_2(x)$ for some $x \in C_2(K)$. Let $[\sigma]$ be any oriented 2-simplex which appears in x. At least one edge of σ, say e, is different from the edges appearing in u and v, and e will also be an edge in some other 2-simplex τ. If we assume that σ and τ have been oriented as before in the computation of $H_2(K)$ and as indicated by the arrows in Fig. 4.6, then the only way that the coefficient of e will vanish in $\partial_2(x)$ is if the coefficients of $[\sigma]$ and $[\tau]$ in x are equal. It follows easily from this that $x = k\Sigma$ for some $k \in \mathbb{Z}$, where $\Sigma \in C_2(K)$ is as above. But then $\partial_2(x) = \partial_2(k\Sigma) = k\partial_2(\Sigma) = k \cdot 0 = 0$ and so $a = b = 0$. Thus,

$$H_1(K) \approx \mathbb{Z} \oplus \mathbb{Z}$$

under the map which sends $(a,b) \in \mathbb{Z} \oplus \mathbb{Z}$ to $a[u] + b[v] \in H_1(K)$.

The only group which remains to be computed is $H_0(K)$, but one can show that

$$H_0(K) \approx \mathbb{Z}$$

just like in Examples 2 and 3. This finishes Example 4.

If we look over the results of the four examples that we have just given, we see that the homology groups of a simplicial complex K do indeed appear to detect "holes" in the underlying space $|K|$. The "holes" correspond in a natural way to the generators of the various groups. For example, generators of the first homology group represent closed edge paths in $|K|$ which do not "bound." The group $H_0(K)$, which happened to be isomorphic to $\mathbb{Z}$ in all of the examples above, also has a nice geometric interpretation. In Problem 4.5 you are asked to show that the rank of $H_0(K)$ equals the number of components of K [and hence of $|K|$ by Problem 2.14(c)]. Note that S^0 consists of two points, so that $H_0(K)$ detects the number of 0-dimensional "spherical holes" in $|K|$, where by a k-dimensional spherical hole we simply want to mean an imbedded k-sphere in $|K|$ which cannot be contracted (or shrunk) to a point on $|K|$. It should be cautioned however that k-cycles or "homological holes" do not necessarily correspond to such geometric spherical holes in general. It is also possible to find a complex K and an imbedded k-sphere in $|K|$ whose associated k-cycle bounds a (k + 1)-chain and yet the sphere neither bounds a (k + 1)-disk in $|K|$ nor can it be contracted to a point in $|K|$. The trouble is that taking the boundary of chains is basically a formal algebraic operation, even though it was motivated by geometric considerations and does give the usual "oriented boundary" when applied to an oriented q-simplex or similar simple configurations.

At any rate, from the evidence that we have so far we should be pretty excited about homology groups because they are invariants which detect geometrical properties. Furthermore, and this is important to emphasize, their computation is a straightforward, though often complicated, combinatorial problem in the sense that one can forget about the geometry of the situation. Homology groups are really a notion that is associated to an abstract simplicial complex, because all that is involved is the definition of an appropriate operator ∂_q on linear combinations of formal symbols $[v_0v_1\cdots v_q]$. One could easily program a computer so that it would be able to compute the homology groups of an arbitrary simplicial complex (see Appendix C). This is the reason that the study of topology which deals with simplicial complexes is called combinatorial topology. In later chapters we shall see just how well homology groups distinguish between spaces.

Finally, let us introduce some standard terminology. If K is a simplicial complex, we know that $C_q(K)$ is a finitely generated free abelian group on as many generators as there are q-simplices in K. Since subgroups and quotient groups of finitely generated groups are again finitely generated, it follows that $H_q(K)$ is finitely generated. Therefore, by the fundamental theorem about such groups (see Theorem 3 in Appendix B), we can write

$$H_q(K) \approx F_q \oplus T_q$$

where F_q is a free group and T_q is the torsion subgroup of $H_q(K)$.

DEFINITION. The rank of F_q, which is also the rank of $H_q(K)$, is called the q-th Betti number of K in honor of E. Betti (1823-1892) and will be denoted by $\beta_q(K)$. The torsion coefficients of T_q are called the q-th torsion coefficients of K.

Clearly, to know the Betti numbers and torsion coefficients is equivalent to knowing the homology groups. The Betti numbers for

Examples 1-4 above are easily determined. So far we have not seen any torsion, but nontrivial torsion coefficients can be found in Problems 4.4(e),(f) and 4.6.

Problems

4.4. Compute the Betti numbers, the torsion coefficients, and the homology groups in all dimensions for a complex K which triangulates

(a) a cylinder

(b) the Möbius strip

(c) two disjoint circles

(d) a circle with a line segment sticking out from it

(e) the projective plane P^2

(f) the Klein bottle

In (c)-(f) let K be the simplicial complex indicated in Fig. 4.7a-d, respectively.

4.5. Let K be a nonempty simplicial complex.

(a) If K is connected, prove that $H_0(K) \approx \mathbb{Z}$.

(b) Suppose that $K = L_1 \cup \cdots \cup L_t$, where L_i is a component of K (see Problem 2.14). Prove that t is equal to $\beta_0(K)$, the 0-th Betti number of K. In fact, show that $H_i(K) \approx H_i(L_1) \oplus \cdots \oplus H_i(L_t)$ for all i.

4.6. Construct a simplicial complex K so that $H_0(K) \approx \mathbb{Z}$, $H_1(K) \approx \mathbb{Z}_3$, and $H_i(K) = 0$ for $i > 1$. Can you generalize your construction to find simplicial complexes K_n, $1 \leq n < \infty$, so that $H_0(K_n) \approx \mathbb{Z}$, $H_1(K_n) \approx \mathbb{Z}_n$, and $H_i(K_n) = 0$ for $i > 1$?

4.7. If K is a simplicial complex, define the q-skeleton of K, K^q, by $K^q = \{ \sigma \in K \mid \dim \overline{\sigma} \leq q \}$. Show that K^q is a simplicial complex. What is the relation between the homology groups of K and those of K^q? If σ is a 3-simplex and $K = \overline{\sigma}$, compute $H_q(K^1)$ for $q \geq 0$.

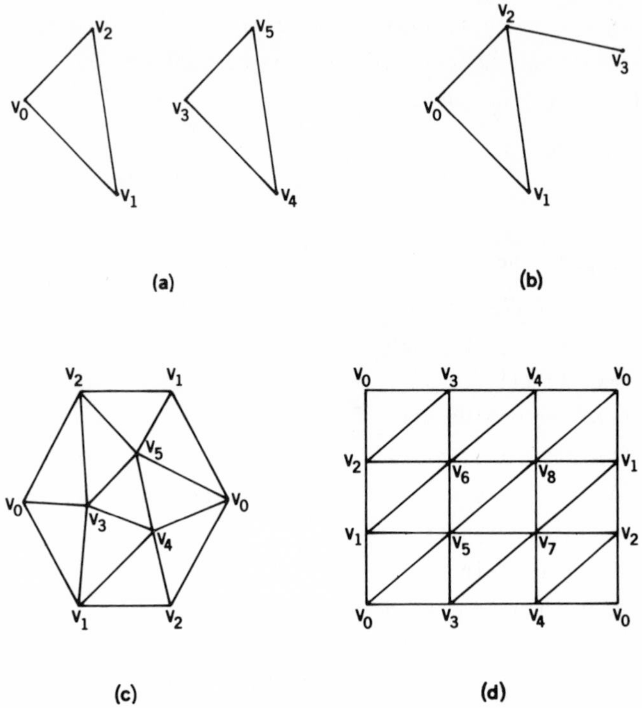

FIGURE 4.7

4.4. THE HOMOLOGY GROUPS OF A CONE

DEFINITION. Let K be a simplicial complex and let v be a point such that, for every $\sigma \in K$, v is linearly independent of the vertices of σ. Define the cone (of K from v), vK, by

$$vK = K \cup \{v\} \cup \{\ vv_0v_1\cdots v_q \mid v_0v_1\cdots v_q \text{ is a simplex of } K\ \}$$

Because of our hypothesis on v it is trivial to check that vK is a simplicial complex, that K is a subcomplex of vK, and that $\dim vK = \dim K + 1$. Figure 4.8 gives some examples of cones vK. Observe that $|vK|$ can also be described as being the space of all line segments from v to points of $|K|$. In fact, if $x \in |vK|$

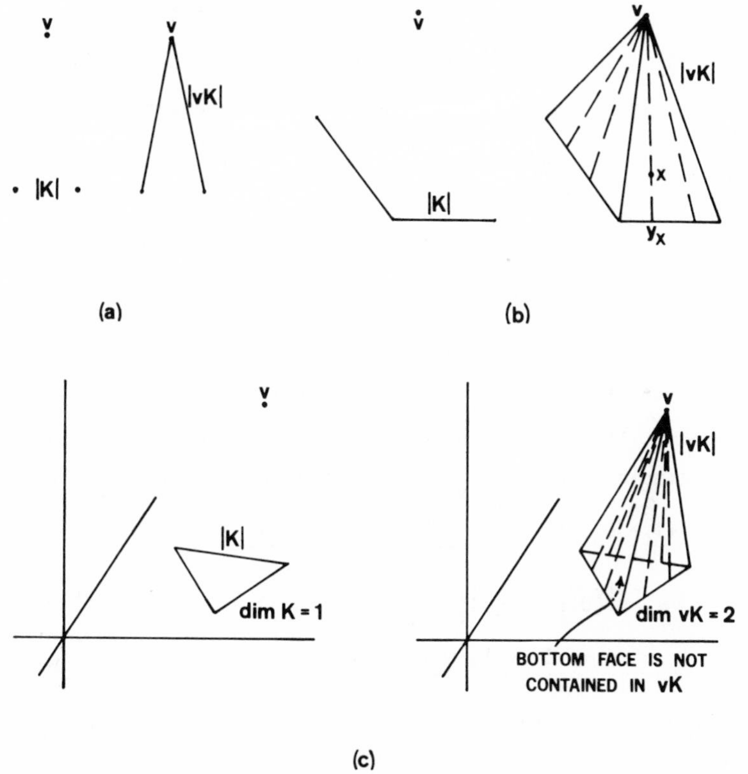

FIGURE 4.8

and $x \neq v$, then there is a unique point $y_x \in |K|$ such that x lies on the line segment from y_x to v, that is, $x = ty_x + (1 - t)v$ for a unique $t \in (0,1]$ (see Fig. 4.8b). Finally, note that if $K = \emptyset$, then $vK = v$.

One of the nice properties of cones is that their homology groups are easy to compute, but first we need some more notation. Suppose that $[v_0v_1\cdots v_q]$ is an oriented q-simplex in a simplicial complex K and that w is a vertex of K. Define a symbol $w[v_0v_1\cdots v_q]$ in two cases: $w[v_0v_1\cdots v_q]$ will denote the oriented $(q + 1)$-simplex $[wv_0v_1\cdots v_q]$ if $wv_0v_1\cdots v_q$ is a $(q + 1)$-simplex of K, and

$w[v_0v_1\cdots v_q] = 0$ if $w = v_i$ for some i. More generally, if $c = \sum_{i=1}^{n} a_i[\sigma_i] \in C_q(K)$, where the $[\sigma_i]$ are oriented q-simplices of K, and if w is a vertex of K, then let wc denote the $(q + 1)$-chain $\sum_{i=1}^{n} a_i(w[\sigma_i])$ whenever $w[\sigma_i]$ is defined for all i. If $c = 0$, we set $wc = 0$.

LEMMA 3. If $c \in C_{q-1}(K)$, $q \geq 2$, and if w is a vertex of K, then

$$\partial_q(wc) = c - w\partial_{q-1}(c)$$

whenever the terms in the equation are defined.

Proof: It suffices to consider the case where c is an oriented $(q - 1)$-simplex $[v_0v_1\cdots v_{q-1}]$. If $w \notin \{v_0,v_1,\cdots,v_{q-1}\}$, then

$$\begin{aligned}\partial_q(wc) &= \partial_q(w[v_0v_1\cdots v_{q-1}]) = \partial_q([wv_0v_1\cdots v_{q-1}])\\ &= [v_0v_1\cdots v_{q-1}] + \sum_{i=1}^{q} (-1)^i[wv_0\cdots\hat{v}_{i-1}\cdots v_{q-1}]\\ &= c + \sum_{i=1}^{q} (-1)^i w[v_0\cdots\hat{v}_{i-1}\cdots v_{q-1}] = c - w\partial_{q-1}(c)\end{aligned}$$

On the other hand, if $w = v_t$ for some t, then $wc = w[v_0v_1\cdots v_{q-1}] = 0$, and so $\partial_q(wc) = \partial_q(0) = 0$. But

$$\begin{aligned}w\partial_{q-1}(c) &= v_t\partial_{q-1}([v_0v_1\cdots v_{q-1}]) = v_t(\sum_{i=0}^{q-1} (-1)^i[v_0\cdots\hat{v}_i\cdots v_{q-1}])\\ &= \sum_{i=0}^{q-1} (-1)^i v_t[v_0\cdots\hat{v}_i\cdots v_{q-1}] = (-1)^t[v_tv_0\cdots\hat{v}_t\cdots v_{q-1}]\\ &= (-1)^{t+t}[v_0v_1\cdots v_{q-1}] = c\end{aligned}$$

This shows that $c - w\partial_{q-1}(c) = c - c = 0$, that is, $\partial_q(wc) = 0 = c - w\partial_{q-1}(c)$. Lemma 3 is proved.

THEOREM 2. If vK is the cone on a simplicial complex K, then

$$H_q(vK) = 0 \text{ if } q \geq 1 \text{ and}$$

$$H_0(vK) \approx \mathbb{Z}$$

Proof: Let $q \geq 1$ and let $z \in Z_q(vK)$. By Lemma 3, $\partial_{q+1}(vz) = z - v\partial_q(z) = z$, since $\partial_q(z) = 0$. Therefore, $Z_q(vK) = B_q(vK)$ and $H_q(vK) = 0$. Finally, $H_0(vK) \approx \mathbb{Z}$ by Problem 4.5(a) because every cone vK is connected. This proves the theorem.

COROLLARY. Let σ be a k-simplex, $k \geq 0$. Then

$$H_q(\overline{\sigma}) = 0 \quad \text{if} \quad q \geq 1 \quad \text{and}$$

$$H_0(\overline{\sigma}) \approx \mathbb{Z}$$

Proof: If $k = 0$, the result follows from Example 1 in Sec. 4.3. If $k \geq 1$ and $\sigma = v_0v_1 \cdots v_k$, then $\overline{\sigma}$ is the cone on $\overline{v_1v_2\cdots v_k}$ $[\overline{\sigma} = v_0(\overline{v_1v_2\cdots v_k})]$ and we have a special case of Theorem 2.

A construction that is closely related to that of taking the cone on a complex is the suspension of a complex.

DEFINITION. Let K be a simplicial complex and suppose that v and w are points so that the cones vK and wK are defined. If $|vK| \cap |wK| = |K|$, define the suspension (of K from v and w), vKw, by $vKw = vK \cup wK$.

Clearly, vKw is a simplicial complex which contains K, vK, and wK as subcomplexes. Examples of vKw can be found in Fig. 4.9. The space $|vKw|$ is formed by taking the union of all line segments from v or w to points of $|K|$. If $K = \emptyset$, then $vKw = \{v,w\}$.

Problem

4.8. Let $X \subset \mathbb{R}^n$. Choose points $v \in \mathbb{R}^{n+1} - \mathbb{R}^n$ and $w \in \mathbb{R}^{n+2} - \mathbb{R}^{n+1}$. Define a cone of X, CX, and a suspension of X, SX, to be a space which is homeomorphic to

$$\{\, tx + (1 - t)v \mid x \in X \text{ and } t \in [0,1] \,\}$$

and

$$\{\, tx + (1 - t)u \mid x \in X,\ t \in [0,1], \text{ and } u = v \text{ or } w\}$$

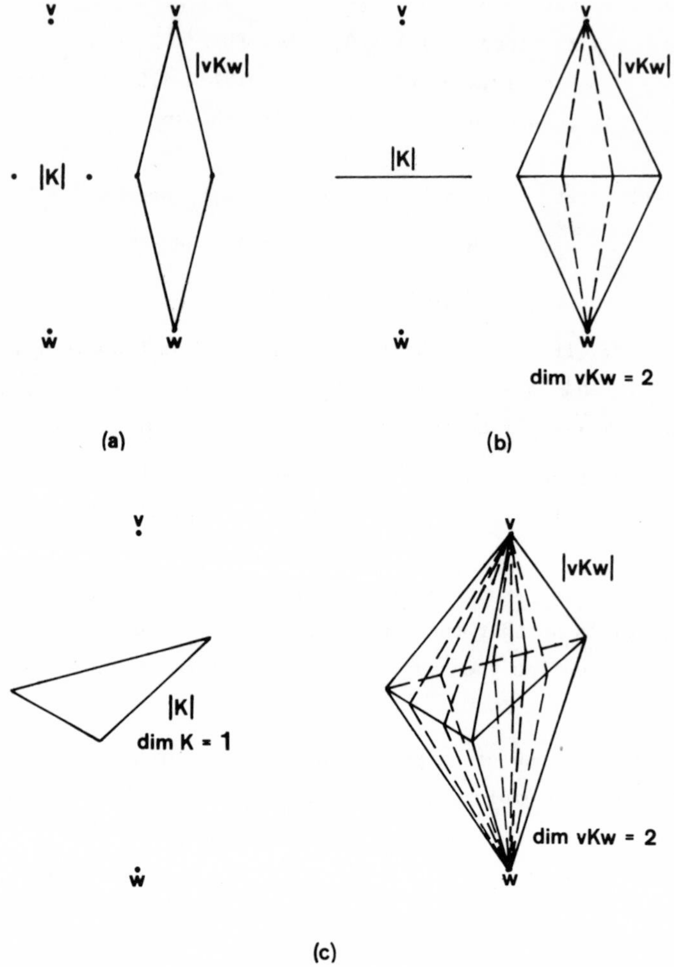

FIGURE 4.9

respectively. (If $X = \emptyset$, then $CX \approx \{v\}$ and $SX \approx \{v,w\}$.)

(a) Show that the spaces CX and SX are uniquely determined (up to homeomorphism), that is, their definition does not depend on a particular choice of v and w.

(b) The cone and suspension of a space are the natural analogs of the cone and suspension of a simplicial complex, respectively. In fact, show that if K is a simplicial complex, then $|vK| \approx C(|K|)$ and $|vKw| \approx S(|K|)$.

(c) Prove that $D^n \approx CS^{n-1}$ and $S^n \approx SS^{n-1}$ for $n \geq 0$.

4.5. THE TOPOLOGICAL INVARIANCE OF HOMOLOGY GROUPS

In the previous two sections we defined the homology groups of a simplicial complex K and actually computed them for several specific examples; however, we are not so much interested in finding an invariant for K, as we are interested in finding one for the underlying space $|K|$. Therefore, the next theorem on the topological invariance of the homology groups is of fundamental importance of the whole theory:

THEOREM 3. Suppose that K and L are simplicial complexes and that $|K|$ is homeomorphic to $|L|$. Then $H_q(K)$ is isomorphic to $H_q(L)$ for all q.

The proof of Theorem 3 is given at the end of Sec. 6.1. The significance of the theorem lies in the fact that it makes the next definition possible.

DEFINITION. Let X be a polyhedron. Choose a triangulation (K,φ) for X and define the q-th homology group of X, $H_q(X)$, by $H_q(X) = H_q(K)$. The q-th Betti number of X, $\beta_q(X)$, and the q-th torsion coefficients of X are defined to be the rank and torsion coefficients of $H_q(X)$, respectively.

Theorem 3 shows that the group $H_q(X)$ is well defined (up to isomorphism), because if (L,ψ) is another triangulation for X, then $H_q(K)$ is isomorphic to $H_q(L)$. Thus, when we were computing homology groups we were really computing topological invariants of spaces and not just properties of particular triangulations. These are exactly the type of computable algebraic invariants we were looking for in Sec. 2.1. In Chapters 6 and 7 we shall show that our theory has a wide range of applications and that the hard work we are putting into it is justified; however, at this point we are lacking one important ingredient of the theory, namely, we do not yet know how homology groups behave with respect to maps. This gap is filled in the next chapter. Right now though let us summarize some of the computations that were made so far in a table. The results in Table 1 are phrased in terms of polyhedra rather than their corresponding simplicial complexes, but this is permissible in view of Theorem 3.

TABLE 1

X	$H_0(X)$	$H_1(X)$	$H_2(X)$	$H_i(X)$, $i > 2$
The cone on a polyhedron; in particular, the disks D^k for $k \geq 0$ (Theorem 2, Problem 4.8)	$\mathbb{Z}$	0	0	0
S^0 (Example 1 in Sec. 4.3, Problem 4.5)	$\mathbb{Z} \oplus \mathbb{Z}$	0	0	0
S^1 (Example 2 in Sec. 4.3)	$\mathbb{Z}$	$\mathbb{Z}$	0	0
S^2 (Example 3 in Sec. 4.3)	$\mathbb{Z}$	0	$\mathbb{Z}$	0
$S^1 \times S^1$ (Example 4 in Sec. 4.3)	$\mathbb{Z}$	$\mathbb{Z} \oplus \mathbb{Z}$	$\mathbb{Z}$	0
P^2 [Problem 4.4(e)]	$\mathbb{Z}$	$\mathbb{Z}_2$	0	0
The Klein bottle [Problem 4.4(f)]	$\mathbb{Z}$	$\mathbb{Z} \oplus \mathbb{Z}_2$	0	0

There is one final observation concerning S^2 and $S^1 \times S^1$ which is worth making at this point, and it has to do with the computation of the top-dimensional homology group $H_2(K)$ in examples 3 and 4 in Sec. 4.3. Since in these cases $H_2(K) = Z_2(K)$, we saw that our only problem was to find the 2-cycles. A little reflection shows that we got a nonzero cycle precisely because it was possible to orient the 2-simplices of K so that adjacent 2-simplices induced opposite orientations on the 1-simplex in their boundary which they had in common. (See Fig. 4.5b and 4.6.) This is not possible in the case of P^2 or the Klein bottle and suggests that the orientability of a surface S is determined by whether or not $H_2(S)$ is zero. The connection between orientability and the group $H_2(S)$ will be studied in greater detail in Sec. 6.5.

4.6. HISTORICAL COMMENTS

By the end of the nineteenth century various topics in topology, such as the theory of surfaces, had been studied quite extensively. Betti [Bet] had introduced some fundamental notions in the study of n-dimensional manifolds in 1871 and further progress in this area was made by V. Eberhardt (1862-1927) [Eb] and Dyck [Dyc 2] in 1890. However, there was still a lack of a unified approach. The man who, to a large degree, was responsible for such a unification and who was the founder of combinatorial topology was H. Poincaré (1854-1912). Poincaré was one of the great mathematicians of all time and made significant contributions to many areas of pure and applied mathematics. His work in topology was an outgrowth of his investigations in the theory of algebraic varieties, that is, the sets of zeros of algebraic functions. Riemann surfaces are a special case of this, and the higher dimensional analogs are manifolds with possible singularities.

In a series of papers [Poin 4-9] beginning with one in 1895, which was followed by five important supplements in 1899-1904, Poincare detailed what he considered to be a systematic approach to the study of n-dimensional spaces. In 1895 in [Poin 4] he still viewed manifolds as algebraic varieties and tried to exploit their analytic structure. He defined the Betti numbers and proved (though somewhat incorrectly) a generalized version of Euler's theorem (see Sec. 6.6). In his first supplement [Poin 5] in 1899 Poincaré abandoned the analytic approach and reformulated his theory in terms of the more fruitful polyhedral definition of manifolds. He also defined the torsion coefficients. Although Poincaré restricted himself to studying manifolds and used curved cells instead of simplices, his basic theory is easily translated into the language of simplicial complexes. The first steps in this direction were taken by Brouwer [Brou 3] in 1910-1911.

Poincaré's 1899 paper quite definitely can be said to have founded combinatorial topology. In this and the second supplement [Poin 6] Poincaré also described a method for computing the Betti numbers and torsion coefficients in terms of "incidence matrices" which are naturally associated to a complex. The incidence matrices are defined in Appendix C and we indicate there how they turn the problem of computing homology groups into a purely mechanical problem that can be solved by any computer. We should emphasize, however, that early topologists worked exclusively with Betti numbers and torsion coefficients. The group-theoretic approach to homology theory which we are presenting did not become fashionable until the 1920s and is due to the influence of the algebraist E. Noether (1882-1935), apparently through conversations with topologists in 1925-1930. The first textbooks ([Se-T 1] and [Alexf-H]) to use homology groups appeared in the middle 1930s. It is interesting to note though that these texts used the group theory mainly as a convenient language, whereas it is quite clear now that algebraical methods in topology can lead to deep results on their own. Homology groups were first called Betti groups (see, for example, [Alex 6] and [Alexf-H]).

One of the really important results Poincaré [Poin 4] introduced in 1895 is his famous Poincaré duality theorem for manifolds, which asserts that if M is a compact oriented n-dimensional manifold without boundary, then the k-th and the (n - k)-th Betti number of M are equal, that is, $\beta_k(M) = \beta_{n-k}(M)$. This result had already been announced by Betti [Bet] in 1871 in the case of 3-dimensional manifolds. However, Betti had defined the "Betti" numbers slightly differently. With his definitions, not only was the duality theorem wrong but also Euler's theorem does not generalize. The modern more complete version of Poincaré duality is fundamental to the theory of manifolds.

Another important notion, introduced by Poincaré [Poin 4] already in 1895, was that of the fundamental group of a space (see Chapter 8). A space X has trivial fundamental group, or is simply-connected, if every closed path in X can be shrunk to a point within X. (This turns out to be a stronger condition than saying that $H_1(X) = 0$.) In [Poin 9] in 1904 Poincaré made his famous conjecture that every compact simply-connected 3-dimensional manifold without boundary is homeomorphic to the 3-sphere S^3. This conjecture is unsolved to this day, even though higher dimensional analogs of it have been proved. The Poincaré conjecture was the outgrowth of an attempt by Poincaré to show that manifolds are characterized by their homology groups and fundamental group. In [Poin 9] Poincaré also gave an example of a 3-dimensional manifold with the same homology groups as S^3 but which was not homeomorphic to S^3 because it had a nontrivial fundamental group. Later in 1919 Alexander [Alex 2] showed that two 3-dimensional manifolds (called lens spaces) could have the same homology and fundamental groups but still not be homeomorphic. (In Alexander's example the fundamental groups are again nontrivial.)

Because Poincaré's definition of homology theory involved cellular decompositions of a space, it was natural to ask about the topological invariance of these definitions. A relatively easy argument shows that they are invariant under subdivision of a given cellular structure. This led Poincaré to make another famous conjecture called the *Hauptvermutung*, which stated that any two cellular decompositions of a polyhedron have isomorphic subdivisions. The conjecture is easily shown to be true for polyhedra of dimension 1 or 2, but in 1961 J. Milnor (1931-) [Mi 2] showed by counterexample that it is false if the dimension of the polyhedron is larger than 5. (The dimension of a polyhedron is the dimension of a simplicial complex which triangulates it. See Sec. 6.3.) A much more difficult result is the celebrated theorem of Kirby and Siebenmann [Ki-S] in 1969 from which it follows that there are actually manifolds of dimension larger than 4 for which the Hauptvermutung also fails. Nevertheless, the Betti numbers and torsion coefficients are topological invariants, and this was proved by Alexander [Alex 1] in 1915.

Chapter 5

MAPS AND HOMOTOPY

5.1. SIMPLICIAL MAPS AGAIN

So far we have been concerned exclusively with spaces, seeking invariants for them which do not change under deformations in the sense of the rubber-sheet geometry alluded to in Sec. 2.1. Now we would like to begin a similar analysis for maps and find invariants also for them which are unchanged under suitable deformations. We shall show how an arbitrary continuous map induces a homomorphism between the corresponding homology groups and develop the basic properties of these homomorphisms, but first we consider the special case of simplicial maps.

DEFINITION. Let K and L be simplicial complexes and let $f : K \to L$ be a simplicial map. Define maps

$$f_{\#q} : C_q(K) \to C_q(L)$$

as follows: if $q < 0$ or $q > \dim K$, then $f_{\#q} = 0$; and if $0 \leq q \leq \dim K$, then $f_{\#q}$ is the unique homomorphism which, on the generators $[v_0v_1\cdots v_q]$ of $C_q(K)$, is given by

$$\begin{aligned} f_{\#q}([v_0v_1\cdots v_q]) &= [f(v_0)f(v_1)\cdots f(v_q)], \quad \text{if } f(v_i) \neq f(v_j) \\ &\quad \text{for } i \neq j, \\ &= 0, \quad \text{otherwise}. \end{aligned}$$

The maps $f_{\#q}$ are clearly well-defined homomorphisms.

LEMMA 1. $\partial_q \circ f_{\#q} = f_{\#q-1} \circ \partial_q$ for all q.

NOTE. In connection with this lemma we would like to describe a convenient terminology which is frequently used in topology and modern algebra. Namely, we shall say that the diagrams

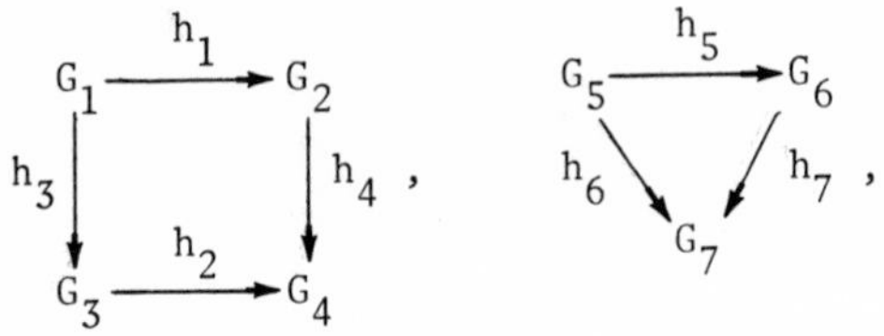

etc.

commute, where the G_i are sets and the h_i maps, provided that $h_4 \circ h_1 = h_2 \circ h_3$, $h_6 = h_7 \circ h_5$, etc. Thus, Lemma 1 could be expressed simply by saying that the diagram

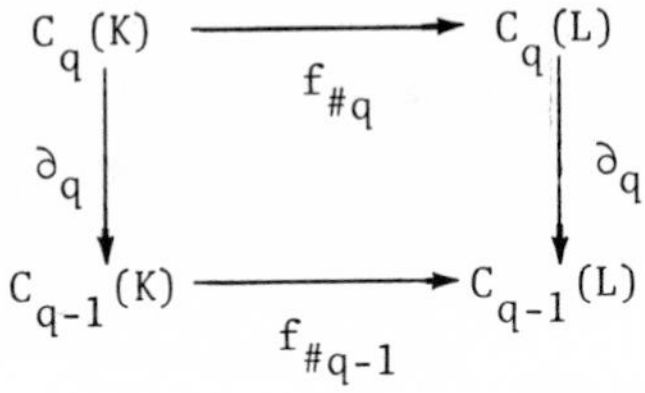

commutes for all q.

Proof of Lemma 1: It clearly suffices to show that

$$(\partial_q \circ f_{\#q})([v_0v_1 \cdots v_q]) = (f_{\#q-1} \circ \partial_q)([v_0v_1 \cdots v_q])$$

for every oriented q-simplex $[v_0v_1 \cdots v_q]$ in K. To do this we merely compute both sides of the equation.

CASE 1. $f(v_0)$, $f(v_1)$, $\cdots$, and $f(v_q)$ are all distinct: Then,

$$
\begin{aligned}
(\partial_q \circ f_{\#q})([v_0v_1\cdots v_q]) &= \partial_q([f(v_0)f(v_1)\cdots f(v_q)]) \\
&= \sum_{i=0}^{q} (-1)^i[f(v_0)\cdots \widehat{f(v_i)}\cdots f(v_q)] \\
&= \sum_{i=0}^{q} (-1)^i f_{\#q-1}([v_0\cdots \hat{v}_i\cdots v_q]) \\
&= f_{\#q-1}\Big(\sum_{i=0}^{q} (-1)^i[v_0\cdots \hat{v}_i\cdots v_q]\Big) \\
&= (f_{\#q-1} \circ \partial_q)([v_0v_1\cdots v_q])
\end{aligned}
$$

CASE 2. $f(v_1), f(v_2), \cdots, f(v_q)$ are distinct and $f(v_0) = f(v_1)$: Then,

$$(\partial_q \circ f_{\#q})([v_0v_1\cdots v_q]) = \partial_q(0) = 0$$

and

$$
\begin{aligned}
(f_{\#q-1} \circ \partial_q)([v_0v_1\cdots v_q]) &= f_{\#q-1}\Big(\sum_{i=0}^{q} (-1)^i[v_0\cdots \hat{v}_i\cdots v_q]\Big) \\
&= \sum_{i=0}^{q} (-1)^i f_{\#q-1}([v_0\cdots \hat{v}_i\cdots v_q]) \\
&= (-1)^0[f(v_1)f(v_2)\cdots f(v_q)] \\
&\quad + (-1)^1[f(v_0)f(v_2)\cdots f(v_q)] \\
&\quad + (-1)^2\cdot 0 + \cdots + (-1)^q\cdot 0 \\
&= [f(v_1)f(v_2)\cdots f(v_q)] - [f(v_0)f(v_2)\cdots \\
&\quad f(v_q)] \\
&= 0
\end{aligned}
$$

CASE 3. $f(v_0), f(v_1), \cdots, \widehat{f(v_i)}, \cdots, f(v_q)$ are distinct and $f(v_i) = f(v_j)$ for some $i < j$: This case follows easily from Case 2 because $[v_0 v_1 \cdots v_q] = \pm[v_i v_j v_0 v_1 \cdots \hat{v}_i \cdots \hat{v}_j \cdots v_q]$.

CASE 4. There exist distinct i,j,k such that $f(v_i) = f(v_j) = f(v_k)$: Then,

$$(\partial_q \circ f_{\#q})([v_0 v_1 \cdots v_q]) = \partial_q(0) = 0$$

and

$$\begin{aligned}(f_{\#q-1} \circ \partial_q)([v_0 v_1 \cdots v_q]) &= f_{\#q-1}\Big(\sum_{t=0}^{q} (-1)^t [v_0 \cdots \hat{v}_t \cdots v_q]\Big) \\ &= \sum_{t=0}^{q} (-1)^t f_{\#q-1}([v_0 \cdots \hat{v}_t \cdots v_q]) \\ &= 0\end{aligned}$$

This finishes the proof of the lemma.

Now the maps $f_{\#q}$ are no more interesting by themselves than were the chain groups $C_q(K)$. What will be important is the maps that they induce on the homology groups and Lemma 1 is essential for that. We shall generalize the construction somewhat because it will be needed again later.

DEFINITION. A chain map $\varphi : C_\#(K) \to C_\#(L)$ is a "vector" $\varphi = (\cdots, \varphi_{-1}, \varphi_0, \varphi_1, \cdots)$ of homomorphisms $\varphi_q : C_q(K) \to C_q(L)$ satisfying $\partial_q \circ \varphi_q = \varphi_{q-1} \circ \partial_q$ for all q. In other words, there is a commutative diagram

$$\begin{array}{ccccccccc}
\cdots & \xrightarrow[\partial_{q+2}]{} & C_{q+1}(K) & \xrightarrow[\partial_{q+1}]{} & C_q(K) & \xrightarrow[\partial_q]{} & C_{q-1}(K) & \xrightarrow[\partial_{q-1}]{} & \cdots \\
 & & \big\downarrow \varphi_{q+1} & & \big\downarrow \varphi_q & & \big\downarrow \varphi_{q-1} & & \\
\cdots & \xrightarrow[\partial_{q+2}]{} & C_{q+1}(L) & \xrightarrow[\partial_{q+1}]{} & C_q(L) & \xrightarrow[\partial_q]{} & C_{q-1}(L) & \xrightarrow[\partial_{q-1}]{} & \cdots
\end{array}$$

It follows from Lemma 1 that $f_{\#} = (\ \cdots, f_{\#-1}, f_{\#0}, f_{\#1}, \cdots)$ is a chain map. Not all chain maps are induced from a simplicial map though, and it is a nontrivial problem to determine just which are.

Suppose that $\varphi : C_{\#}(K) \to C_{\#}(L)$ is an arbitrary chain map. Define maps

$$\varphi_{*q} : H_q(K) \to H_q(L)$$

as follows: Let $a \in H_q(K)$. Then a is a coset of $B_q(K)$ in $Z_q(K)$, that is, $a = [z] = z + B_q(K)$ for some q-cycle $z \in Z_q(K)$. Set $\varphi_{*q}(a) = \varphi_q(z) + B_q(L) \in H_q(L)$.

LEMMA 2. φ_{*q} is a well-defined homomorphism.

Proof: First, let us show that φ_{*q} is well defined. Suppose that $a = z + B_q(K) = z' + B_q(K) \in H_q(K)$ for some $z, z' \in Z_q(K)$. Then $z - z' \in B_q(K)$. In other words, there is a $u \in C_{q+1}(K)$ such that $z - z' = \partial_{q+1}(u)$. It follows that $\varphi_q(z) - \varphi_q(z') = \varphi_q(z - z') = \varphi_q(\partial_{q+1}(u)) = \partial_{q+1}(\varphi_{q+1}(u))$, where the last equality holds because φ is a chain map. Therefore, $\varphi_q(z) - \varphi_q(z') \in B_q(L)$ and $\varphi_q(z)$ determines the same coset of $B_q(L)$ in $C_q(L)$ as does $\varphi_q(z')$. It remains to show that $\varphi_q(z) + B_q(L)$ is in fact a coset of $B_q(L)$ in $Z_q(L)$, that is, that $\varphi_q(z) \in Z_q(L)$. But $\partial_q(\varphi_q(z)) = \varphi_{q-1}(\partial_q(z)) = \varphi_{q-1}(0) = 0$ since φ is a chain map and z is a q-cycle in K. This proves that φ_{*q} is well defined.

Next, let $[z_1] = z_1 + B_q(K)$, $[z_2] = z_2 + B_q(K) \in H_q(K)$. Then

$$\begin{aligned}
\varphi_{*q}([z_1] + [z_2]) &= \varphi_{*q}((z_1 + z_2) + B_q(K)) \\
&= \varphi_q(z_1 + z_2) + B_q(L) \\
&= (\varphi_q(z_1) + \varphi_q(z_2)) + B_q(L) \\
&= (\varphi_q(z_1) + B_q(L)) + (\varphi_q(z_2) + B_q(L)) \\
&= \varphi_{*q}([z_1]) + \varphi_{*q}([z_2])
\end{aligned}$$

Thus φ_{*q} is a homomorphism and the proof of Lemma 2 is finished.

DEFINITION. The maps φ_{*q} are called the maps on homology induced by the chain map φ. In particular, if $f : K \to L$ is a simplicial map, we shall let $f_{*q} : H_q(K) \to H_q(L)$ denote the map on homology induced by the chain map $f_{\#}$.

EXAMPLES

1. Let $K = \partial(\overline{v_0v_1v_2})$ and let $f : K \to K$ be the simplicial map defined by $f(v_i) = v_0$. Then $|f| : |K| \to |K|$ is a constant map, namely, $|f|(x) = v_0$ for every $x \in |K|$. We also know from Sec. 4.3 that $H_0(K) \approx \mathbb{Z} \approx H_1(K)$ and $H_q(K) = 0$ for $q > 1$. Thus, the only interesting dimensions are 0 and 1. Clearly, $f_{\#1} : C_1(K) \to C_1(K)$ is the zero map by definition, and hence $f_{*1} = 0 : H_1(K) \to H_1(K)$. Next, recall that $v_0 + B_0(K)$ is a generator of $H_0(K)$. Therefore, $f_{*0}(v_0 + B_0(K)) = f_{\#0}(v_0) + B_0(K) = v_0 + B_0(K)$ implies that f_{*0} is the identity map.

2. Let K be as in Example 1 and define a simplicial map $f : K \to K$ by $f(v_0) = v_0$, $f(v_1) = v_2$, and $f(v_2) = v_1$. It follows as above that $f_{*q} = 0$ if $q > 1$ and f_{*0} is the identity map. The only possible change can come in f_{*1}. Now, it was shown in Sec. 4.3, Example 2, that $a = ([v_0v_1] + [v_1v_2] + [v_2v_0]) + B_1(K)$ is a generator of $H_1(K)$. Therefore,

$$
\begin{aligned}
f_{*1}(a) &= f_{\#1}([v_0v_1] + [v_1v_2] + [v_2v_0]) + B_1(K) \\
&= ([f(v_0)f(v_1)] + [f(v_1)f(v_2)] + [f(v_2)f(v_0)]) + B_1(K) \\
&= ([v_0v_2] + [v_2v_1] + [v_1v_0]) + B_1(K) \\
&= -a
\end{aligned}
$$

and so $f_{*1} = -1_{H_1(K)}$.

The next lemma lists some basic properties of $f_{\#q}$ and f_{*q}. The proofs are straightforward and left as an exercise for the reader.

LEMMA 3. Let $f : K \to L$ and $g : L \to M$ be simplicial maps between simplicial complexes. Then

(a) $(g \circ f)_{\#q} = g_{\#q} \circ f_{\#q} : C_q(K) \to C_q(M)$
(b) $(g \circ f)_{*q} = g_{*q} \circ f_{*q} : H_q(K) \to H_q(M)$
(c) if $K = L$ and $f = 1_K$, then $f_{\#q}$ and f_{*q} are also the identity homomorphisms

Let us finish this section with some more notation which will be useful in the future.

DEFINITION. Suppose that K, L, and M are simplicial complexes and that $\varphi : C_\#(K) \to C_\#(L)$ and $\psi : C_\#(L) \to C_\#(M)$ are chain maps. Define the composite of φ and ψ, $\psi \circ \varphi$, by

$$\psi \circ \varphi = (\cdots, \mu_{-1}, \mu_0, \mu_1, \cdots)$$

where $\mu_i = \psi_i \circ \varphi_i : C_i(K) \to C_i(M)$. Also, let $1_{C_\#(K)}$ denote the identity chain map

$$(\cdots, 1_{C_{-1}(K)}, 1_{C_0(K)}, 1_{C_1(K)}, \cdots) : C_\#(K) \to C_\#(K)$$

LEMMA 4. Let $\varphi : C_\#(K) \to C_\#(L)$ and $\psi : C_\#(L) \to C_\#(M)$ be chain maps. Then

(a) the composite $\psi \circ \varphi : C_\#(K) \to C_\#(M)$ is a chain map
(b) $(\psi \circ \varphi)_{*q} = \psi_{*q} \circ \varphi_{*q} : H_q(K) \to H_q(M)$
(c) if $K = L$ and $\varphi = 1_{C_\#(K)}$, then $\varphi_{*q} = 1_{H_q(K)}$

Proof: Easy (Problem 5.5).

Problems

5.1. Compute f_{*q}, $q \geq 0$, for the following simplicial maps f:

(a) $K = \partial(\overline{v_0v_1v_2})$ and $f : K \to K$, where $f(v_0) = v_1$, $f(v_1) = v_2$, and $f(v_2) = v_0$.
(b) $K = \partial(\overline{v_0v_1v_2v_3})$ and $f : K \to K$, where $f(v_0) = v_0$, $f(v_1) = v_1$, $f(v_2) = v_3$, and $f(v_3) = v_2$.
(c) Let K and L be the 1-dimensional simplicial complexes in Fig. 5.1 with vertices v_i and w_i, respectively. Define $f : K \to L$ by $f(v_i) = w_i$, $0 \leq i \leq 4$, and $f(v_5) = w_0$.

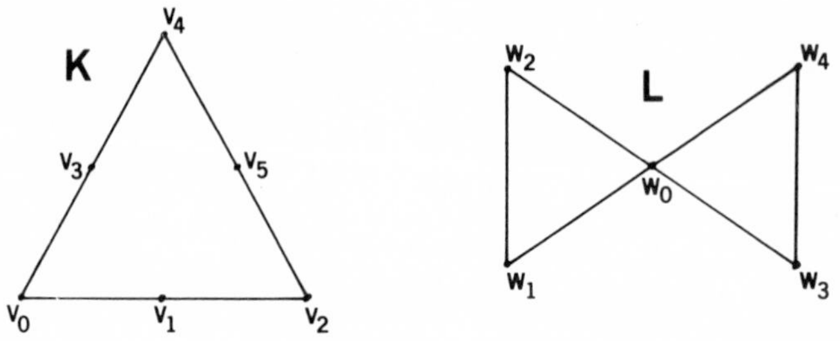

FIGURE 5.1

5.2. In Example 4 of Sec. 4.3 we defined a triangulation K for the torus $S^1 \times S^1$ and found two generators $[u]$ and $[v]$ for the group $H_1(K) \approx \mathbb{Z} \oplus \mathbb{Z}$.

(a) Define a simplicial map $f : K \to K$ such that $f_{*1}([u]) = [v]$ and $f_{*1}([v]) = [u]$. Is such an f unique? If not, construct another. To what simple continuous map $\varphi : S^1 \times S^1 \to S^1 \times S^1$ does f correspond?

(b) Define simplicial maps $f,g : K \to K$ such that $f_{*1}(a[u] + b[v]) = a[u]$ and $g_{*1}(a[u] + b[v]) = (a + b)[u]$. To what continuous maps $\varphi,\psi : S^1 \times S^1 \to S^1 \times S^1$ do f, g, respectively, correspond?

5.3. Let K and L be nonempty connected simplicial complexes and let $f : K \to L$ be any simplicial map. Prove that $f_{*0} : H_0(K) \to H_0(L)$ is always an isomorphism. In fact, if $L = K$, then $f_{*0} = 1_{H_0(K)}$.

5.4. Prove Lemma 3.

5.5. Prove Lemma 4.

5.2. HOMOTOPY

Suppose that K and L are simplicial complexes. Although the set of maps $\{|f| \mid f : K \to L \text{ is a simplicial map}\}$ comprises only a small fraction of the set of all continuous maps $g : |K| \to |L|$, we shall show that any such g can be "approximated" by simplicial maps.

In order to do this we need the idea of taking a map and deforming it a little. This is made precise in the next definition.

DEFINITION. Suppose that $f, g : X \to Y$ are continuous maps. We shall say that f and g are homotopic, denoted by $f \simeq g$, if there exists a continuous map $h : X \times [0,1] \to Y$ such that $h(x,0) = f(x)$ and $h(x,1) = g(x)$ for all $x \in X$. The map h itself is called a homotopy between f and g.

It is easy to check that $\simeq$ is an equivalence relation in the set of all continuous maps from X to Y.

DEFINITION. The equivalence class of a continuous map $f : X \to Y$ under $\simeq$, called the homotopy class of f, is denoted by $[f]$. Let $[X,Y]$ denote the set of homotopy classes of continuous maps from X to Y, that is,

$$[X,Y] = \{ [f] \mid f : X \to Y \text{ is a continuous map} \}$$

We shall soon see that $\simeq$ is the "right" equivalence relation on maps for topology in the sense that the invariants which we shall define for maps depend only on the homotopy class of the map, but right now suppose that h is a homotopy between two maps $f, g : X \to Y$. Observe that if we define $h_t : X \to Y$ by $h_t(x) = h(x,t)$, then h_t is a family of maps parametrized by $[0,1]$ such that $h_0 = f$ and $h_1 = g$. Conversely, given a one-parameter family of maps $h_t : X \to Y$, $t \in [0,1]$, with $h_0 = f$ and $h_1 = g$ which "varies continuously in t," then defining $h(x,t) = h_t(x)$ shows that $f \simeq g$. (By "varies continuously in t" we mean that if s is close to t, then the maps h_s and h_t are close, that is, $h_s(x)$ is uniformly close to $h_t(x)$ for all $x \in X$.) Therefore, $f \simeq g$ is equivalent to being able to deform f into g by a continuously varying family of maps h_t. The maps h_t will also be called a homotopy between f and g, and in the future we shall let the nature of the particular problem at hand determine whether to interpret a homotopy as a map h or as a family of maps h_t.

Next, it will be worthwhile to consider an example which points out the need to be careful when dealing with the question of whether two maps are homotopic. There are certain subtle points which are easily overlooked and not fully appreciated by someone who sees the definition of homotopy for the first time. Consider the two maps $i,c : S^1 \to S^1$, where $i = 1_{S^1}$ and c is the constant map $c(p) = (1,0)$ for all $p \in S^1$. Define $h_t : S^1 \to \mathbb{R}^2$, $t \in [0,1]$, by $h_t(x,y) = (1 + t(x - 1),ty)$ for $(x,y) \in S^1$. Then $h_0 = c$ and $h_1 = i$ and so h_t is a homotopy in $\mathbb{R}^2$ between c and i (see Fig. 5.2). This shows, of course, that i is homotopic to c when both are considered as maps of S^1 into $\mathbb{R}^2$; however, it is not true that i is homotopic to c when they are considered as maps into S^1. This may sound confusing, but we can clear up the problem as follows: Suppose that $\varphi : Y \to Z$ is a fixed continuous map. Define $\varphi_\# : [X,Y] \to [X,Z]$ by $\varphi_\#([f]) = [\varphi \circ f]$. In Problem 5.7(a) you are asked to prove that if $f,g : X \to Y$ and $f \simeq g$, then $\varphi \circ f \simeq \varphi \circ g$. Therefore, $\varphi_\#$ is well defined. If $j : S^1 \to \mathbb{R}^2$ is the

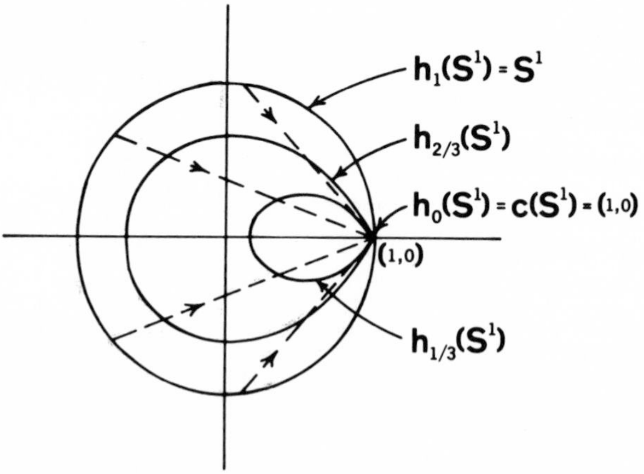

FIGURE 5.2

inclusion map, then $j_\# : [S^1,S^1] \to [S^1,\mathbb{R}^2]$. Furthermore, $[i],[c] \in [S^1,S^1]$, and what we showed above was that $j_\#([i]) = j_\#([c])$. Nevertheless, $[i] \neq [c]$. The point is that the homotopy h_t wanders outside of S^1 (see Fig. 5.2). If we could find a homotopy h_t between i and c such that $h_t(S^1) \subset S^1$ for all $t \in [0,1]$, then i would be homotopic to c in S^1. We claim that there is no such homotopy, although we shall not be able to prove this fact until Sec. 7.1. Intuitively, we can see that for a fixed $p \in S^1$, $h_t(p)$ would be a path σ_p from p to $(1,0)$ in S^1 as t ranges from 0 to 1 and it is impossible to find a continuous collection of such paths σ_p. For example, we could let σ_p correspond to the shortest arc on S^1 from p to $(1,0)$. This would work fine for all $p \in S^1$ except $(-1,0)$. There we have to choose between two possible paths and whichever one we picked as $\sigma_{(-1,0)}$ would destroy the continuity in the sense that there would be points $q \in S^1$ arbitrarily close to $(-1,0)$ for which the paths σ_q and $\sigma_{(-1,0)}$ would be far apart.

The study of the sets $[X,Y]$ is a problem in what is properly called homotopy theory which is an important part of algebraic topology; but since we are concentrating on homology theory in this book, we shall therefore limit our discussion of homotopy to those few aspects which are relevant to our goals. One fact that is essential to remember when dealing with homotopies between maps is that the domain and range of the map are important. (This is what the discussion concerning the example in Fig. 5.2 was all about.)

Here is one more definition.

DEFINITION. Two imbeddings $f,g : X \to Y$ are said to be isotopic if there is a homotopy $h_t : X \to Y$ between f and g such that h_t is an imbedding for all $t \in [0,1]$.

Knot theory is the study of isotopy classes of imbeddings $f : S^1 \to \mathbb{R}^3$ (or S^3).

Problems

5.6. Prove that $\simeq$ is an equivalence relation in the set of all continuous maps from a space X to a space Y.

5.7. Suppose that we have continuous maps $f,g : X \to Y$ and $\varphi,\psi : Y \to Z$.

(a) If $f \simeq g$, prove that $\varphi \circ f \simeq \varphi \circ g$.

(b) If $\varphi \simeq \psi$, prove that $\varphi \circ f \simeq \psi \circ f$.

(c) If $f \simeq g$ and $\varphi \simeq \psi$, prove that $\varphi \circ f \simeq \psi \circ g$.

5.8. Prove that a continuous map $f : S^1 \to X$ is homotopic to a constant map $c : S^1 \to X$, if and only if f extends to D^2, that is, there exists a continuous map $g : D^2 \to X$ such that $g|S^1 = f$.

5.9. Define $i,f : S^1 \times D^1 \to S^1 \times D^1$ by $i(x,t) = (x,t)$ and $f(x,t) = (x,0)$ for all $(x,t) \in S^1 \times D^1$. Show that $f \simeq i$.

5.10. Show that any continuous map $f : X \to \mathbb{R}^n$ is homotopic to a constant map.

5.3. SIMPLICIAL APPROXIMATIONS

In this section we shall define what we mean by a simplicial approximation to a continuous map and develop some basic properties which are needed to prove the main existence theorem (Theorem 2).

DEFINITION. Let K be a simplicial complex and let v be a vertex of K. The star of v, denoted by $st(v)$, is defined by

$$st(v) = st(v,K) = \{\, p \in \operatorname{int} \sigma \mid \sigma \text{ is a simplex of } K \text{ which has } v \text{ as a vertex} \,\}$$

NOTE. v is the only vertex of K in $st(v)$.

Some examples of stars of vertices are given in Fig. 5.3.

LEMMA 5. Let $v_0, v_1, \cdots,$ and v_q be vertices in a simplicial complex K. Then $v_0, v_1, \cdots,$ and v_q are the vertices of a simplex of K if and only if $\bigcap_{i=0}^{q} st(v_i) \neq \emptyset$.

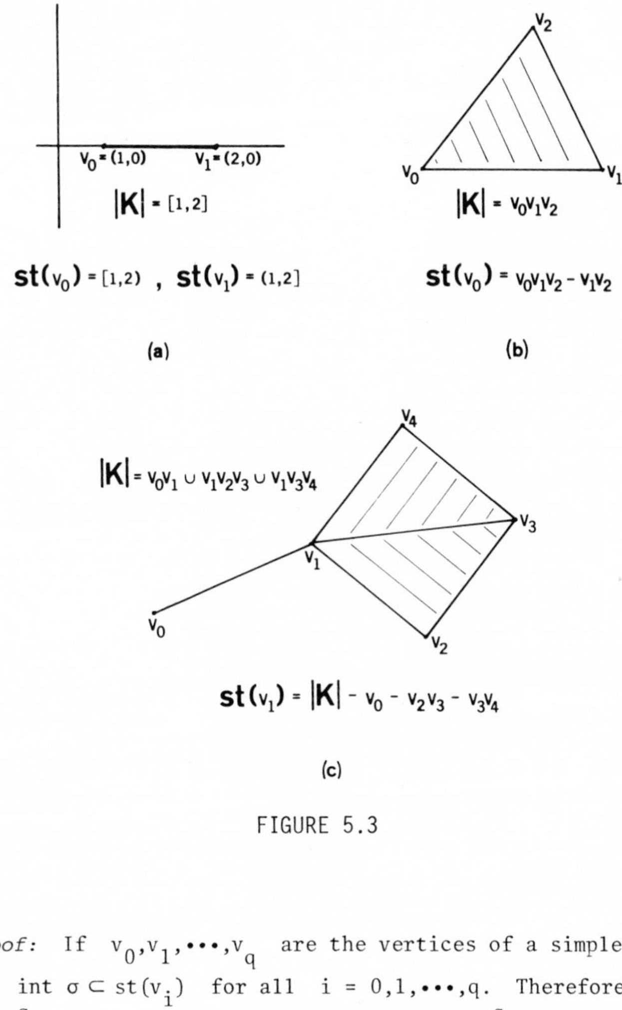

FIGURE 5.3

Proof: If $v_0, v_1, \cdots, v_q$ are the vertices of a simplex σ in K, then $\text{int}\,\sigma \subset st(v_i)$ for all $i = 0,1,\cdots,q$. Therefore, $\emptyset \neq \text{int}\,\sigma \subset \bigcap_{i=0}^{q} st(v_i)$. Conversely, suppose that $\bigcap_{i=0}^{q} st(v_i) \neq \emptyset$ and let $x \in \bigcap_{i=0}^{q} st(v_i)$. Then, for each $i = 0,1,\cdots,q$, there is a simplex σ_i in K which has v_i as a vertex and $x \in \text{int}\,\sigma_i$. By Problem 2.10, $\sigma_0 = \sigma_1 = \cdots = \sigma_q$ and so each v_i is a vertex of σ_0. Since each face of a simplex of K is a simplex in K, the lemma is proved.

DEFINITION. Let K and L be simplicial complexes and suppose that $f : |K| \to |L|$ is a continuous map. A simplicial approximation to f is a simplicial map $\varphi : K \to L$ with the property that if $f(x) \in \sigma$ for some $x \in |K|$ and simplex $\sigma \in L$, then $|\varphi|(x) \in \sigma$.

The main properties of simplicial approximations are summarized in the next two lemmas.

LEMMA 6. Let $f : |K| \to |L|$ be a continuous map and suppose that $\varphi : K \to L$ is a simplicial approximation to f.

(a) The map $|\varphi| : |K| \to |L|$ is homotopic to f.

(b) If $f = |\psi|$, where $\psi : K \to L$ is a simplicial map, then $\varphi = \psi$. This can be interpreted as saying that the only simplicial approximation to a simplicial map is the map itself. It is not true, however, that an arbitrary continuous map has a unique simplicial approximation (see Problem 5.11).

Proof: (a) Define a homotopy $h : |K| \times [0,1] \to |L|$ between f and $|\varphi|$ by $h(x,t) = t|\varphi|(x) + (1 - t)f(x)$, for $x \in |K|$ and $t \in [0,1]$. The map h is a map into $|L|$ because by hypothesis $|\varphi|(x)$ and $f(x)$ lie in a simplex of L and therefore the line segment from $|\varphi|(x)$ to $f(x)$ is contained in $|L|$.

(b) Let v be a vertex of K. Then $w = f(v)$ is a vertex of L. Since a vertex is also a 0-simplex, it follows from the definition that $\varphi(v) = w$. This proves Lemma 6.

LEMMA 7. Let $f : |K| \to |L|$ be a continuous map and suppose that φ is a map from the vertices of K to those of L. Then φ is a simplicial approximation to f if and only if $f(\mathrm{st}(v)) \subset \mathrm{st}(\varphi(v))$ for every vertex v in K.

Proof: Suppose that φ is a simplicial approximation to f and let $x \in \mathrm{st}(v)$. Then $f(x) \in \mathrm{int}\ \sigma$ for some simplex σ in L. Now $x \in \mathrm{int}\ \tau$, where τ is a simplex in K which has v as a vertex. Since φ is a simplicial map, $|\varphi|$ maps τ onto a simplex

σ_1 in L which has $\varphi(v)$ as a vertex and $|\varphi|(x) \in \text{int}\,\sigma_1$. By hypothesis $|\varphi|(x) \in \sigma$. It follows from the definition of a simplicial complex that σ_1 is a face of σ. Therefore, $f(x) \in \text{int}\,\sigma \subset \text{st}(\varphi(v))$. This proves that $f(\text{st}(v)) \subset \text{st}(\varphi(v))$ because x was an arbitrary point of st(v).

Conversely, assume that $f(\text{st}(v)) \subset \text{st}(\varphi(v))$ for every vertex v in K, where φ is a map from the vertices of K to those of L. First we show that φ is a simplicial map. Let $v_0v_1\cdots v_q$ be a q-simplex in K. By Lemma 5, $\bigcap_{i=0}^{q} \text{st}(v_i) \neq \emptyset$; and so

$$\emptyset \neq f\Big(\bigcap_{i=0}^{q} \text{st}(v_i)\Big) \subset \bigcap_{i=0}^{q} f(\text{st}(v_i)) \subset \bigcap_{i=0}^{q} \text{st}(\varphi(v_i))$$

It follows again from Lemma 5 that $\varphi(v_0),\varphi(v_1),\cdots,\varphi(v_q)$ are the vertices of a simplex in L, proving that φ is a simplicial map. Next, let $x \in |K|$. Then $x \in \text{int}\,\tau$ and $f(x) \in \text{int}\,\sigma$, where τ and σ are simplices of K and L, respectively. If v is any vertex of τ, then $x \in \text{st}(v)$ and $f(x) \in \text{st}(\varphi(v))$ by hypothesis. Hence $\varphi(v)$ is a vertex of σ for each vertex v of τ. Since φ is a simplicial map, $|\varphi|(\tau) \subset \sigma$. In particular, $|\varphi|(x) \in \sigma$, which shows that φ is a simplicial approximation to f. This proves the lemma.

The characterization of simplicial approximations given in Lemma 7 brings us a step closer to proving their existence because we have the following corollary:

LEMMA 8. A continuous map $f : |K| \to |L|$ admits a simplicial approximation if and only if for each vertex $v \in K$, $\text{st}(v) \subset f^{-1}(\text{st}(w))$ for some vertex w in L.

Proof: If we have a simplicial approximation φ, set $w = \varphi(v)$ and use Lemma 7. Conversely, define a vertex map φ by setting $\varphi(v) = w$, where we have chosen one of the w's for which $\text{st}(v) \subset f^{-1}(\text{st}(w))$. Lemma 7 implies that φ will be a simplicial approximation. This proves Lemma 8.

We can see from Lemma 8, that the way to prove that a continuous map f has a simplicial approximation is to subdivide K so that the stars, st(v), become sufficiently small and will always be contained in one of the sets from the cover $\{f^{-1}(st(w)) | w$ is a vertex of $L\}$ of K. The map f may not have a simplicial approximation with respect to K itself because the simplices of K could be too "large" (see Problem 5.12).

Problems

5.11. Let $v_0 = (0,0)$, $v_1 = (\pi/2,0)$, $v_2 = (\pi/2,\pi/2)$, and $v_3 = (\pi/2,-\pi/2)$ be points in $\mathbb{R}^2$. Consider the simplicial complex $K = \overline{v_0v_1v_2} \cup \overline{v_0v_1v_3}$ and define $f : |K| \to |K|$ by $f(x,y) = (x, \sin x)$, where $(x,y) \in |K|$ (see Fig. 5.4). Find two distinct simplicial approximations $\varphi : K \to K$ to f.

5.12. Define simplicial complexes K and L in $\mathbb{R}$ by $K = \{0,5,[0,5]\}$ and $L = \{0,1,5,[0,1],[1,5]\}$, so that $|K| = |L|$. Show that the identity map $1_{|K|} : |K| \to |L|$ does not admit a simplicial approximation $\varphi : K \to L$.

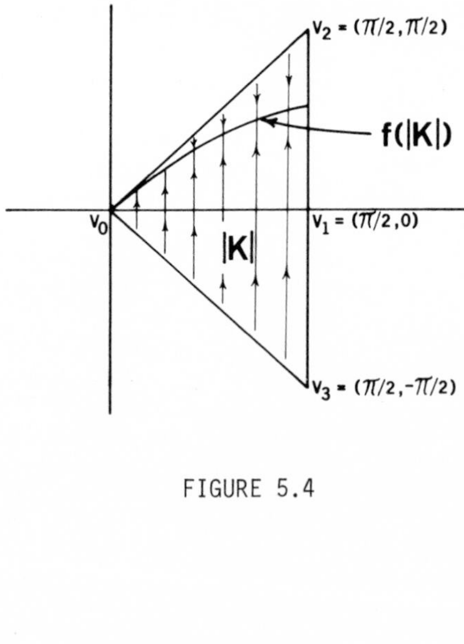

FIGURE 5.4

5.4. THE BARYCENTRIC SUBDIVISION

DEFINITION. If $\sigma = v_0 v_1 \cdots v_q$ is a q-simplex in $\mathbb{R}^n$, define the barycenter, $\hat{\sigma}$, of σ to be the point

$$\hat{\sigma} = \sum_{i=0}^{q} \frac{1}{q+1} v_i$$

The barycenter of a 1-simplex is its midpoint and the barycenter of an arbitrary simplex is its "weighted" center.

Now let K be a simplicial complex and define a new complex sd K as follows: The vertices of sd K are the barycenters $\hat{\sigma}$ of the simplices σ in K; a collection $\hat{\sigma}_0, \hat{\sigma}_1, \cdots, \hat{\sigma}_q$ of vertices in sd K will form a q-simplex $\hat{\sigma}_0 \hat{\sigma}_1 \cdots \hat{\sigma}_q$ in sd K if and only if the σ_i are all distinct and $\sigma_0 < \sigma_1 < \cdots < \sigma_q$. It is easy to show that sd K is a simplicial complex and $|\text{sd } K| = |K|$.

DEFINITION. sd K is called the first barycentric subdivision of K. Furthermore, define the n-th barycentric subdivision of K, $\text{sd}^n K$, inductively by $\text{sd}^0 K = K$ and $\text{sd}^n K = \text{sd}(\text{sd}^{n-1} K)$ for $n \geq 1$.

Figure 5.5 gives an example of a first barycentric subdivision. Intuitively it should be obvious that the simplices of $\text{sd}^n K$ get smaller and smaller as $n \to \infty$. This fact (which was the reason for defining sd K in the first place) will be very useful in light of the discussion at the end of Sec. 5.3 and we make it precise in the following lemma.

DEFINITION. If K is a simplicial complex in $\mathbb{R}^n$, define the mesh of K by

$$\text{mesh } K = \max \{ \text{diam } \sigma \mid \sigma \text{ is a simplex of } K \}$$

LEMMA 9. Suppose that K is an m-dimensional complex in $\mathbb{R}^n$, then

$$\text{mesh (sd } K) \leq \frac{m}{m+1} \text{ mesh } K$$

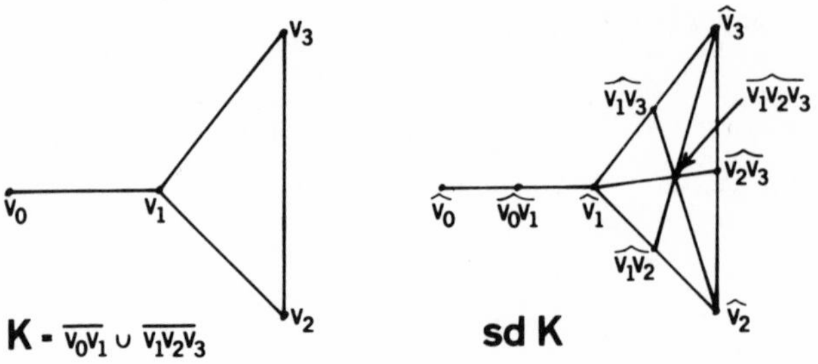

FIGURE 5.5

Proof: Let us begin by proving two facts from which the lemma will follow easily.

CLAIM 1. If σ is a simplex, then diam $\sigma = |v - w|$, where v and w are vertices of σ.

Let $\sigma = v_0v_1\cdots v_q$ and let $x,y \in \sigma$. Suppose that $y = \sum_{i=0}^{q} b_iv_i$, where the b_i are the barycentric coordinates of y. Then

$$|x - y| = \left| \left(\sum_{i=0}^{q} b_i\right)x - \sum_{i=0}^{q} b_iv_i \right|$$

$$= \left| \sum_{i=0}^{q} b_i(x - v_i) \right|$$

$$\leq \sum_{i=0}^{q} b_i|x - v_i|$$

$$\leq \max \{ |x - v_i| : 0 \leq i \leq q \}$$

By symmetry,

$$|x - v_i| \leq \max \{ |v_j - v_i| : 0 \leq j \leq q \}$$

If follows that

$$|x - y| \leq \max \{ |v_j - v_i| : 0 \leq i,j \leq q \}$$

This clearly implies that

$$\begin{aligned}\operatorname{diam}\sigma &= \sup\{\ |x-y| : x,y \in \sigma\ \}\\ &= \max\{\ |v_j - v_i| : 0 \le i,j \le q\ \}\\ &= |v_s - v_t|, \text{ for some } s \text{ and } t\end{aligned}$$

and proves Claim 1.

CLAIM 2. If $\sigma = v_0v_1\cdots v_q$ is a q-simplex, then

$$\operatorname{diam}\tau \le \frac{q}{q+1}\operatorname{diam}\sigma \quad \text{for every} \quad \tau \in \operatorname{sd}\overline{\sigma}$$

By Claim 1, $\operatorname{diam}\tau = |v' - w'|$, where v' and w' are vertices of τ. By definition of $\operatorname{sd}\overline{\sigma}$, there is no loss in generality if we assume that $v' = \sum_{i=0}^{r} \frac{1}{r+1} v_i$ and $w' = \sum_{j=0}^{s} \frac{1}{s+1} v_j$ with $0 \le r \le s \le q$. Then, as in the proof of Claim 1 we get that

$$|v' - w'| \le \max\{\ |v_i - w'| : 0 \le i \le r\ \}$$

and

$$\begin{aligned}|v_i - w'| &= \left|\ v_i - \sum_{j=0}^{s} \frac{1}{s+1} v_j\ \right|\\ &= \frac{1}{s+1}\left(\sum_{j=0}^{s} |v_i - v_j|\right)\\ &\le \frac{1}{s+1}\max\{\ |v_i - v_j| : 0 \le j \le s\ \}\end{aligned}$$

Putting these two inequalities together gives us

$$\begin{aligned}|v' - w'| &\le \frac{s}{s+1}\max\{\ |v_i - v_j| : 0 \le i \le r,\ 0 \le j \le s\ \}\\ &\le \frac{s}{s+1}\operatorname{diam}\sigma\end{aligned}$$

This shows that $\operatorname{diam}\tau \le \frac{s}{s+1}\operatorname{diam}\sigma$. But $\frac{s}{s+1} \le \frac{q}{q+1}$ since $s \le q$; and so Claim 2 is proved.

Using Claim 2 we see that

$$\begin{aligned}\text{mesh}\ (\text{sd } K) &= \max \{ \text{diam } \tau \mid \tau \in \text{sd } K \} \\ &\leq \max \{ \frac{q}{q+1} \text{ diam } \sigma \mid \sigma \in K \} \\ &= \frac{q}{q+1} \text{ mesh } K\end{aligned}$$

which proves Lemma 9.

Next, we define natural maps between the homology groups of K and those of $\text{sd } K$. Recall the notation from Sec. 4.4.

DEFINITION. Define a homomorphism $\text{sd}_{\#q} : C_q(K) \to C_q(\text{sd } K)$ inductively as follows: As usual, it suffices to describe the map on the oriented q-simplices. If v is a vertex of K, let $\text{sd}_{\#0}(v) = \hat{v}$. If $[v_0v_1]$ is an oriented 1-simplex of K, let $\text{sd}_{\#1}([v_0v_1]) = [\hat{v}_0\widehat{v_0v_1}] + [\widehat{v_0v_1}\hat{v}_1]$. Assume that $\text{sd}_{\#q-1}$ has been defined for $q \geq 2$ and let $[\sigma]$ be an oriented q-simplex of K. Define $\text{sd}_{\#q}$ by

$$\text{sd}_{\#q}([\sigma]) = \hat{\sigma}\ \text{sd}_{\#q-1}(\partial_q([\sigma]))$$

Finally, we set $\text{sd}_{\#q} = 0$ if $q < 0$ or $q > \dim K$.

In the example in Fig. 5.5, $\text{sd}_{\#2}([v_1v_2v_3]) = [\hat{v}_1\widehat{v_1v_2}\widehat{v_1v_2v_3}] + [\widehat{v_1v_2}\hat{v}_2\widehat{v_1v_2v_3}] + [\hat{v}_2\widehat{v_2v_3}\widehat{v_1v_2v_3}] + [\widehat{v_2v_3}\hat{v}_3\widehat{v_1v_2v_3}] + [\hat{v}_3\widehat{v_3v_1}\widehat{v_1v_2v_3}] + [\widehat{v_3v_1}\hat{v}_1\widehat{v_1v_2v_3}]$. In other words, $\text{sd}_{\#q}([v_0v_1 \cdots v_q])$ is the q-chain which is the sum of the q-simplices into which $v_0v_1 \cdots v_q$ is decomposed by the first barycentric subdivision with the induced orientations.

LEMMA 10. $\partial_q \circ \text{sd}_{\#q} = \text{sd}_{\#q-1} \circ \partial_q$, that is,

$$\text{sd}_{\#} = (\ \cdots, \text{sd}_{\#-1}, \text{sd}_{\#0}, \text{sd}_{\#1}, \cdots) : C_{\#}(K) \to C_{\#}(\text{sd } K)$$

is a chain map.

Proof: Now

$$(\partial_1 \circ \text{sd}_{\#1})([v_0v_1]) = \partial_1(\ [\hat{v}_0\widehat{v_0v_1}] + [\widehat{v_0v_1}\hat{v}_1]\ )$$

$$= \widehat{v_0v_1} - \hat{v}_0 + \hat{v}_1 - \widehat{v_0v_1} = \hat{v}_1 - \hat{v}_0$$
$$= sd_{\#0}(v_1 - v_0) = (sd_{\#0} \circ \partial_1)([v_0v_1])$$

Assume inductively that $\partial_{q-1} \circ sd_{\#q-1} = sd_{\#q-2} \circ \partial_{q-1}$ for $2 \le q <$ dim K. We shall prove that $\partial_q \circ sd_{\#q} = sd_{\#q-1} \circ \partial_q$. Let $[\sigma]$ be an oriented q-simplex of K. Then

$$\begin{aligned}(\partial_q \circ sd_{\#q})([\sigma]) &= \partial_q(\hat{\sigma}\, sd_{\#q-1}(\partial_q([\sigma])))\\ &= sd_{\#q-1}(\partial_q([\sigma])) - \hat{\sigma}\, \partial_{q-1}(sd_{\#q-1}(\partial_q([\sigma])))\\ &= (sd_{\#q-1} \circ \partial_q)([\sigma])\end{aligned}$$

where the second equality sign follows from Lemma 3 in Sec. 4.4 and the third from our inductive hypothesis which implies that

$$\partial_{q-1} \circ sd_{\#q-1} \circ \partial_q = sd_{\#q-2} \circ \partial_{q-1} \circ \partial_q = 0$$

Hence the lemma is proved by induction since it is trivially true for all other values of q.

If follows from Lemma 10 and Lemma 2 that $sd_{\#}$ induces homomorphisms $sd_{*q} : H_q(K) \to H_q(\text{sd } K)$.

DEFINITION. Define homomorphisms $sd^n_{\#q} : C_q(K) \to C_q(sd^n\, K)$ and $sd^n_{*q} : H_q(K) \to H_q(sd^n\, K)$ inductively by $sd^0_{\#q} = 1_{C_q(K)}$, $sd^0_{*q} = 1_{H_q(K)}$ and $sd^n_{\#q} = sd_{\#q} \circ sd^{n-1}_{\#q}$, $sd^n_{*q} = sd_{*q} \circ sd^{n-1}_{*q}$ for $n \ge 1$.

One can show that sd_{*q}, and hence sd^n_{*q}, are isomorphisms for all $n \ge 0$. Since this fact will not be needed in what follows, we shall leave it as an exercise for the reader (Problem 5.18). It is one of the many consequences of the next theorem which is a very important algebraic result and on which much of the next section is based.

THEOREM 1. Let K and L be simplicial complexes and suppose that $\varphi, \psi : C_{\#}(K) \to C_{\#}(L)$ are chain maps. Assume further that for every simplex $\sigma \in K$ we can find a subcomplex $L_\sigma \subset L$ with the property that

1. $L_\sigma = \overline{\tau_\sigma}$ or $\mathrm{sd}\,\overline{\tau_\sigma}$ for some simplex τ_σ
2. $L_\tau \subset L_\sigma$ whenever $\tau < \sigma$
3. $\varphi_q([\sigma]), \psi_q([\sigma]) \in C_q(L_\sigma) \subset C_q(L)$ for each oriented q-simplex $[\sigma]$ in K

If $\varphi_{*0} = \psi_{*0}$, then $\varphi_{*q} = \psi_{*q}$ for all q.

Proof: First observe that $\varphi_{*q} = \psi_{*q}$ if and only if $(\varphi_q - \psi_q)(z)$ is a boundary in $C_q(L)$ for every $z \in Z_q(K)$. This suggests that in order to show that $\varphi_{*q} = \psi_{*q}$ we should try to construct a map $\mu : Z_q(K) \to C_{q+1}(L)$ such that $\varphi_q - \psi_q = \partial_{q+1} \circ \mu$. The obvious way to define such a μ would be by induction on q, but one would soon encounter certain difficulties. By changing the approach slightly, induction will be possible.

DEFINITION. Suppose that $\varphi, \psi : C_\#(K) \to C_\#(L)$ are chain maps. A chain homotopy from φ to ψ is a "vector" $D = (\cdots, D_{-1}, D_0, D_1, \cdots)$ of homomorphisms $D_q : C_q(K) \to C_{q+1}(L)$ with the property that

$$\varphi_q - \psi_q = \partial_{q+1} \circ D_q + D_{q-1} \circ \partial_q$$

for all q. Whenever there is a chain homotopy between φ and ψ we shall call φ and ψ chain homotopic and write $\varphi \simeq \psi$.

REMARK. The definition of a chain homotopy is motivated by what happens when we have a homotopy $h : \sigma \times [0,1] \to X$ between two maps $f,g : \sigma \to X$, where σ is a q-simplex (see Fig. 5.6).

It is easy to see that $\simeq$ is an equivalence relation among the set of chain maps from $C_\#(K)$ to $C_\#(L)$.

CLAIM. If two chain maps $\varphi, \psi : C_\#(K) \to C_\#(L)$ are chain homotopic, then $\varphi_{*q} = \psi_{*q}$ for all q.

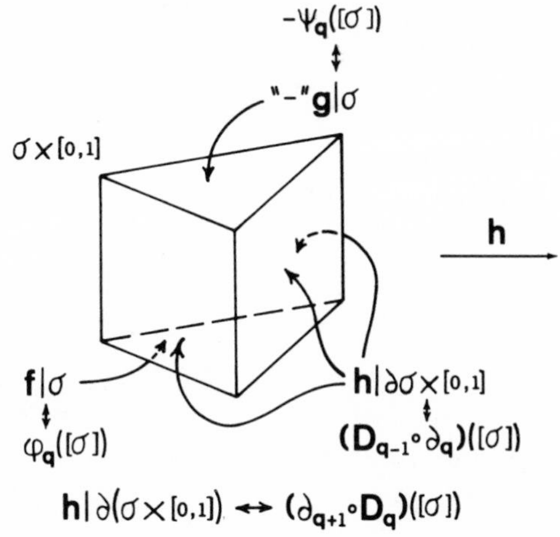

FIGURE 5.6

The claim is obvious because if $z \in Z_q(K)$, then $(\varphi_q - \psi_q)(z) = (\partial_{q+1} \circ D_q)(z) + (D_{q-1} \circ \partial_q)(z) = \partial_{q+1}(D_q(z)) + D_{q-1}(0) = \partial_{q+1}(D_q(z))$, where D is a chain homotopy from φ to ψ. Therefore, to prove Theorem 1 we shall construct a chain homotopy D between our two given maps φ and ψ. The maps $D_i : C_i(K) \to C_{i+1}(L)$ clearly have to be the zero map if $i \leq -1$. Assume inductively that we have already defined a homomorphism $D_{q-1} : C_{q-1}(K) \to C_q(L)$ for $q \geq 0$ satisfying

(i) $\varphi_{q-1} - \psi_{q-1} = \partial_q \circ D_{q-1} + D_{q-2} \circ \partial_{q-1}$

(ii) $D_{q-1}([\tau]) \subset C_q(L_\tau)$ for each oriented $(q-1)$-simplex $[\tau]$ in K

We shall now show how to define $D_q : C_q(K) \to C_{q+1}(L)$. If we choose a fixed orientation for each q-simplex σ in K and let $[\sigma]$ denote the corresponding oriented simplex, then $C_q(K)$ will be a free abelian group on the set $S_q^+ = \{ [\sigma] \mid \sigma$ is a q-simplex of $K \}$. Therefore, it suffices to define D_q on each element of S_q^+ because we can then extend D_q linearly.

CASE 1. $q = 0$: Let $[\sigma] \in S_0^+$ and consider the diagram

$$\begin{array}{ccccc} \varphi_0([\sigma]),\psi_0([\sigma]) \in C_0(L_\sigma) = & Z_0(L_\sigma) & \subset & Z_0(L) & = C_0(L) \\ & \downarrow & & \downarrow & \\ & H_0(L_\sigma) & \xrightarrow[i_*]{} & H_0(L) & \end{array}$$

where the vertical maps are the natural projections and i_* is induced from the inclusion $L_\sigma \subset L$. It is easy to see that i_* is injective. From this and the fact that $\varphi_{*0} = \psi_{*0}$ it follows that $\varphi_0([\sigma])$ and $\psi_0([\sigma])$ determine the same homology class in $H_0(L_\sigma)$. In other words, there is some $c_{[\sigma]} \in C_1(L_\sigma)$ such that $(\varphi_0 - \psi_0)([\sigma]) = \partial_1(c_{[\sigma]})$. If we define $D_0([\sigma]) = c_{[\sigma]}$, then D_0 will satisfy (i) and (ii) because $(\partial_1 \circ D_0 + D_{-1} \circ \partial_0)([\sigma]) = (\partial_1 \circ D_0)([\sigma]) = \partial_1(c_{[\sigma]}) = (\varphi_0 - \psi_0)([\sigma])$.

CASE 2. $q > 0$: Let $[\sigma] \in S_q^+$. The inductive hypothesis (ii) and conditions (2) and (3) in Theorem 1 imply that $(\varphi_q - \psi_q - D_{q-1} \circ \partial_q)([\sigma]) \in C_q(L_\sigma)$. Furthermore, it follows from (i) that

$$\begin{aligned} \partial_q \circ (\varphi_q - \psi_q - D_{q-1} \circ \partial_q) &= \partial_q \circ \varphi_q - \partial_q \circ \psi_q - \partial_q \circ D_{q-1} \circ \partial_q \\ &= \varphi_{q-1} \circ \partial_q - \psi_{q-1} \circ \partial_q \\ &\quad - (\varphi_{q-1} - \psi_{q-1} - D_{q-2} \circ \partial_{q-1}) \circ \partial_q \\ &= 0 \end{aligned}$$

This means that $(\varphi_q - \psi_q - D_{q-1} \circ \partial_q)([\sigma])$ actually belongs to $Z_q(L_\sigma)$. But by Theorem 2 in Sec. 4.4 $H_q(L_\sigma) = 0$, so that for each $[\sigma] \in S_q^+$ we can choose a $c_{[\sigma]} \in C_{q+1}(L_\sigma)$ such that $(\varphi_q - \psi_q - D_{q-1} \circ \partial_q)([\sigma]) = \partial_{q+1}(c_{[\sigma]})$. Define $D_q([\sigma]) = c_{[\sigma]}$. Then D_q will satisfy (i) and (ii).

This finishes our inductive definition of the chain homotopy D from φ to ψ and Theorem 1 now follows from the claim.

REMARK. The proof of Theorem 1 is based on the "method of acyclic models" which is an important general method in algebraic topology for constructing maps, in particular, chain homotopies. We have considered only a very special case, but there is much more to Theorem 1 than meets the eye. In our case the "acyclic models" are the L_σ, and the meaning of the term "acyclic" is that $H_q(L_\sigma) = 0$ for $q \neq 0$. It is really this last property of the complexes L_σ which makes the proof of Theorem 1 work.

Problems

5.13. Let K be a simplicial complex.

(a) Prove that sd K is a simplicial complex and that $|\text{sd } K| = |K|$.

(b) If L is a subcomplex of K, prove that sd L is a subcomplex of sd K.

5.14. If σ is a simplex, prove that $\text{sd } \bar{\sigma} = \hat{\sigma} \text{ sd } (\partial\bar{\sigma})$.

5.15. (a) Prove that $\simeq$ is an equivalence relation in the set of chain maps from one simplicial complex to another.

(b) Suppose that $\varphi^1, \varphi^2 : C_\#(K) \to C_\#(L)$ and $\psi^1, \psi^2 : C_\#(L) \to C_\#(M)$ are chain maps such that $\varphi^1 \simeq \varphi^2$ and $\psi^1 \simeq \psi^2$. Prove that $\psi^1 \circ \varphi^1 \simeq \psi^2 \circ \varphi^2 : C_\#(K) \to C_\#(M)$. In other words, composites of chain homotopic chain maps are chain homotopic.

5.16. Give an alternate proof of Theorem 2 in Sec. 4.4 based on the following observations:

(a) $c = (\cdots, c_{-1}, c_0, c_1, \cdots) : C_\#(vK) \to C_\#(vK)$ is a chain map, where $c_q = 0$ for all $q \neq 0$ and $c_0(\sum_{i=1}^{k} a_i v_i) = (\sum_{i=1}^{k} a_i)v$ for all vertices v_i in vK and $a_i \in \mathbb{Z}$; and

(b) $c \simeq 1_{C_\#(vK)} : C_\#(vK) \to C_\#(vK)$.

5.17. Define the chain complexes $C_\#(K)$ and $C_\#(L)$ to be chain equivalent if there exist chain maps $\varphi : C_\#(K) \to C_\#(L)$ and $\psi : C_\#(L) \to C_\#(K)$ such that $\psi \circ \varphi \simeq 1_{C_\#(K)}$ and $\varphi \circ \psi \simeq 1_{C_\#(L)}$. Prove that chain equivalent chain complexes have isomorphic homology groups.

5.18. Let K be a simplicial complex. Prove that $sd^n_{*q} : H_q(K) \to H_q(sd^n K)$ is an isomorphism for all $n \geq 0$. (*Hint:* It suffices to consider the case $n = 1$. Define a simplicial map $\varphi : sd\, K \to K$ by letting $\varphi(\hat{\sigma})$ be any vertex of the simplex $\sigma \in K$. Show that $\varphi_\# \circ sd_\# = 1_{C_\#(K)}$ by computing $(\varphi_{\#q} \circ sd_{\#q})([v_0 v_1 \cdots v_q])$ and using induction. Even though $sd_{\#q} \circ \varphi_{\#q}$ may not equal $1_{C_\#(sd\, K)}$, show that $sd_{*q} \circ \varphi_{*q} = 1_{H_q(sd\, K)}$ by using Theorem 1. Namely, if $\tau = \hat{\sigma}_0 \hat{\sigma}_1 \cdots \hat{\sigma}_q \in sd\, K$, then let $L_\tau = sd\, \overline{\sigma_q}$. Thus, φ_{*q} is an inverse of sd_{*q}.)

5.5. THE SIMPLICIAL APPROXIMATION THEOREM; INDUCED HOMOMORPHISMS

We are finally ready to show that continuous maps can be approximated by simplicial maps.

THEOREM 2 (The Simplicial Approximation Theorem). Let K and L be simplicial complexes and suppose that $f : |K| \to |L|$ is a continuous map. Then there is an integer $N \geq 0$ such that for each $n \geq N$, f admits a simplicial approximation $\varphi : sd^n K \to L$.

Proof: Consider the open cover $\mathcal{U} = \{ f^{-1}(st(w)) \mid w$ is a vertex of $L \}$ of $|K|$. Let $\varepsilon > 0$ be a Lebesgue number for $\mathcal{U}$ (see Appendix A for the definition of "Lebesgue number"), which exists because $|K|$ is compact (see Problem 2.19). If $m = \dim K$, we can choose $N \geq 0$ such that $(m/(m + 1))^N$ mesh $K \leq \varepsilon/2$ because $m/(m + 1) < 1$. It follows from Lemma 9 and induction that mesh $(sd^n K) \leq (m/(m + 1))^n$ mesh $K \leq \varepsilon/2$ whenever $n \geq N$. In particular, if v is a vertex of $sd^n K$, then diam $(st(v, sd^n K)) \leq 2(\text{mesh } (sd^n K)) \leq \varepsilon$ and therefore, by definition of a Lebesgue number, $st(v) \subset f^{-1}(st(w))$ for some vertex w of L. By Lemma 8, f admits a simplicial approximation $\varphi : sd^n K \to L$ and Theorem 2 is proved.

Using Theorem 2 we can now show how continuous maps induce homomorphisms on the homology groups. Let $f : |K| \to |L|$ be a continuous map, where K and L are simplicial complexes. By the simplicial approximation theorem there is an $n \geq 0$ such that f admits a simplicial approximation $\varphi : sd^n K \to L$.

DEFINITION. The homomorphism

$$f_{*q} : H_q(K) \to H_q(L)$$

given by $f_{*q} = \varphi_{*q} \circ sd^n_{*q}$ is called the homomorphism induced on the q-th homology group by the continuous map f.

LEMMA 11. (a) f_{*q} is a well-defined homomorphism.

(b) Suppose that K, L, and M are simplicial complexes and $f : |K| \to |L|$ and $g : |L| \to |M|$ are continuous maps. Then $(g \circ f)_{*q} = g_{*q} \circ f_{*q}$. If $K = L$ and $f = 1_{|K|}$, then $f_{*q} = 1_{H_q(K)}$.

Proof: (a) Suppose that $\varphi' : sd^n K \to L$ is another simplicial approximation to f. Let $\sigma = v_0 v_1 \cdots v_s \in sd^n K$. By Lemma 5, $\bigcap_{j=0}^{s} st(v_j, sd^n K) \neq \emptyset$, and so, by Lemma 7, $\emptyset \neq f(\bigcap_{j=0}^{s} st(v_j, sd^n K)) \subset \bigcap_{j=0}^{s} f(st(v_j, sd^n K)) \subset \bigcap_{j=0}^{s} (\, st(\varphi(v_j)) \cap st(\varphi'(v_j))\,)$. Using Lemma 5 again, this shows that the vertices $\varphi(v_j)$ and $\varphi'(v_j)$ belong to a simplex of L. Therefore, $\varphi_{\#q}$ and $\varphi'_{\#q}$ satisfy the hypotheses of Theorem 1, because for every $\sigma \in sd^n K$ we can set $L_\sigma = \overline{\tau}$, where τ is the smallest simplex in L which contains $|\varphi|(\sigma)$ and $|\varphi'|(\sigma)$. It follows that $\varphi_{*q} = \varphi'_{*q}$ and the definition of f_{*q} is independent of the choice of φ.

Next, assume $\psi : sd^k K \to L$ is a simplicial approximation to f. Without loss of generality we may suppose that $n \geq k$. Since the vertices of $sd^k K$ are vertices of $sd^n K$, we may extend ψ to a simplicial map $\varphi' : sd^n K \to L$ as follows: If v is a vertex of $sd^n K$ and $v \in \text{int}\, \sigma$, where σ is a simplex of $sd^k K$, let $\varphi'(v)$ be any vertex of $|\psi|(\sigma)$. It is easy to check that φ' is also a simplicial approximation to f and hence $\varphi'_{*q} = \varphi_{*q}$ by what we just showed above. Furthermore, a simple induction on $n - k$ shows that $\psi_{*q} = \varphi'_{*q} \circ sd^{n-k}_{*q}$. (If $n - k = 1$ and $\sigma \in sd^k K$, then let

$L_\sigma = \overline{|\psi|(\sigma)}$ and apply Theorem 1 to $\psi_\#$ and $\varphi'_\# \circ sd_\#$.) Therefore, $f_{*q} = \varphi_{*q} \circ sd^n_{*q} = \varphi'_{*q} \circ sd^{n-k}_{*q} \circ sd^k_{*q} = \psi_{*q} \circ sd^k_{*q}$ which proves (a).

(b) Let $\psi : sd^k\, L \to M$ be a simplicial approximation to g and choose n such that f admits a simplicial approximation $\varphi : sd^n\, K \to sd^k\, L$. Let v be a vertex of $sd^n\, K$ and define a map φ' from the vertices of $sd^n\, K$ to the vertices of L by $\varphi'(v) = w$, where w is any vertex of the unique simplex $\tau \in L$ such that $\varphi(v) \in \text{int}\, \tau$. It is easy to check that φ' is a simplicial approximation to f. We have a diagram

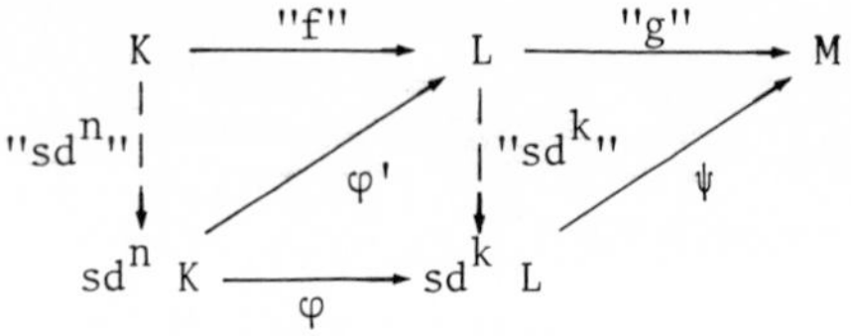

An easy application of Theorem 1 shows that $\varphi_{*q} = sd^k_{*q} \circ \varphi'_{*q}$. It is also easy to see that $\psi \circ \varphi : sd^n\, K \to M$ is a simplicial approximation to $g \circ f$. From these two facts, Lemma 3, and the definitions it follows that $(g \circ f)_{*q} = (\psi \circ \varphi)_{*q} \circ sd^n_{*q} = \psi_{*q} \circ \varphi_{*q} \circ sd^n_{*q} = \psi_{*q} \circ (sd^k_{*q} \circ \varphi'_{*q}) \circ sd^n_{*q} = g_{*q} \circ f_{*q}$. This proves the first part of (b). The second part is obvious from Lemma 3(c), and so Lemma 11 is proved.

We shall finish this chapter with another important property of the map f_{*q}, namely, that it depends only on the homotopy class of f. Intuitively, this should not be a surprising result because if h is a homotopy between two maps f and g and if z is a q-cycle, then the q-cycles $f_{\#q}(z)$ and $g_{\#q}(z)$ are homologous via the "(q + 1)-chain" $h(z \times [0,1])$.

THEOREM 3. Let K and L be simplicial complexes and suppose that $f,g : |K| \to |L|$ are continuous maps which are homotopic. Then $f_{*q} = g_{*q} : H_q(K) \to H_q(L)$ for all q.

Proof: Let $h : |K| \times [0,1] \to |L|$ be a homotopy between f and g and let $\epsilon > 0$ be a Lebesgue number for the open cover $\mathcal{U} = \{ h^{-1}(st(w)) \mid w$ is a vertex of $L \}$ of $|K| \times [0,1]$. Choose a partition $0 = a_0 < a_1 < \cdots < a_r = 1$ of $[0,1]$ such that $0 < a_i - a_{i-1} < \epsilon$, for $i = 1,2,\cdots,r$. Then for each $x \in |K|$ and $i = 1,2,\cdots,r$ there is a vertex $w \in L$ such that both $h(x,a_{i-1})$ and $h(x,a_i)$ belong to $st(w)$. This means that if we define $f_i : |K| \to |L|$ by $f_i(x) = h(x,a_i)$ and let $\mathcal{U}_i = \{ f_i^{-1}(st(w)) \cap f_{i-1}^{-1}(st(w)) \mid w$ is a vertex of $L \}$, then $\mathcal{U}_i$ will be an open cover of $|K|$. Let ϵ_i be a Lebesgue number for $\mathcal{U}_i$ and set $\delta = \min \{\epsilon_1,\epsilon_2,\cdots,\epsilon_r\}$. Next, choose n so large that mesh $(sd^n K) \leq \delta/2$. It follows that diam $(st(v,sd^n K)) \leq \delta$ for all vertices v in $sd^n K$ and that for each i and each vertex $v \in sd^n K$ there is a vertex w in L with $st(v,sd^n K) \subset f_{i-1}^{-1}(st(w)) \cap f_i^{-1}(st(w))$. Therefore, by Lemma 8, we can find a simplicial map $\varphi_i : sd^n K \to L$ which is a simplicial approximation to both f_{i-1} and f_i. (We can set $\varphi_i(v) = w$.) Consider φ_i and φ_{i+1}. They are simplicial approximations to the same continuous map f_i, and so, as in the proof of Lemma 11(a), φ_i and φ_{i+1} satisfy the hypotheses of Theorem 1. (If σ is a simplex of $sd^n K$, then let $L_\sigma = \overline{\tau}$ where τ is the smallest simplex of L which contains $|\varphi_i|(\sigma)$ and $|\varphi_{i+1}|(\sigma)$.) Thus, $(\varphi_i)_{*q} = (\varphi_{i+1})_{*q}$ and it follows from the definitions that $(f_i)_{*q} = (f_{i+1})_{*q}$. Since i was arbitrary, we get that $f_{*q} = (f_0)_{*q} = (f_1)_{*q} = \cdots = (f_r)_{*q} = g_{*q}$ and Theorem 3 is proved.

Problems

5.19. Let K and L be as in Problem 5.12. Find a simplicial approximation $\varphi : sd^n K \to L$ to $1_{|K|}$ for some $n \geq 0$. What is the smallest n for which φ exists?

5.20. Let X and Y be polyhedra. Use the simplicial approximation theorem to show that $[X,Y]$ is countable.

5.6. HISTORICAL COMMENTS

The definition of simplicial maps and the simplicial approximation theorem are due to Brouwer [Brou 3] in 1910-1911. Brouwer's work will be discussed in more detail later, but its significance lay in the fact that it established the usefulness of algebraic topology in solving geometric problems. The concept of homotopy and deformations can be traced back at least as far as 1866 to the paper [J 2] by Jordan which essentially solved the question of when two curves on an orientable surface are homotopic. Poincaré's definition of the fundamental group of a manifold in [Poin 4] in 1895 generalized these ideas and gave them new importance. The word "homotopy" was introduced by [De-H].

The algebraic concepts in this section, such as the various homomorphisms induced on homology groups, developed quite naturally as soon as topologists had adopted the language of groups. In particular, most of them can already be found in the 1935 text by Alexandroff and Hopf [Alexf-H]. The "method of acyclic models" was formulated by S. Eilenberg (1913-) and S. MacLane (1909-) in a joint paper [Eil-M] in 1953.

Chapter 6

FIRST APPLICATIONS OF HOMOLOGY THEORY

6.1. A QUICK REVIEW

To get to this stage we have had to work our way through the proofs of quite a few lemmas and theorems. There is always a danger in such a case that the many details may cause one to lose track of the overall objective which is to discover and/or prove concrete theorems about geometrical objects. However, we know enough now to begin making some significant applications. First, here is a summary of what our labors have accomplished:

I. To every simplicial complex K and to every integer q we have associated an abelian group $H_q(K)$ called the q-th homology group of K.

II. To every continuous map $f : |K| \to |L|$ between the underlying spaces of two simplicial complexes K and L and to each integer q we have associated a homomorphism

$$f_{*q} : H_q(K) \to H_q(L)$$

These homomorphisms have certain natural properties which are perhaps described most simply by saying that every diagram

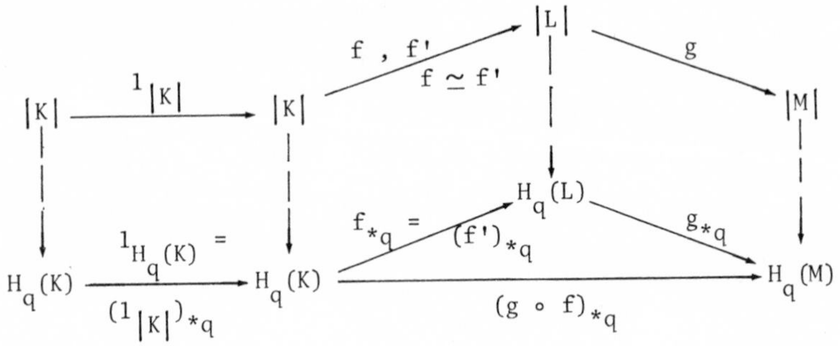

"commutes." To be precise, let K, L, and M be simplicial complexes and suppose that $f, f' : |K| \to |L|$ and $g : |L| \to |M|$ are continuous maps. Then, for all q,

1. $(1_{|K|})_{*q} = 1_{H_q(K)}$
2. $(g \circ f)_{*q} = g_{*q} \circ f_{*q}$
3. If $f \simeq f'$, then $f_{*q} = (f')_{*q}$

Everything that we need to know from the last two chapters (other than the basic definitions which are necessary for some specific computations) are contained in (I) and (II). From an abstract point of view, the rest could be forgotten. If we wanted, then we could simply take (I) and (II) as axioms about certain groups $H_q(K)$ and homomorphisms f_{*q} and assume their existence. The point is that we really only needed Chapters 4 and 5 to show that such groups and homomorphisms exist and that we are not working with a vacuous theory. We shall not emphasize the axiomatic approach in this book; nevertheless, it is worth pointing out that much of what we shall prove will follow from the purely formal properties of $H_q(K)$ and f_{*q} listed in (I) and (II). The fact that they are defined geometrically will be irrelevant. In Sec. 2.1 we attempted in vague terms to give a definition of algebraic topology. Now we can see in much clearer terms what was meant by "algebraic invariants" associated to spaces. As we discuss various applications we shall see how theorems concerning geometry are proved by abstract algebraic methods dealing with these algebraic invariants.

Our first application of properties (I) and (II) will be to give a proof of Theorem 3 in Chapter 4. In fact, one can prove more.

DEFINITION. Two spaces X and Y are said to have the same homotopy type, denoted by $X \simeq Y$, if there exist continuous maps $f : X \to Y$ and $g : Y \to X$ such that $g \circ f \simeq 1_X$ and $f \circ g \simeq 1_Y$. In that case, f is called a homotopy equivalence.

It is obvious that every homeomorphism is a homotopy equivalence. Furthermore, "$\simeq$" is an equivalence relation in the set of spaces (Problem 6.1).

LEMMA 1. $\mathbb{R}^n \simeq D^n \simeq$ point.

Proof: Let $X = \mathbb{R}^n$ or D^n and let $f : X \to 0$ and $g : 0 \to X$ be the constant map and inclusion map, respectively. Clearly, $f \circ g = 1_0$. Consider the continuous map $h : X \times [0,1] \to X$ defined by $h(x,t) = tx$. Since $h(x,0) = 0 = (g \circ f)(x)$ and $h(x,1) = x = 1_X(x)$, we see that h is a homotopy between $g \circ f$ and 1_X. This proves that X and 0 have the same homotopy type and proves the lemma.

Some further examples of spaces which have the same homotopy type are given in the problems for this section. Very loosely speaking, two spaces will have the same homotopy type if one can squash, thicken, or deform one into the other without eliminating or introducing any "holes." For example, S^n does not have the same homotopy type as a point. This follows from the next theorem and Lemma 2 in Sec. 6.2.

The main theorem about homotopy equivalent spaces, which also generalizes Theorem 3 of Chapter 4, is the following:

THEOREM 1. Homotopy equivalent polyhedra have isomorphic homology groups.

Proof: It clearly suffices to prove that if K and L are simplicial complexes and $|K| \simeq |L|$, then $H_q(K) \approx H_q(L)$ for all q. Therefore, let $f : |K| \to |L|$ and $g : |L| \to |K|$ be continuous maps such that $g \circ f \simeq 1_{|K|}$ and $f \circ g \simeq 1_{|L|}$. Then (II) implies that $g_{*q} \circ f_{*q} = 1_{H_q(K)}$ and $f_{*q} \circ g_{*q} = 1_{H_q(L)}$ for all q. Hence $f_{*q} : H_q(K) \to H_q(L)$ is an isomorphism, proving the theorem.

Since homeomorphisms are homotopy equivalences we have

COROLLARY 1. Theorem 3 in Sec. 4.5.

Also, the computation in Example 1 of Sec. 4.3 gives

COROLLARY 2. If a polyhedron X has the homotopy type of a point, then $H_0(X) \approx \mathbb{Z}$ and $H_i(X) = 0$ for $i \neq 0$.

Because of Theorem 1 we should not expect homology theory to be too helpful in general with regard to the problem of showing that two spaces are actually homeomorphic. For example, combining Lemma 1 and Corollary 2 we see that D^n has the same homology groups as a point. The best that we could hope for is that our theory would differentiate between distinct homotopy types. Unfortunately, it fails to do even that except in special cases. (There exist spaces that have the same homology groups but which are not homotopy equivalent.) It does, however, often present us with a computable means for obtaining negative results. In other words, if we can show that two polyhedra have different homology groups, then we can conclude that they do not have the same homotopy type.

Important Reminder for the Future: From now on we shall frequently drop the subscript "q" on maps in expressions such as "$f_{*q}(c)$" and write simply "$f_*(c)$." There will not be any confusion because it will be clear from the context which homomorphism we are talking about. Also, since homeomorphic spaces are indistinguishable from the point of view of the topologist we shall at all times feel free to replace given spaces and maps by homeomorphic spaces and the corresponding maps if it is convenient to do so. We shall leave it to the reader to justify this step in the particular case at hand. The point is that in a diagram

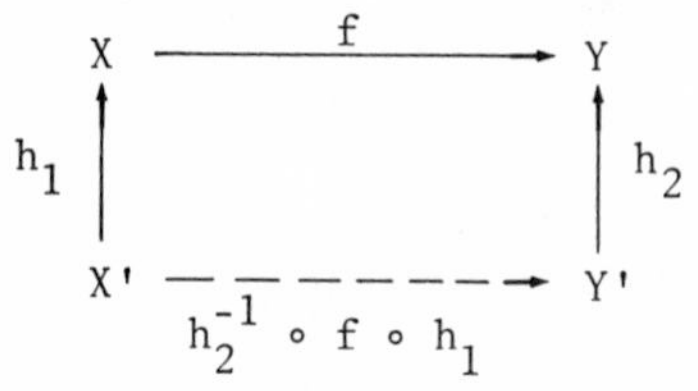

where h_1 and h_2 are homeomorphisms, everything that is true about X, Y, and f is true about X', Y', and $h_2^{-1} \circ f \circ h_1$, and vice versa. For example, in proofs we invariably replace a space by a triangulation of it because the theory we have developed works best for simplicial complexes, that is, in the diagram above usually $X' = |K|$ and $Y' = |L|$, where (K,h_1) and (L,h_2) are triangulations.

Problems

6.1. Prove that "$\simeq$" is an equivalence relation in the set of spaces.

6.2. Prove that the following pairs of spaces are homotopy equivalent:

(a) S^1 and $S^1 \times D^1$;

(b) X and $X \times D^n$ for any space X;

(c) $S^1 \times D^1$ and the Möbius strip; and

(d) $X \times Y$ and $X \times Z$, where $Y \simeq Z$ and X is arbitrary.

6.3. (a) Let $X \subset \mathbb{R}^n$. We shall say that X is star-shaped if there is some point $x_0 \in X$ such that for every $x \in X$ the line segment from x_0 to x belongs to X (see Fig. 6.1). Prove that every star-shaped region has the homotopy type of a point.

(b) In Problem 4.8 we defined the cone CX of a space X. Prove that CX always has the homotopy type of a point. [This actually extends the result in (a).]

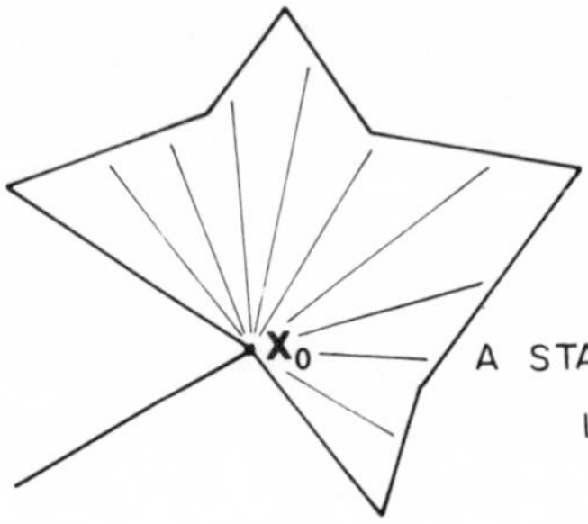

FIGURE 6.1

6.2. LOCAL HOMOLOGY GROUPS

Let X be a polyhedron. The homology groups of X which we defined in Sec. 4.5 are global invariants in the sense that their definition depends on knowledge of all of X. We shall now define some invariants whose definition depends only on the local structure of a space.

DEFINITION. Let X be a space. A local triangulation for $x \in X$ is a pair (K,h), where (K,h) is a triangulation of some neighborhood U of x in X. We say that X is locally triangulable if every $x \in X$ admits a local triangulation.

Clearly a polyhedron is locally triangulable because one can take U to be the whole space and (K,h) a triangulation. An example of a space that is not is the space $X \subset \mathbb{R}^2$ in Fig. 2.9b. The points $(0,t) \in X$ for $t \in [-1,1]$ do not have a polyhedral neighborhood. Suppose that a space X is locally triangulable. Does it follow that X itself is a polyhedron? In other words, can we piece the local triangulation together to form a global triangulation? Since $\mathbb{R}^n$ is locally triangulable but not a polyhedron, we see that the answer is no, but this is only a superficial answer. The answer in this case was no because we are restricting triangulations to have a finite number of simplices. However, we could either allow infinite triangulations or assume X is compact and then we would have an interesting and extremely difficult problem before us. The results of Kirby and Siebenmann [Ki-S] on triangulations of n-dimensional topological manifolds to which we referred earlier (see Sec. 3.1) suggest that local triangulability does not imply triangulability in general.

DEFINITION. Given a local triangulation (K,h) for $x \in X$, define the closed star of x, $st^c(x) = st^c(x,K,h)$, and the link of x, $\ell k(x) = \ell k(x,K,h)$, by

$$st^c(x) = \{\sigma \in K \mid \text{there is a } \tau \in K \text{ with } h^{-1}(x) \in \tau \text{ and } \sigma < \tau\}$$

and

$$\ell k(x) = \{\sigma \in st^c(x) \mid h^{-1}(x) \notin \sigma\}$$

It is easy to see that $st^c(x)$ and $\ell k(x)$ are simplicial complexes which depend on the local triangulation (K,h). Furthermore, if $h^{-1}(x)$ is a vertex of K, then $|st^c(x)|$ is just the closure of $st(h^{-1}(x))$ in $|K|$ as defined in Sec. 5.3. Let K be as in Fig. 5.3c. If $X = |K|$, $h = 1_X$, and $x = v_1$, then $|st^c(x)| = X$ and $\ell k(x) = \{v_0,v_2,v_3,v_4,v_2v_3,v_3v_4\}$.

DEFINITION. Let $x \in X$ and $k \geq 0$. Define the k-th local homology group of x in X, $LH_k(x;X)$, by $LH_k(x;X) = H_k(\ell k(x,K,h))$, where (K,h) is a local triangulation for $x \in X$.

Although it is not obvious from the definition, the local homology groups $LH_k(x;X)$ are well defined up to isomorphism and do not depend on the local triangulation (K,h). Before we prove this, however, let us consider two important special cases in detail. We assume that (K,h) is a local triangulation for a point x in some space X.

EXAMPLE 1. $\dim K = n \geq 1$ and $h^{-1}(x) \in \operatorname{int} \sigma$, where σ is an n-simplex of K: In this case, $st^c(x) = \bar{\sigma}$ and $\ell k(x) = \partial\bar{\sigma}$. Therefore, $|\ell k(x)| \approx S^{n-1}$ and the local homology groups $LH_k(x;X) \approx H_k(S^{n-1})$ are completely determined by the next lemma.

LEMMA 2. $H_k(S^{n-1}) \approx \mathbb{Z}$, if $k = 0$ or $n - 1$ and $n \geq 2$,
$\approx \mathbb{Z} \oplus \mathbb{Z}$, if $k = 0$ and $n = 1$,
$= 0$, if $k \neq 0$ or $n - 1$.

Proof: Let $\tau = v_0v_1 \cdots v_n$ be any n-simplex. Let $M = \overline{\tau}$ and $N = \partial\overline{\tau}$. Since S^{n-1} is homeomorphic to $\partial\tau = |N|$, it suffices to compute $H_k(N)$. We shall assume that $n \geq 2$, because the case $n = 1$ is trivial. Recalling the definition of the simplicial homology groups we see that $B_k(N) = B_k(M)$ for $0 \leq k \leq n - 2$ or $k \geq n$ and $Z_k(N) = Z_k(M)$ for all k. Therefore, $H_k(N) = H_k(M)$ for $0 \leq k \leq n - 2$ or $k \geq n$, and the corollary to Theorem 2 in Sec. 4.4 proves Lemma 2 for these values of k. It only remains to compute $H_{n-1}(N)$.

Now we know that $(\partial_{n-1} \circ \partial_n)([v_0v_1 \cdots v_n]) = 0$, and so $\Sigma = \partial_n([v_0v_1 \cdots v_n]) \in Z_{n-1}(N)$. We want to show that Σ is in fact a generator of $Z_{n-1}(N)$. Choose any $z = \sum_{i=0}^{n} a_i[v_0v_1 \cdots \hat{v}_i \cdots v_n] \in Z_{n-1}(N)$ where $a_i \in \mathbb{Z}$. Then

$$0 = \partial_{n-1}(z) = \sum_{i=0}^{n} a_i\partial_{n-1}([v_0v_1 \cdots \hat{v}_i \cdots v_n])$$

$$= \sum_{i=0}^{n} a_i\Big(\sum_{j=0}^{i-1} (-1)^j [v_0v_1 \cdots \hat{v}_j \cdots \hat{v}_i \cdots v_n]$$

$$+ \sum_{j=i+1}^{n} (-1)^{j-1} [v_0v_1 \cdots \hat{v}_i \cdots \hat{v}_j \cdots v_n] \Big)$$

Collecting coefficients of the typical oriented (n - 2)-simplex $[v_0v_1 \cdots \hat{v}_s \cdots \hat{v}_t \cdots v_n]$, where $s < t$, and setting them equal to 0 we get that $a_s = (-1)^{s+t}a_t$, that is, $z = a_0\Sigma$. Hence, $Z_{n-1}(N) = \mathbb{Z}\Sigma$. Since $B_{n-1}(N) = 0$ and $Z_{n-1}(N)$ has no elements of finite order, $H_{n-1}(N) = Z_{n-1}(N) \approx \mathbb{Z}$ and the lemma is proved.

EXAMPLE 2. $\dim K = n \geq 1$ and $h^{-1}(x) \in \operatorname{int} \sigma$, where σ is an $(n - 1)$-simplex of K which is a face of precisely t n-simplices $\tau_1, \tau_2, \cdots, \tau_t$ of K with $t \geq 1$: If $n = 1$, then $\ell k(x)$ consists of t points and so $LH_k(x;X) = 0$ for $k \neq 0$ and $LH_0(x;X)$ is a direct sum of t copies of $\mathbb{Z}$. On the other hand, suppose that $n \geq 2$ (see Fig. 6.2).

LEMMA 3. $LH_k(x;X) \approx \mathbb{Z}$, if $k = 0$,
$\approx$ direct sum of $t - 1$ copies of $\mathbb{Z}$, if $k = n - 1$,
$= 0$, if $k \neq 0$ or $n - 1$.

Proof: Let $\sigma = w_0 w_1 \cdots w_{n-1}$ and let v_j be the vertex of τ_j which is not in σ. The following facts are easily checked:

(a) $C_k(\ell k(x)) = C_k(st^c(x))$ if $0 \leq k < n - 1$.

(b) $B_{n-2}(\ell k(x)) = B_{n-2}(st^c(x))$. [[Clearly $B_{n-2}(\ell k(x)) \subset B_{n-2}(st^c(x))$ and the only real problem about showing the reverse inclusion is showing that $\partial_{n-1}([w_0 w_1 \cdots w_{n-1}])$ belongs to $B_{n-2}(\ell k(x))$, but this is implied by the equation

$$0 = \partial_{n-1}(\partial_n([v_1 w_0 w_1 \cdots w_{n-1}]))$$

$$= \partial_{n-1}([w_0 w_1 \cdots w_{n-1}]) + \sum_{i=1}^{n} (-1)^i [v_1 w_0 w_1 \cdots \hat{w}_{i-1} \cdots w_{n-1}].]]$$

(c) $st^c(x)$ is a cone. In fact, $st^c(x) = w_{n-1}M$, where $M = st^c(x) - \{\ \tau \in st^c(x) \mid \tau$ has w_{n-1} as a vertex $\}$.

It follows from (a) and (b) that $LH_k(x;X) = H_k(\ell k(x)) = H_k(st^c(x))$ for $0 \leq k \leq n - 2$. Therefore, (c) and Theorem 2 in Sec. 4.4 imply

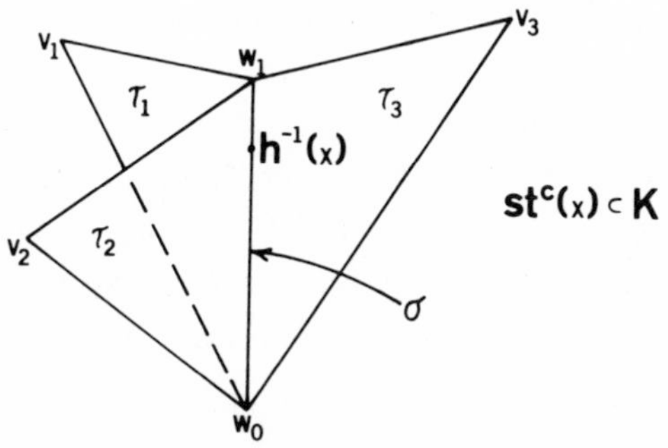

FIGURE 6.2

that $LH_0(x;X) \approx \mathbb{Z}$ and $LH_k(x;X) = 0$ if $1 \leq k \leq n - 2$. The only group that remains to be computed is $LH_{n-1}(x;X)$.

Define $a_j \in Z_{n-1}(\ell k(x)) = H_{n-1}(\ell k(x)) = LH_{n-1}(x;X)$ for $1 \leq j \leq t$ by $a_j = \partial_n([v_1 w_0 w_1 \cdots w_{n-1}] - [v_j w_0 w_1 \cdots w_{n-1}])$. Note that $a_1 = 0$. Let L_j denote the simplicial complex $\partial(\overline{\tau_1} \cup \overline{\tau_2} \cup \cdots \cup \overline{\tau_j}) - \{\sigma\}$, so that $L_1 \subset L_2 \subset \cdots \subset L_t = \ell k(x)$. We shall show that $LH_{n-1}(x;X) = H_{n-1}(L_t) = \mathbb{Z}a_1 \oplus \mathbb{Z}a_2 \oplus \cdots \oplus \mathbb{Z}a_t$ by induction on t. If $t = 1$, then $L_1 = v_1(\partial\overline{\sigma})$ is a cone and we can use Theorem 2 in Sec. 4.4 to conclude that $H_{n-1}(L_1) = 0 = \mathbb{Z}a_1$. Assume now that $H_{n-1}(L_s) = \mathbb{Z}a_1 \oplus \mathbb{Z}a_2 \oplus \cdots \oplus \mathbb{Z}a_s$ whenever $s \leq t - 1$, where $2 \leq t$, and let $s = t$. Let $a \in H_{n-1}(L_t)$. Since $H_{n-1}(st^c(x)) = 0$ by (c) above and Theorem 2 in Sec. 4.4, $a = \partial_n(\sum_{i=1}^{t} c_i[v_i w_0 w_1 \cdots w_{n-1}])$ for some $c_i \in \mathbb{Z}$. It follows that $a + c_t a_t \in H_{n-1}(L_{t-1})$. By the inductive hypothesis $a + c_t a_t = \sum_{i=1}^{t-1} c_i' a_i$ for some $c_i' \in \mathbb{Z}$. This shows that $H_{n-1}(L_t) = \mathbb{Z}a_1 + \mathbb{Z}a_2 + \cdots + \mathbb{Z}a_t$. Now suppose that $\sum_{i=1}^{t} c_i a_i = 0$ for some $c_i \in \mathbb{Z}$. But $a_t = [v_t w_1 w_2 \cdots w_{n-1}] + \cdots$ and $[v_t w_1 w_2 \cdots w_{n-1}]$ does not appear in any other a_i for $i \neq t$. Hence c_t must equal 0 so that we can use the inductive hypothesis to conclude that $c_i a_i = 0$ for $1 \leq i \leq t - 1$. Thus, $H_{n-1}(L_t) = \mathbb{Z}a_1 \oplus \mathbb{Z}a_2 \oplus \cdots \oplus \mathbb{Z}a_t$ and the proof of Lemma 3 is finished.

Next, we show that the local groups $LH_k(x;X)$ are well defined. Although we cannot show that the underlying spaces of the links $\ell k(x,K,h)$ are homeomorphic for different local triangulations we can show that they are homotopy equivalent, which is sufficient.

THEOREM 2. If (K_i,h_i), $i = 1,2$, are local triangulations for a point x in a space X, then $|\ell k(x,K_1,h_1)| \simeq |\ell k(x,K_2,h_2)|$. In particular, $H_k(\ell k(x,K_1,h_1)) \approx H_k(\ell k(x,K_2,h_2))$, so that the local homology groups $LH_k(x;X)$ are well defined up to isomorphism and independent of the local triangulation of X.

Proof: Let (K,h) be a local triangulation for $x \in X$ and let $y = h^{-1}(x) \in |K|$. It is easy to see that $st^c(x,K,h)$ and the cone $y\ell k(x,K,h)$ have the same underlying space. (If we wanted to, we could replace the local simplicial structure of K around y, $st^c(x, K,h)$, by the cone $y\ell k(x,K,h)$. See Fig. 6.3a and b.) This means that every point in $|st^c(x,K,h)|$ other than y can be written uniquely in the form $y + t(u - y)$, where $u \in |\ell k(x,K,h)|$ and $0 < t \leq 1$ (see Sec. 4.4). Set

$$A_{t,K,h} = \{ y + s(u - y) : u \in |\ell k(x,K,h)| \text{ and } 0 \leq s \leq t \}$$

and

$$B_{t,K,h} = \{ y + t(u - y) : u \in |\ell k(x,K,h)| \}$$

Then $A_{t,K,h}$ is just $|st^c(x,K,h)|$ contracted by a factor of t and $B_{t,K,h}$ is a "translation" of $|\ell k(x,K,h)|$ (see Fig. 6.3c).

After this preliminary discussion we are now ready to prove Theorem 2. Let (K_i,h_i), $i = 1,2$, be local triangulations for

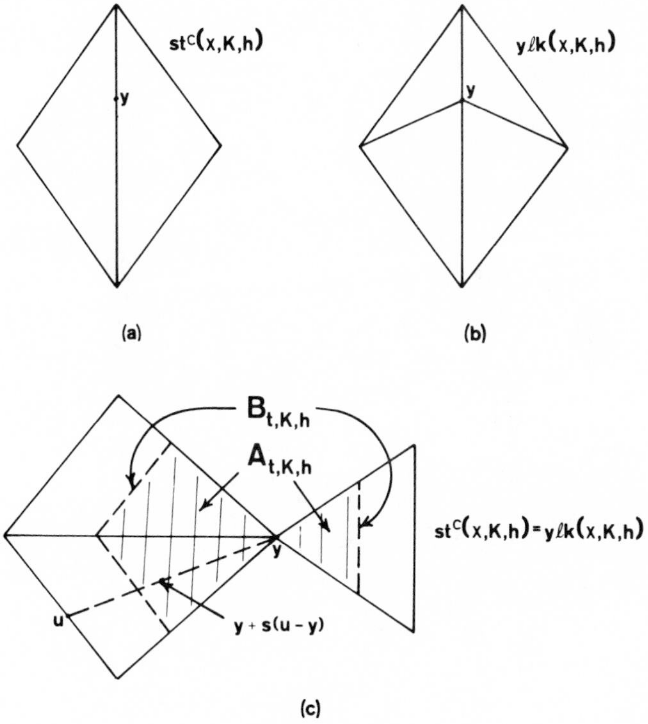

FIGURE 6.3

$x \in X$ and assume that $v_i = h_i^{-1}(x)$ is a vertex of K_i. For $c_i \in (0,1]$ set $A_i = A_{c_i,K_1,h_1}$, $A_i' = (h_1^{-1} \circ h_2)(A_{c_i,K_2,h_2})$, $B_i = B_{c_i,K_1,h_1}$, and $B_i' = (h_1^{-1} \circ h_2)(B_{c_i,K_2,h_2})$. Choose real numbers a_i and a_i', $1 \le i \le 6$, with $0 < a_6 < a_5 < \cdots < a_1 = 1$ and $0 < a_6' < a_5' < \cdots < a_1' \le 1$ such that $A_{2j} \supset A_j'$ and $A_{2j}' \supset A_j$ for $1 \le j \le 3$ (see Fig. 6.4). This is always possible because $A_{t,K,h}$ is compact and can be made as small as desired by choosing a small enough t. Finally, if $u \in \ell k(x,K_1,h_1)$, $u' \in \ell k(x,K_2,h_2)$, and $0 < t,t' \le 1$, then we shall let (u,t) and $(u',t')'$ denote the points $v_1 + t(u - v_1)$ and $(h_1^{-1} \circ h_2)(v_2 + t'(u' - v_2))$, respectively.

Now $\ell k(x,K_1,h_1) = B_1$ is homeomorphic to B_2 via the map which sends $u = (u,1)$ to (u,a_2). Similarly, $\ell k(x,K_2,h_2)$ is homeomorphic to $(h_2^{-1} \circ h_1)(B_2')$, which in turn is homeomorphic to B_2'. Therefore, in order to show that $\ell k(x,K_1,h_1) \simeq \ell k(x,K_2,h_2)$ it suffices to show that $B_2 \simeq B_2'$. In other words, we only need to define maps $\varphi : B_2 \to B_2'$ and $\psi : B_2' \to B_2$ such that $\psi \circ \varphi \simeq 1_{B_2}$ and $\varphi \circ \psi \simeq 1_{B_2'}$. Observe that B_2 lies strictly between B_1' and B_2' and that

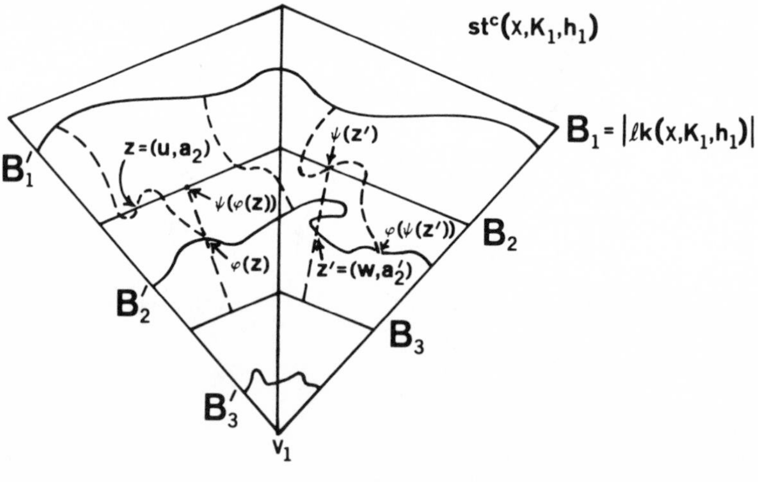

FIGURE 6.4

B_2' lies strictly between B_2 and B_3. Since the parts between B_1' and B_2' and between B_2 and B_3 are naturally homeomorphic to $B_2' \times [a_2',a_1']$ and $B_2 \times [a_3,a_2]$, respectively, the maps φ and ψ are defined to be the obvious projections. (In Fig. 6.4, the curved dotted lines correspond to the lines $p \times [a_2',a_1']$, $p \in B_2'$, and the straight dotted lines correspond to $q \times [a_3,a_2]$, $q \in B_2$.) More precisely, if $(u,a_2) = (u',t')'$ and $(w',a_2')' = (w,t)$, then we define $\varphi((u,a_2)) = (u',a_2')'$ and $\psi((w',a_2')') = (w,a_2)$. Next, define a homotopy λ_s between 1_{B_2} and $\psi \circ \varphi$ by $\lambda_s((u,a_2)) = (u_s,a_2)$, where u_s is defined by the equation $(u',t' + s(a_2' - t'))' = (u_s, t_s)$. Then $\lambda_0 = 1_{B_2}$, $\lambda_1 = \psi \circ \varphi$, and what λ_s does is move the point (u,a_2) along the projection on B_2 of the dotted line from z to $\varphi(z)$ in Fig. 6.4. A homotopy μ_s between $1_{B_2'}$ and $\varphi \circ \psi$ can be defined analogously by $\mu_s((w',a_2')') = (w_s',a_2')'$, where w_s' satisfies $(w,t + s(a_2 - t)) = (w_s',t_2')'$. This finishes the proof of Theorem 2.

If follows from Theorem 2 that the local homology groups are local topological invariants of a space.

Problems

6.4. Let $X = |K|$, where K is the simplicial complex in Fig. 6.5. Compute the local homology groups $LH_k(x;X)$ for $x = v_0$ or v_1.

6.5. Lemma 3 has an alternate proof. First, let X and Y be two nonempty polyhedra and define the wedge of X and Y, denoted by $X \vee Y$, to be any space C with the property that $C = A \cup B$, where $A \approx X$, $B \approx Y$, and $A \cap B$ consists of a single point. It is easy to show that $X \vee Y$ always exists and that, up to homeomorphism, it is a well-defined polyhedron. Intuitively, $X \vee Y$ is obtained by pasting X and Y together along a point from each. Clearly, $(X \vee Y) \vee Z$ and $X \vee (Y \vee Z)$ are homeomorphic so that one can drop the parantheses when considering iterated wedges.

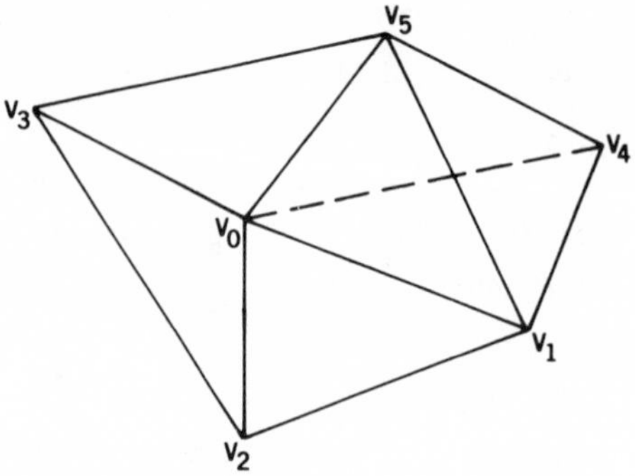

$\mathbf{K} = \overline{v_0v_2v_5} \cup \overline{v_0v_1v_2} \cup \overline{v_0v_1v_3} \cup \partial(\overline{v_0v_3v_4v_5})$

FIGURE 6.5

(a) Prove that $H_k(X \vee Y) \approx H_k(X) \oplus H_k(Y)$ for all $k \neq 0$.

(b) In Example 2 prove that $\ell k(x) \simeq S^{n-1} \vee S^{n-1} \vee \cdots \vee S^{n-1}$ [(t - 1)-fold wedge].

(c) Prove Lemma 3 using (a), (b), and Lemma 2.

6.6. Prove that an open subset of $\mathbb{R}^n$ is never homeomorphic to an open subset of $\mathbb{R}^m$ if $n \neq m$.

6.3. SOME INVARIANCE THEOREMS

We have said that the topologist's main goal is to classify spaces up to homeomorphism using computable invariants. This goal is justified by the belief that the relation of being "homeomorphic" reflects the intuitive notion of equivalence of spaces in the sense of the rubber sheet geometry discussed in Sec. 2.1. The objective of this section is to use homology theory to show that certain obvious geometric notions associated with simplicial complexes are indeed homeomorphism (or topological) invariants. If they were not, we would have a very unsatisfactory theory.

The first standard result along these lines is the following:

THEOREM 3. (a) S^n and S^m do not have the same homotopy type unless $n = m$. In particular, S^n is homeomorphic to S^m only if $n = m$.

(b) $\mathbb{R}^n$ is not homeomorphic to $\mathbb{R}^m$ unless $n = m$.

Proof: Part (a) follows from Lemma 2 and Theorem 1. Part (b) already follows from Problem 6.6 which is an easy application of local homology groups, but there is also a more direct proof using part (a) that does not need the results from Sec. 6.2. Let $e_i = (0,0,\cdots,0,1) \in \mathbb{R}^i$ and define the stereographic projection

$$P_n : S^n - e_{n+1} \to \mathbb{R}^n$$

by

$$P_n((x_1,x_2,\cdots,x_{n+1})) = (x_1/(1 - x_{n+1}),x_2/(1 - x_{n+1}),\cdots,x_n/(1 - x_{n+1}))$$

The map P_n can also be described geometrically as follows: If $x \in S^n - e_{n+1}$ and if L_x is the ray which starts at e_{n+1} and passes through x, then $P_n(x)$ is the point which is the intersection of L_x with $\mathbb{R}^n$ (see Fig. 6.6 for the case $n = 2$). It is easy to check that P_n is a homeomorphism, so that we may think of S^n as $\mathbb{R}^n \cup \infty$, where ∞ is the "point of $\mathbb{R}^n$ at infinity." Also, $P_n|S^{n-1} = 1_{S^{n-1}}$.

Suppose now that $h : \mathbb{R}^n \to \mathbb{R}^m$ is a homeomorphism. Define $H : S^n \to S^m$ by $H(x) = (P_m^{-1} \circ h \circ P_n)(x)$ if $x \neq e_{n+1}$ and $H(e_{n+1}) = e_{m+1}$. The map H will be a homeomorphism and therefore $n = m$ by part (a).

Theorem 3(b) asserts the topological invariance of the dimension of Euclidean space. We want to show next that the dimension of a simplicial complex is also a topological invariant. This is a less trivial fact.

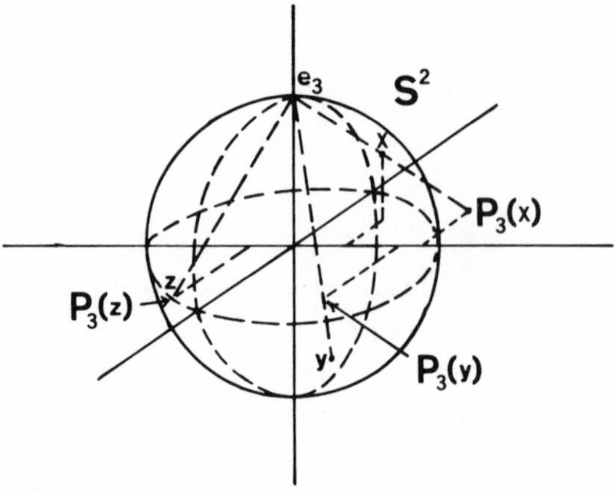

FIGURE 6.6

THEOREM 4 (Invariance of Dimension). If K and L are simplicial complexes such that $|K| \approx |L|$, then $\dim K = \dim L$.

Proof: If K is a 0-dimensional complex, then $|K|$ consists of a finite set of points and the theorem is obvious. Suppose now that $\dim K = n \geq 1$. Let σ be an n-dimensional simplex of K and choose $x \in \operatorname{int} \sigma$. It follows from the computation of the local homology groups in Example 1 of Sec. 6.2 that $LH_{n-1}(x;|K|) \approx \mathbb{Z}$ if $n \geq 2$, $LH_{n-1}(x;|K|) \approx \mathbb{Z} \oplus \mathbb{Z}$ if $n = 1$, and $LH_k(x;|K|) = 0$ if $k \geq n$. Furthermore, if y is an arbitrary point of $|K|$, then $LH_k(y;|K|) = 0$ for $k \geq n$, because $\dim \ell k(y,K,1_{|K|}) \leq n - 1$. Therefore, $\dim K$ can be characterized by the equation $\dim K = \min \{ k \mid LH_t(y;|K|) = 0 \text{ for all } t \geq k \text{ and all } y \in |K| \}$. Since the right-hand side of this equation is a topological invariant by Theorem 2 in Sec. 6.2, we have proved Theorem 4.

DEFINITION. Let X be a polyhedron and let (K,φ) be a triangulation of X. Define the dimension of X, $\dim X$, by $\dim X = \dim K$.

Theorem 4 implies that the dimension of a polyhedron is well defined so that the notion of dimension is an intrinsic property of spaces as it ought to be and not an accidental property of some triangulation. Thus, one could use the dimensionality of a space to distinguish between homeomorphism types. For example, since $\dim S^k = k$, we see again that S^n cannot be homeomorphic to S^m unless $n = m$.

There is a relative version of Theorem 4 which we shall need later.

DEFINITION. A simplicial pair (K,L) is a pair of simplicial complexes K and L such that L is a subcomplex of K. If $K \neq L$, then define $\dim (K,L)$ to be the largest dimension of the simplices in $K - L$. If $K = L$, then set $\dim (K,L) = -1$.

Note that $\dim (K,\emptyset) = \dim K$. It is possible to have $\dim L > \dim (K,L)$.

ADDENDUM TO THEOREM 4 (Invariance of Relative Dimension). Let (K,L) and (K',L') be simplicial pairs and suppose $h : |K| \to |K'|$ is a homeomorphism such that $h(|L|) = |L'|$. Then $\dim (K,L) = \dim (K',L')$.

Proof: Although $K - L$ is not a simplicial complex for an arbitrary simplicial pair (K,L), $\overline{K - L} = \{ \sigma \in K \mid \text{there is a } \tau \in K - L \text{ and } \sigma < \tau \}$ is. Furthermore, $\dim \overline{K - L} = \dim (K,L)$. Since the h in the Addendum clearly maps $|\overline{K - L}|$ homeomorphically onto $|\overline{K' - L'}|$, it follows from Theorem 4 that $\dim (K,L) = \dim \overline{K - L} = \dim \overline{K' - L'} = \dim (K',L')$ and we are done.

DEFINITION. A polyhedral pair (X,Y) is a pair of polyhedra X and Y such that $Y \subset X$ and such that X admits a triangulation (K,φ) with $Y = \varphi(|L|)$ for some subcomplex L of K. The pair $((K,L),\varphi)$ is called a triangulation for (X,Y). Set $\dim (X,Y) = \dim (K,L)$.

If (X,Y) is a polyhedral pair, then $\dim (X,Y)$ is well defined by the Addendum to Theorem 4. Note that (X,Y) is more than just a pair of polyhedra because the definition also implies that Y is "nicely" imbedded in X. This last property will be important later in Sec. 7.5 when we consider the problem of extending maps.

Next, if K is a simplicial complex, then define a subspace $\delta_s^r K$ of $|K|$ by

$$\delta_s^r K = \{ x \in \sigma \mid \sigma \text{ is an s-simplex of } K \text{ that is a face of exactly } r \text{ distinct } (s+1)\text{-simplices of } K \text{ but not a face on any } (s+2)\text{-simplex} \}$$

Observe that $\delta_s^r K$ may be empty.

LEMMA 4. If K and L are simplicial complexes and $h : |K| \to |L|$ is a homeomorphism, then $h(\delta_s^0 K) = \delta_s^0 L$ for all $s \geq 0$.

Proof: Since $\delta_0^0 K$ is just the collection of isolated points of $|K|$, the lemma is clearly true if $s = 0$. Therefore, assume that $s \geq 1$. Define a subspace $\eta_s K$ of $|K|$ by

$$\eta_s K = \{ x \in |K| : \text{there is a neighborhood } U \text{ of } x \text{ in } |K| \text{ such that if } y \in U, \text{ then } LH_{s-1}(x;|K|) \approx \mathbb{Z} \oplus \mathbb{Z}, \text{ if } s = 1, \text{ and } \approx \mathbb{Z}, \text{ if } s > 1 \}$$

CLAIM. $\delta_s^0 K = c\ell(\eta_s K)$ for $s \geq 1$.

To prove the claim, suppose that σ is an s-simplex of K and $\sigma \subset \delta_s^0 K$. If $x \in \operatorname{int} \sigma$, then it follows from Example 1 in Sec. 6.2 that $x \in \eta_s K$. This obviously implies that $\delta_s^0 K \subset c\ell(\eta_s K)$. Conversely, let $x \in \eta_s K$ and choose a simplex $\sigma \in K$ of maximal dimension k such that $x \in \sigma$. Then $k \geq s$ because the maximality of k guarantees that $\dim \ell k(x,K,1_{|K|}) \leq k - 1$. On the other hand, if $k > s$, then, again by Example 1 in Sec. 6.2, there would exist points $y \in \sigma$ arbitrarily close to x with $LH_{s-1}(y;|K|) \approx \mathbb{Z}$ or 0 depending on whether $s > 1$ or $s = 1$. This contradicts the existence of the neighborhood U in the definition of $\eta_s K$. Therefore $k = s$ and $\sigma \subset \delta_s^0 K$, which shows that $\eta_s K \subset \delta_s^0 K$. Since $\delta_s^0 K$ is a closed set, it follows that $c\ell(\eta_s K) \subset \delta_s^0 K$ and the claim is proved.

Now let $h : |K| \to |L|$ be a homeomorphism between the underlying spaces of two simplicial complexes. From Theorem 2 it is clear that $h(\eta_s K) = \eta_s L$. Hence $h(\delta_s^0 K) = h(c\ell(\eta_s K)) = c\ell(h(\eta_s K)) = c\ell(\eta_s L) = \delta_s^0 L$ and Lemma 4 is proved.

We can rephrase part of Lemma 4 using some common terminology.

DEFINITION. A simplicial complex K is said to be homogeneously n-dimensional if every simplex of K is a face of some n-simplex in K.

THEOREM 5 (Invariance of "homogeneously n-dimensional"). If K and L are simplicial complexes and $|K| \approx |L|$, then K is homogeneously n-dimensional if and only if L is.

Proof: This is an immediate consequence of Lemma 4, Theorem 4, and the observation that the property of being homogeneously n-dimensional is characterized by the fact that $\dim K = n$ and $\delta_0^0 K = \delta_1^0 K = \cdots = \delta_{n-1}^0 K = \emptyset$.

Theorem 5 justifies

DEFINITION. Let X be a polyhedron and let $h : |K| \to X$ be a triangulation. We shall say that X is homogeneously n-dimensional if K is.

In order to prove the invariance of the boundary of a complex we shall need

LEMMA 5. If K and L are simplicial complexes and $h : |K| \to |L|$ is a homeomorphism, then $h(\delta_s^r K) = \delta_s^r L$ for $s,r \geq 0$ and $r \neq 2$.

Proof: We begin by considering a special case. Suppose that K is an n-dimensional complex and let us give an alternate description of the set $\delta_{n-1}^r K$ for $r,n \geq 1$ and $r \neq 2$. Define subspaces $\mu^r K$ of $|K|$ by

$$\mu^r K = \{\, x \in |K| : LH_{n-1}(x;|K|) \text{ is isomorphic to a direct sum of } r \text{ or } r-1 \text{ copies of } \mathbb{Z} \text{ depending on whether } n = 1 \text{ or } n > 1, \text{ respectively} \,\}$$

Let σ be an $(n - 1)$-simplex of K and suppose that $\sigma \subset \delta_{n-1}^r K$. If $x \in \operatorname{int} \sigma$, it follows from Example 2 in Sec. 6.2 that $x \in \mu^r K$.

This implies that $\delta^r_{n-1}K \subset c\ell(\mu^r K)$. Although we shall not be able to prove that $\delta^r_{n-1}K$ and $c\ell(\mu^r K)$ are always the same set (see Fig. 6.7), we shall be able to use $c\ell(\mu^r K)$ to give a topological characterization of $\delta^r_{n-1}K$.

CLAIM 1. $c\ell(\mu^r K)$ is either empty or the underlying space of a k-dimensional subcomplex $\lambda^r K$ of K, where $k \leq n - 1$.

Suppose that $x \in \mu^r K$ and let σ be the unique simplex of K such that $x \in \text{int}\ \sigma$. If $y \in \text{int}\ \sigma$, then $\ell k(y,K,1_{|K|}) = \ell k(x,K,1_{|K|})$, which means that $LH_{n-1}(y;|K|) \approx LH_{n-1}(x;|K|)$ and $y \in \mu^r K$. Therefore, $\text{int}\ \sigma \subset \mu^r K$ and $\sigma \subset c\ell(\mu^r K)$, so that $c\ell(\mu^r K) = |\lambda^r K|$, where

$$\lambda^r K = \{\ \tau \in K \mid \text{there exists a } \sigma \in K \text{ and an } x \in \mu^r K \text{ with } x \in \text{int}\ \sigma \text{ and } \tau < \sigma\ \}$$

This proves the first part of Claim 1. On the other hand, assume that $\lambda^r K$ contains an n-simplex τ. If $x \in \text{int}\ \tau$, then it is easy to see that $x \in \mu^r K$. Since $r \neq 2$, the definition of $\mu^r K$ and the computation in Example 1 of Sec. 6.2 imply contradictory results about the local homology groups of x in $|K|$. Thus, $\lambda^r K$ cannot contain any n-simplices and Claim 1 is proved.

Next, suppose that σ is an $(n - 1)$-simplex in $\lambda^r K$. It follows from Example 2 in Sec. 6.2 that the only way that $x \in \text{int}\ \sigma$ can

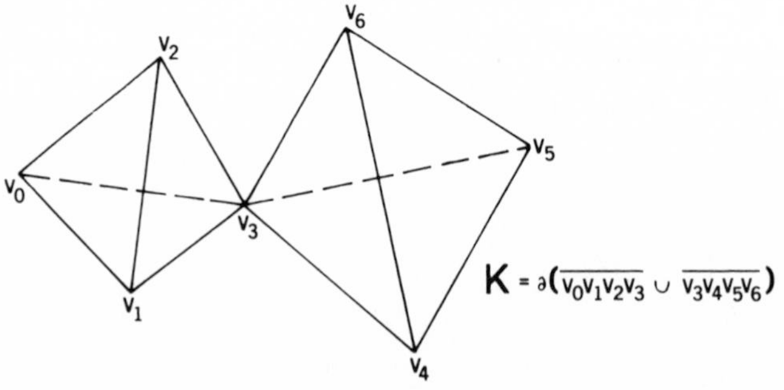

$\delta^3_1 K = \phi$ BUT $c\ell(\mu^3 K) = \{v_3\}$

FIGURE 6.7

have the correct local homology groups in $|K|$ is if σ is a face of r distinct n-simplices of K. In other words, $\sigma \subset \delta^r_{n-1}K$ and we have proved

CLAIM 2. $\delta^r_{n-1}K = \delta^0_{n-1}(\lambda^r K)$.

Now we are ready to prove Lemma 5. If $r = 0$, then we are done by Lemma 4. Assume that $r > 0$. Since $\delta^r_s K \subset \delta^0_{s+1}K$ holds for any simplicial complex K whenever $r > 0$ and $s \geq 0$, and since $\delta^0_{s+1}K$ is clearly the underlying space of a subcomplex of K, we may assume without loss of generality by Lemma 4 and Theorem 4 that $|K| = \delta^0_{s+1}K$ and $|L| = \delta^0_{s+1}L$. This means that if $r \neq 2$, then the discussion leading up to Claim 1 and 2 is applicable (set $n = s + 1$). By Theorem 2, $c\ell(\mu^r K)$ is a topological invariant, that is, $h(c\ell(\mu^r K)) = c\ell(\mu^r L)$, so that we can make an additional reduction and assume $K = \lambda^r K$ and $L = \lambda^r L$. By Claim 2, $\delta^r_s K = \delta^0_s K$ and $\delta^r_s L = \delta^0_s L$ if $r \neq 2$, and Lemma 4 implies that $h(\delta^r_s K) = \delta^r_s L$. Lemma 5 is proved.

NOTE. There is a good reason for the assumption $r \neq 2$ in Lemma 5. To have Lemma 5 hold for the case $r = 2$ would be almost like saying that there is an essentially unique way to triangulate a space. For example, if $K = \{0,2,[0,2]\}$ and $L = \{0,1,2,[0,1],[1,2]\}$, then $\delta^2_0 K = \emptyset \neq \{1\} = \delta^2_0 L$.

If K is an n-dimensional simplicial complex, then $|\partial K| = \delta^1_0 K \cup \delta^1_1 K \cup \cdots \cup \delta^1_{n-1}K$. Therefore, Lemma 5 clearly implies

THEOREM 6 (Invariance of Boundary). If K and L are simplicial complexes and $h : |K| \to |L|$ is a homeomorphism, then $h(|\partial K|) = |\partial L|$. In particular, the boundary of a polyhedron is a well-defined subset.

Finally, let us state a frequently encountered theorem that is closely related to the results of this section and which is an extension of the usual inverse function theorem in analysis:

THEOREM C (Invariance of Domain). If U and V are homeomorphic subsets of $\mathbb{R}^n$ and if U is open, then so is V.

Theorem C is proved in most books on algebraic topology. See, for example, [Hoc-Y, p. 277]. The theorem is an easy consequence of the generalized Jordan curve theorem (see Theorem G in Sec. 7.6).

Problems

6.7. Prove that the cylinder $S^1 \times D^1$ is not homeomorphic to the Möbius strip.

6.8. Consider the space X in Fig. 6.8 consisting of the 2-sphere with two circles attached at a point. Show that X has the same homology groups as the torus $S^1 \times S^1$ but that X is not homeomorphic to $S^1 \times S^1$.

6.9. Assume that each point of a space X has a neighborhood which is homeomorphic to $\mathbb{R}^2$. Prove that every triangulation of X is proper.

6.4. HOMOLOGY WITH ARBITRARY COEFFICIENTS; THE MOD 2 HOMOLOGY GROUPS

When we defined the chain group $C_q(K)$ of a simplicial complex K, it amounted to taking formal integer sums of oriented q-simplices of K. From an abstract point of view it is easy to generalize this definition.

Let G be an arbitrary abelian group and let S_q again denote the set of oriented q-simplices of K.

DEFINITION. The group $C_q(K;G)$ of q-chains of K with coefficients in G is defined as follows:

(1) If $0 \leq q \leq \dim K$, then

$$C_q(K;G) = \{ f : S_q \to G \mid \text{if } q \geq 1, \text{ then } f(-\alpha) = -f(\alpha) \text{ for all } \alpha \in S_q \}$$

(2) If $q < 0$ or $q > \dim K$, then

$$C_q(K;G) = 0$$

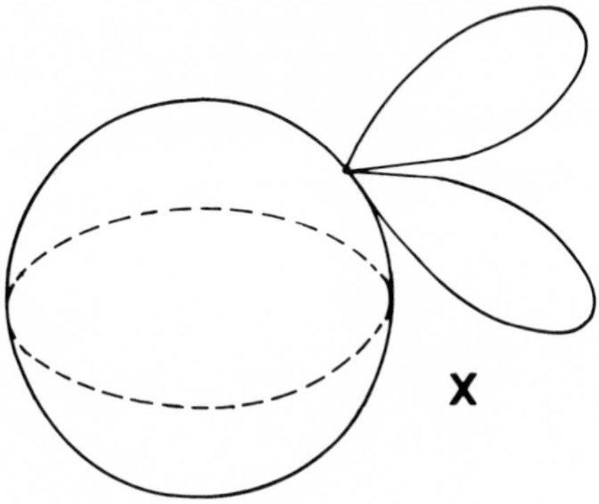

FIGURE 6.8

Note that $C_q(K;\mathbb{Z}) = C_q(K)$. The operation $+$ which makes $C_q(K;G)$ into a group is defined as in the case of $C_q(K)$ and is just pointwise addition of functions. If $g \in G$ and $\alpha \in S_q$, then define a function $g\cdot\alpha : S_q \to G$ by $(g\cdot\alpha)(\beta) = 0$, if $\beta \neq \pm\alpha$, and $(g\cdot\alpha)(\varepsilon\alpha) = \varepsilon g$, where $\varepsilon = \pm 1$. Clearly, $g\cdot\alpha \in C_q(K;G)$. Let S_q^+ be a subset of S_q as in Sec. 4.3 and observe that if $f \in C_q(K;G)$, then

$$f = \sum_{\alpha\in S_q^+} f(\alpha)\cdot\alpha$$

In other words, every element of $C_q(K;G)$ can be written in the form

$$\sum_{\alpha\in S_q^+} g_\alpha\cdot\alpha$$

where $g_\alpha \in G$, and this representation is also unique. This explains the term "q-chains with coefficients in G." The addition of q-chains corresponds to collecting the coefficients and adding, that is,

$$\sum_{\alpha\in S_q^+} g_\alpha\cdot\alpha + \sum_{\alpha\in S_q^+} g'_\alpha\cdot\alpha = \sum_{\alpha\in S_q^+} (g_\alpha + g'_\alpha)\cdot\alpha$$

The element $0\cdot[\sigma^q]$ is the zero element in $C_q(K;G)$.

All the constructions and lemmas of Chapters 4 and 5 generalize in a straightforward manner to give us a homology theory with coefficients in G. Essentially the only changes that need to be made is to replace any reference to $\mathbb{Z}$ by G. We shall go over the main steps very quickly and leave it to the reader to fill in the details.

First of all, the boundary homomorphisms

$$\partial_q : C_q(K;G) \to C_{q-1}(K;G)$$

are defined basically as before. They are the zero map when $q \leq 0$ or $q > \dim K$ and when $1 \leq q \leq \dim K$ they are determined by the condition that

$$\partial_q(g\cdot[v_0v_1\cdots v_q]) = \sum_{i=0}^{q} (-1)^i g\cdot[v_0v_1\cdots\hat{v}_i\cdots v_q]$$

for all oriented q-simplices $[v_0v_1\cdots v_q]$ of K and $g \in G$. One can prove that $\partial_{q-1} \circ \partial_q = 0$ for all q so that $B_q(K;G) = \operatorname{Im} \partial_{q+1} \subset Z_q(K;G) = \operatorname{Ker} \partial_q$. This means that it makes sense to define $H_q(K;G)$, the q-th homology group of K with coefficients in G, by

$$H_q(K;G) = Z_q(K;G)/B_q(K;G)$$

Next, let K and L be simplicial complexes and $f : K \to L$ a simplicial map. Define homomorphisms

$$f_{\#q} : C_q(K;G) \to C_q(L;G)$$

as before. If $q < 0$ or $q > \dim K$, then $f_{\#q} = 0$; and if $0 \leq q \leq \dim K$, then, for all $g \in G$ and $[v_0v_1\cdots v_q] \in S_q$,

$$\begin{aligned} f_{\#q}(g\cdot[v_0v_1\cdots v_q]) &= g\cdot[f(v_0)f(v_1)\cdots f(v_q)], \text{ if } f(v_i) \neq f(v_j) \text{ for } i \neq j, \text{ and} \\ &= 0, \text{ otherwise} \end{aligned}$$

The maps $f_{\#q}$ again commute with the boundary maps ∂_q and hence define homomorphisms

$$f_{*q} : H_q(K;G) \to H_q(L;G)$$

for all q. The groups $H_q(K;G)$ and the homomorphisms f_{*q} satisfy properties (I) and (II) in Sec. 6.1 because all the lemmas and theorems needed to prove them are true in this situation also, after obvious changes are made to account for the coefficients G rather than $\mathbb{Z}$. In particular, the homology groups with coefficients in G are topological invariants. Therefore, if X is a polyhedron, then we can define

$$H_q(X;G) = H_q(K;G)$$

where K is a simplicial complex which triangulates X, and this group will be well defined up to isomorphism.

The groups $H_q(X;G)$ are important invariants associated to a space, even though it turns out that they are already determined by the ordinary homology groups $H_q(X;\mathbb{Z})$ and $H_{q-1}(X;\mathbb{Z})$ by means of the so-called "universal coefficient theorem." However, we shall stop discussing the general theory now and concentrate on the special case $G = \mathbb{Z}_2$. The groups $H_q(X;\mathbb{Z}_2)$ are usually called the mod 2 homology groups of X. This case is the most geometrically interesting after the case $G = \mathbb{Z}$ and it will be worthwhile to rework the definition of $H_q(X;\mathbb{Z}_2)$ in detail by itself, just as we did for $H_q(X)$.

Let K be a simplicial complex and let $\mathbb{Z}_2$ be the set $\{0,1\}$ with the standard addition and multiplication modulo 2. Since $+1 = -1$ in $\mathbb{Z}_2$, it is unnecessary to orient the simplices of K to define $C_q(K;\mathbb{Z}_2)$. The point is that if $f \in C_q(K;\mathbb{Z}_2)$, then $f(-[\sigma]) = -f([\sigma]) = f([\sigma])$ for every oriented q-simplex $[\sigma] \in S_q$, so that f determines a well-defined function g_f from the set T_q of q-simplices of K to $\mathbb{Z}_2$ given by $g_f(\sigma) = f([\sigma])$. In other words, we can identify $C_q(K;\mathbb{Z}_2)$ with $\{\ g : T_q \to \mathbb{Z}_2\ \}$ because the correspondence $f \to g_f$ is an isomorphism. Next, if $\sigma \in T_q$, define $\sigma_F : T_q \to \mathbb{Z}_2$ by $\sigma_F(\tau) = 0$, if $\tau \neq \sigma$, and $\sigma_F(\sigma) = 1$. Thus, any function $g : T_q \to \mathbb{Z}_2$ satisfies $g = \sum_{\sigma \in T_q} g(\sigma)\sigma_F$. This shows that we can think of $C_q(K;\mathbb{Z}_2)$ as consisting of linear sums of q-simplices of K, that is, if we drop the subscript "F" on σ_F (which we shall do from now on), a typical element of $C_q(K;\mathbb{Z}_2)$ will look like $\sigma_1 + \sigma_2 + \cdots + \sigma_k$ for $\sigma_i \in T_q$. To add elements of $C_q(K;\mathbb{Z}_2)$ we merely collect the coefficients of like simplices and add, but we must remember that $\sigma + \sigma = 0$ for every $\sigma \in T_q$ because $2 = 0$ in $\mathbb{Z}_2$. For example, if $K = \overline{v_0v_1v_2v_3}$, then $v_0v_1 + v_1v_2 + v_2v_0$, $v_1v_2 + v_2v_3 + v_3v_1 \in C_1(K;\mathbb{Z}_2)$ and $(v_0v_1 + v_1v_2 + v_2v_0) + (v_1v_2 + v_2v_3 + v_3v_1) = v_0v_1 + v_3v_1 + v_2v_3 + v_2v_0$.

The boundary map $\partial_q : C_q(K;\mathbb{Z}_2) \to C_{q-1}(K;\mathbb{Z}_2)$ has a simple description:

$$\partial_q(v_0v_1 \cdots v_q) = \sum_{i=0}^{q} v_0v_1 \cdots \hat{v}_i \cdots v_q$$

In other words, the boundary of a q-simplex is the sum of all of its (q - 1)-dimensional faces. The fact that $\partial_{q-1} \circ \partial_q = 0$ is easy to see in this case because each (q - 2)-dimensional face of a q-simplex belongs to precisely two (q - 1)-dimensional faces of the simplex and $2 = 0$ in $\mathbb{Z}_2$. The mod 2 homology groups of K, $H_q(K;\mathbb{Z}_2)$, are now defined to be the usual quotient group of the kernel of ∂_q by the image of ∂_{q+1}.

REMARK. We could represent a chain $c \in C_q(K;\mathbb{Z}_2)$ by the subset $|c|$ of $|K|$ which is the union of all the q-simplices appearing in c. (The set $|c|$ is usually called the support of c.) Then $|c + d|$ is the closure of the symmetric difference of $|c|$ and $|d|$, that is, $|c + d| = c\ell(\,(|c| \cup |d|) - (|c| \cap |d|)\,)$. Also, $|\partial_q(c)|$ is the union of all the (q - 1)-simplices which are the face of an odd number of q-simplices appearing in c. Recalling our intuitive discussion of "holes" in Sec. 4.1, we see that it is the mod 2 homology theory which makes precise the first approach in that section.

Since $\mathbb{Z}_2$ is a field, $H_q(K;\mathbb{Z}_2)$ is actually a vector space over $\mathbb{Z}_2$ [as are $C_q(K;\mathbb{Z}_2)$, $Z_q(K;\mathbb{Z}_2)$, and $B_q(K;\mathbb{Z}_2)$].

DEFINITION. The q-th connectivity number of K, $\varkappa_q(K)$, is defined to be the dimension of the vector space $H_q(K;\mathbb{Z}_2)$ over $\mathbb{Z}_2$. If X is a polyhedron, the q-th connectivity number of X, $\varkappa_q(X)$, is defined to be the q-th connectivity number of K, where (K,φ) is a triangulation of X.

Connectivity numbers are the mod 2 analogs of Betti numbers. They are well defined in the case of a polyhedron because of the topological invariance of the groups $H_q(X;\mathbb{Z}_2)$. The motivation behind the terminology can be found in Sec. 6.6 and Problem 6.11(b). In particular, we shall see that if S is a surface, then $\varkappa_1(S)$ is just what we called the connection number of S in Problem 3.16(b).

Problems

6.10. Fill in the details that were omitted in our discussion of the homology theory with coefficients in G. In particular, show that the relevant lemmas and theorems in Chapters 4 and 5 have analogs in this theory and prove them.

6.11. (a) If X is a point, prove that $H_0(X;G) \approx G$ and $H_q(X;G) = 0$ for $q \neq 0$.

(b) Let X be a polyhedron. Prove that $H_0(X;G)$ is isomorphic to a direct sum of as many copies of G as there are components of X. In particular, the 0-th connectivity number of X, $\varkappa_0(X)$, is nothing but the number of components of X.

(c) Compute $H_q(S^n;G)$ for all q.

6.12. Compute $H_q(X;\mathbb{Z}_2)$ for all q, where

(a) $X = S^1 \times S^1$

(b) $X = P^2$

(c) X is the Klein bottle

6.5. PSEUDOMANIFOLDS; ORIENTABILITY

In this section we shall use the mod 2 homology groups to study pseudomanifolds.

DEFINITION. A connected polyhedron X is said to be an n-dimensional pseudomanifold if it admits a triangulation (K,φ) satisfying

1. K is homogeneously n-dimensional
2. Every (n - 1)-simplex of K is a face of at most two n-simplices of K

3. If σ and σ' are two distinct n-simplices of K, then there exists a sequence $\sigma_1, \sigma_2, \cdots, \sigma_k$ of n-simplices in K such that $\sigma_1 = \sigma$, $\sigma_k = \sigma'$, and σ_i meets σ_{i+1} in an (n - 1)-dimensional face for $1 \leq i < k$

If $\partial X = \emptyset$, then we shall call X a closed n-dimensional pseudomanifold.

Every surface is clearly a closed 2-dimensional pseudomanifold, and so are all the spheres S^n. An n-simplex σ is an n-dimensional pseudomanifold with boundary $\partial\sigma$. Not every closed n-dimensional pseudomanifold is an n-dimensional manifold (see Sec. 3.1 for a definition). The basic difficulty that may occur can be seen by means of the simple example in Fig. 6.9 which shows a 2-dimensional pseudomanifold that is not a bordered surface because it is not locally Euclidean at p. On the other hand, every triangulable n-dimensional manifold is an n-dimensional pseudomanifold (Problem 6.16). Figure 6.9 also shows that the boundary of an n-dimensional pseudomanifold need not be an (n - 1)-dimensional pseudomanifold. Nevertheless, pseudomanifolds have nice enough properties to warrant a separate study since one meets them frequently.

If the properties in the definition of a pseudomanifold were not topological invariants, we would have an awkward definition. We shall prove their invariance shortly, but first let us prove

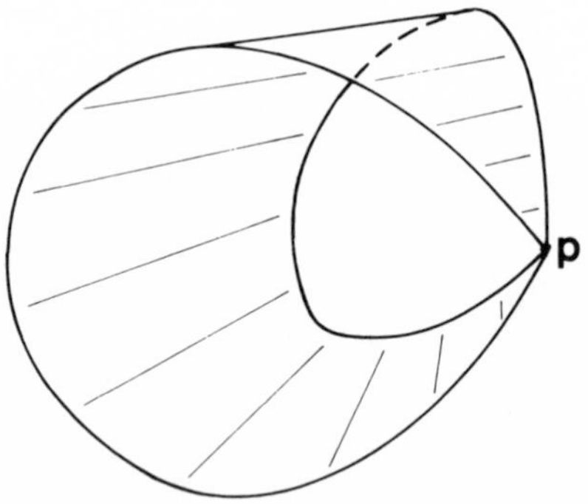

FIGURE 6.9

LEMMA 6. Let X be an n-dimensional pseudomanifold.

(a) If $\partial X = \emptyset$, then $H_n(X;\mathbb{Z}_2) \approx \mathbb{Z}_2$.

(b) If $\partial X \neq \emptyset$, then $H_n(X;\mathbb{Z}_2) = 0$.

Proof: If $n = 0$, then X is a point by Problem 2.13, and Lemma 6 is a special case of Problem 6.11. Assume therefore that $n \geq 1$ and let (K,φ) be any triangulation for X satisfying (1)-(3) in the definition of an n-dimensional pseudomanifold.

(a) Assume that $\partial X = \emptyset$. Since the mod 2 homology groups are topological invariants, it suffices to show that $H_n(K;\mathbb{Z}_2) \approx \mathbb{Z}_2$. Let $\Sigma \in C_n(K;\mathbb{Z}_2)$ be the n-chain which is the sum of all the n-simplices of K.

CLAIM. $\partial_n(\Sigma) = 0$.

First, observe that $|\partial K|$ is the union of all $(n - 1)$-simplices of K which belong to only one n-simplex of K because K is homogeneously n-dimensional. Since $\partial X = \emptyset$, Theorem 6 implies that $\partial K = \emptyset$. It follows from this and conditions (1) and (2) that every $(n - 1)$-simplex of K belongs to precisely two n-simplices of K. Therefore, $\partial_n(\Sigma) = 2\cdot(\text{sum of all } (n - 1)\text{-simplices of } K) = 0$, since $2 = 0$ in $\mathbb{Z}_2$, and the claim is proved.

On the other hand, let $c \in Z_n(K;\mathbb{Z}_2)$. If $c \neq 0$, then let σ be any n-simplex of K which appears in c. Let σ' be any other n-simplex of K. By condition (3) we can choose a sequence $\sigma_1,\sigma_2,\cdots,\sigma_s$ of n-simplices of K such that $\sigma_1 = \sigma$, $\sigma_s = \sigma'$, and σ_j meets σ_{j+1} in an $(n - 1)$-dimensional face for $1 \leq j < s$. A simple induction on j shows that σ_j appears in c for every j; otherwise, we would not have $\partial_n(c) = 0$. This shows that $c = 0$ or Σ and $H_n(K;\mathbb{Z}_2) \approx \mathbb{Z}_2$. Lemma 6(a) is proved.

(b) Assume that $\partial X \neq \emptyset$. It suffices to show that $H_n(K;\mathbb{Z}_2) = 0$ in this case. Let $c \in Z_n(K;\mathbb{Z}_2) = H_n(K;\mathbb{Z}_2)$. If $c \neq 0$, suppose that the n-simplex σ of K appears in c. Let σ' be any n-simplex of K and choose a chain of n-simplices $\sigma_1,\sigma_2,\cdots,\sigma_s$ in K from σ to σ' as in part (a). The only way that the $(n - 1)$-simplex $\sigma_j \cap \sigma_{j+1}$, $1 \leq j < s$, does not appear in $\partial_n(c)$ is if σ_j appears in c for all j. It follows that c is the sum of all the n-simplices of K, but then $\partial_n(c)$ will contain the (n - 1)-simplices of ∂K, so that $\partial_n(c) \neq 0$. We arrived at this contradiction by assuming that $c \neq 0$. It follows that $Z_n(K;\mathbb{Z}_2) = H_n(K;\mathbb{Z}_2) = 0$, proving part (b) and finishing the proof of Lemma 6.

Lemma 6 leads us to an important observation. If we were to compute the mod 2 homology groups for some examples such as points, spheres, or orientable surfaces, we would soon be tempted to conclude that one can get the mod 2 groups $H_q(X;\mathbb{Z}_2)$ from $H_q(X)$ by simply replacing every occurrence of $\mathbb{Z}$ in $H_q(X)$ by $\mathbb{Z}_2$ and making a similar mod 2 reduction on the torsion part of $H_q(X)$. After all, this is essentially what is happening on the chain level. Unfortunately, this procedure is not quite correct when passing to homology. The simplest example that shows this is $X = P^2$. By Lemma 6(a), $H_2(P^2;\mathbb{Z}_2) \approx \mathbb{Z}_2$, but we know from previous calculations that $H_2(P^2) = 0$. What has happened is that the $\mathbb{Z}_2$ in $H_1(P^2) \approx \mathbb{Z}_2$ has "shifted up" a dimension in the mod 2 homology. There is a good explanation for this phenomenon, but it is too complicated to discuss in this book. We only want to put the reader on guard when making computations mod 2.

THEOREM 7 (Invariance of Pseudomanifolds). Let X be an n-dimensional pseudomanifold and let (L,ψ) be any triangulation of X. Then L satisfies properties (1)-(3) in the definition of a pseudomanifold.

Proof: Theorem 7 follows from Problem 2.13 if n = 0. Assume that $n \geq 1$. Property (1) follows from Theorem 5. Property (2) holds because by Lemma 5 we must have $\delta^r_{n-1}L = \emptyset$ for $r \geq 3$. It remains to demonstrate property (3) for L. By dividing up the n-simplices of L into maximal collections satisfying (3), it is easy to see that $L = L_1 \cup L_2 \cup \cdots \cup L_k$, where the L_i are nonempty subcomplexes of L satisfying (1)-(3) and such that $\dim (L_i \cap L_j) \leq n - 2$ for $i \neq j$. The theorem will be proved once we show that k must equal 1. Note that each $|L_i|$ is a pseudomanifold and $|\partial L| = |\partial L_1| \cup |\partial L_2| \cup$

$\cdots \cup |\partial L_k|$. (The last equation follows from the fact that $\dim (L_i \cap L_j) \leq n - 2$ for $i \neq j$.)

CASE 1. $\partial X = \emptyset$: In this case $\partial L = \partial L_i = \emptyset$, so that $H_n(L;\mathbb{Z}_2) \approx H_n(L_i;\mathbb{Z}_2) \approx \mathbb{Z}_2$ by Lemma 6(a). Let $\Sigma_i \in Z_n(L_i;\mathbb{Z}_2) = H_n(L_i;\mathbb{Z}_2)$ be the unique nonzero element. Clearly, there is a natural inclusion $Z_n(L_i;\mathbb{Z}_2) \subset Z_n(L;\mathbb{Z}_2)$, and the proof of Lemma 6(a) shows that Σ_i is just the sum of the n-simplices in L_i. Therefore, $\Sigma_i \neq \Sigma_j$ in $Z_n(L;\mathbb{Z}_2)$ unless $i = j$. This implies that $Z_n(L;\mathbb{Z}_2) = H_n(L;\mathbb{Z}_2)$ contains at least $k + 1$ distinct elements $\Sigma_1,\Sigma_2,\cdots,\Sigma_k$, and 0. If $k > 1$, then we have a contradiction to the fact that $H_n(L;\mathbb{Z}_2)$ is a group with only two elements. Thus $k = 1$ and L satisfies (3).

CASE 2. $\partial X \neq \emptyset$: There is a standard way to reduce Case 2 to Case 1. The idea is to imbed X in a pseudomanifold without boundary, and the easiest way to do this is to form the "double of X," DX. Intuitively, DX is obtained by taking two disjoint copies of X and pasting them together along ∂X (see Fig. 6.10).

We shall define DX first on the simplicial level. Therefore, let K be a simplicial complex in $\mathbb{R}^n$. For each vertex $v \in K$, let $v^* = (v,\text{dist}(v,|\partial K|)) \in \mathbb{R}^n \times \mathbb{R}$, if $\partial K \neq \emptyset$, and $v^* = (v,1) \in \mathbb{R}^n \times \mathbb{R}$, if $\partial K = \emptyset$. Identifying $\mathbb{R}^n \times \mathbb{R}$ with $\mathbb{R}^{n+1}$ in the obvious way, define a simplicial complex K^* in $\mathbb{R}^{n+1}$ by

$$K^* = \{ \sigma^* = v_0^*v_1^*\cdots v_k^* \mid \sigma = v_0v_1\cdots v_k \in K \}$$

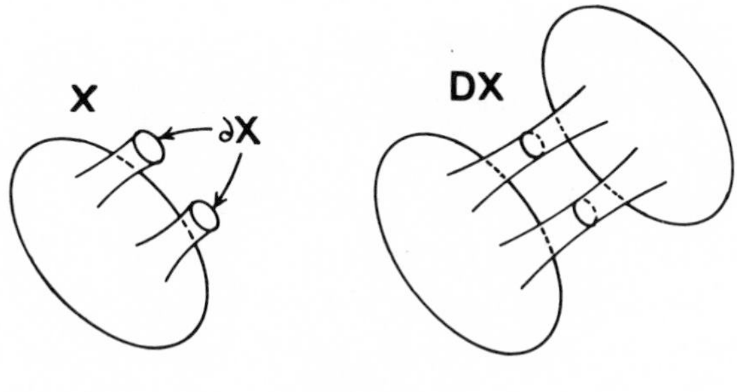

FIGURE 6.10

The complex K^* is clearly isomorphic to K via the simplicial map which sends v to v^*. Only one technicality keeps us from defining DK to be the complex $K \cup K^*$. We want $K \cap K^* = \partial K$, but unfortunately there are a few cases such as $K = \overline{\sigma}$, where σ is a simplex, in which $K^* = K$. This forces us to impose the following condition on K:

(*) If σ is a simplex of K whose vertices belong to ∂K, then $\sigma \in \partial K$.

It is easy to check that if K satisfies (*), then $K \cap K^* = \partial K$. Furthermore, for any complex K, the first barycentric subdivision of K, sd K, always satisfies (*) even if K does not.

DEFINITION. Let K be a simplicial complex in $\mathbb{R}^n$ satisfying (*). Define the double of K, DK, by $DK = K \cup K^*$, where K^* is defined as above.

It may be helpful to summarize the main properties of DK:

1. DK is a simplicial complex which depends only on K.
2. $DK = K \cup K^*$, where $K^* \approx K$ and $K \cap K^* = \partial K$; in particular, $K \subset DK$.
3. $\partial(DK) = \emptyset$.

(Property (3) follows from the definition of ∂ and DK.)

DEFINITION. Let X be a polyhedron. Define the double of X, DX, to be any space which is homeomorphic to $|D(\text{sd } K)|$, where (K,φ) is any triangulation of X.

Note that if $\partial X = \emptyset$, then DX consists of two disjoint copies of X.

Now we return to the proof of Case 2 in Theorem 7. There is no loss of generality if we assume that L satisfies property (*) above, because we can always replace L by sd L. This means that the doubles DL and DL_i are defined. If $\partial L_i = \emptyset$ for some i, then, by Lemma 6(a), $\mathbb{Z}_2 \approx H_n(L_i;\mathbb{Z}_2) = Z_n(L_i;\mathbb{Z}_2) \subset Z_n(L;\mathbb{Z}_2) = H_n(L;\mathbb{Z}_2) = 0$, which is a contradiction. Therefore, $\partial L_i \neq \emptyset$ for all i and $|DL_i|$ is a closed pseudomanifold contained in the closed pseudomanifold $|DL|$ [see Problem 6.14(c)]. The proof of Case 1 implies that $k = 1$ and Case 2 is proved. This finishes the proof of Theorem 7.

Next, we want to show that it makes sense to talk about the orientability of a pseudomanifold.

DEFINITION. Let X be a pseudomanifold and let (K,φ) be a triangulation of X. We shall say that X is orientable if the n-simplices of K can be oriented coherently, that is, the n-simplices of K can be oriented simultaneously in such a way that any two n-simplices which meet in a common $(n - 1)$-dimensional face induce opposite orientations on that face. Otherwise, we shall call X nonorientable.

The next theorem shows that the definition is well defined and independent of the particular triangulation that is chosen. It also gives a nice criterion for orientability.

THEOREM 8. (a) A closed n-dimensional pseudomanifold X is orientable if and only if $H_n(X) \approx \mathbb{Z}$.

(b) A pseudomanifold X with $\partial X \neq \emptyset$ is orientable if and only if the double of X, DX, is orientable.

Proof: Let X be an n-dimensional pseudomanifold and let (K,φ) be a triangulation of X.

(a) $\partial X = \emptyset$: Assume that X is orientable and that the n-simplices $\sigma \in K$ can be oriented coherently. Let $[\sigma]$ denote σ with this particular coherent orientation and let

$$\Sigma = \sum_{\text{n-simplices } \sigma\in K} [\sigma] \in C_n(K)$$

It follows from the definition of a coherent orientation that $\partial_n(\Sigma) = 0$. Furthermore, any n-cycle of K must be a multiple of Σ. (This follows from a standard argument repeatedly given in the past.) Therefore, $H_n(K) = Z_n(K) \approx \mathbb{Z}$, that is, $H_n(X) \approx \mathbb{Z}$. Conversely, if $H_n(X) \approx \mathbb{Z}$, then $Z_n(K) = H_n(K) \approx \mathbb{Z}$. Pick an orientation for each n-simplex σ in K and let $[\sigma]$ denote σ with this particular orientation. Suppose now that $z \in Z_n(K)$ is a nonzero n-cycle. Then

$$z = \sum_{\text{n-simplices } \sigma \in K} a_\sigma[\sigma]$$

where $a_\sigma \in \mathbb{Z}$ and $a_\tau \neq 0$ for some n-simplex τ. The fact that $\partial_n(z) = 0$ implies that $|a_\sigma| = |a_\tau|$ for all n-simplices $\sigma \in K$. (Otherwise, the coefficients of the (n - 1)-simplices in $\partial_n(z)$ do not vanish.) It is now easy to check that the oriented n-simplices $(a_\sigma/|a_\sigma|)[\sigma]$ are oriented coherently. In other words, X is orientable and part (a) of the theorem is proved.

(b) $\partial X \neq \emptyset$: Suppose that X is orientable and that the n-simplices of K can be oriented coherently. We may assume that K satisfies property (*), so that DK is defined. (Otherwise, replace K by sd K. It is not hard to see that the n-simplices of sd K can also be oriented coherently.) Let K* be as in the definition of DK. Let $[\sigma]$ denote the coherently oriented n-simplex $\sigma \in K$. If $[\sigma] = [v_0v_1\cdots v_k]$, then define the corresponding oriented n-simplex $[\sigma^*]$ of K* by $[\sigma^*] = [v_0^*v_1^*\cdots v_k^*]$. It follows that the $[\sigma]$'s and $-[\sigma^*]$'s are coherently oriented n-simplices for DK, and DX is therefore orientable. Alternatively, it is straightforward to check that

$$D\Sigma = \sum_{\text{n-simplices } \sigma \in K} [\sigma] + \sum_{\text{n-simplices } \sigma^* \in K^*} -[\sigma^*]$$

is an n-cycle of DK and that every other n-cycle of DK is a multiple of $D\Sigma$, that is, $H_n(DK) \approx \mathbb{Z}$, and DX is orientable by part (a). Conversely, assume that DX is orientable. Choose a triangulation (K,φ) of X such that K satisfies property (*). By part (a) the n-simplices of DK can be oriented coherently. This clearly induces a coherent orientation on the n-simplices of K and proves that X is orientable. Thus, (b) is proved, finishing the proof of Theorem 8.

If we specialize to the case of surfaces, we have our long-awaited precise definition of orientability together with a means for detecting the property in a direct manner via Theorem 8(a). In Problem 6.17 the reader is asked to show the compatibility of this definition with our use of the term in Sec. 1.2 and Chapter 3. It is also a good thought-provoking exercise to compare our definition with the one proposed in Sec. 4.2, where we talked about covering a surface with disks that are oriented at every point in a compatible way. The main point here would be to make the latter precise (see Problem 6.18).

Problems

6.13. Let the simplicial complexes L and L_i be as in the first paragraph of the proof of Theorem 7. Prove that

$$H_n(L;\mathbb{Z}_2) \approx H_n(L_1;\mathbb{Z}_2) \oplus H_n(L_2;\mathbb{Z}_2) \oplus \cdots \oplus H_n(L_k;\mathbb{Z}_2)$$

It follows that if $\partial L = \emptyset$, then the order of $H_n(L;\mathbb{Z}_2)$ is 2^k.

6.14. (a) If K is any simplicial complex, prove that sd K satisfies property (*) in this section.

(b) Following the definition of DK we listed three properties which DK possessed. Prove them.

(c) Prove that if X is a pseudomanifold with $\partial X \neq \emptyset$, then DX is a closed pseudomanifold.

6.15. In the proof of Case 2 of Theorem 7 we made the claim that L may be assumed to satisfy property (*). Prove this.

6.16. Prove that if X is an n-dimensional manifold which admits a triangulation, then X is an n-dimensional closed pseudomanifold.

6.17. By computing $H_2(S)$ for all the surfaces S in Table 2, Sec. 3.4, show [via Theorem 8(a)] that the definition of orientability given in this section is compatible with our previous intuitive use of the word in Table 2.

6.18. Let S be a surface and (K,φ) a fixed triangulation of S.

(a) Show that (i) $|st^c(x)| \approx D^2$, (ii) $|\ell k(x)| \approx S^1$, (iii) for each $y \in |st^c(x)| - |\ell k(x)|$ there is a map $r'_y : (|st^c(x)| - y) \to |\ell k(x)|$ which leaves the points of $|\ell k(x)|$ fixed, and (iv) any two such maps r'_y are homotopic (see Fig. 6.11, where the dotted arrows are intended to suggest how r'_y could be defined). Let $r_y : |\ell k(y)| \to |\ell k(x)|$ be the restriction of r'_y.

(b) Define a local orientation for S at $x \in S$ to be a choice of generators $\mu_x \in H_1(\ell k(x)) \approx \mathbb{Z}$. Define an orientation for S to be a collection $\{\mu_x\}_{x \in S}$ of local orientations μ_x for S at x satisfying the following compatibility (or continuity) condition: For each $x \in S$ and $y \in |st^c(x)| - |\ell k(x)|$, $(r_y)_*(\mu_y) = \mu_x$. [Note that, although r_y is not unique, it is unique up to homotopy by (a.iv). Therefore, $(r_y)_*$ is a well-defined homomorphism from $H_1(\ell k(y))$ to $H_1(\ell k(x))$.] Call S orientable if S admits an orientation. Prove that a surface S is orientable if and only if $H_2(S) \approx \mathbb{Z}$. In other words, this new notion of orientability agrees with

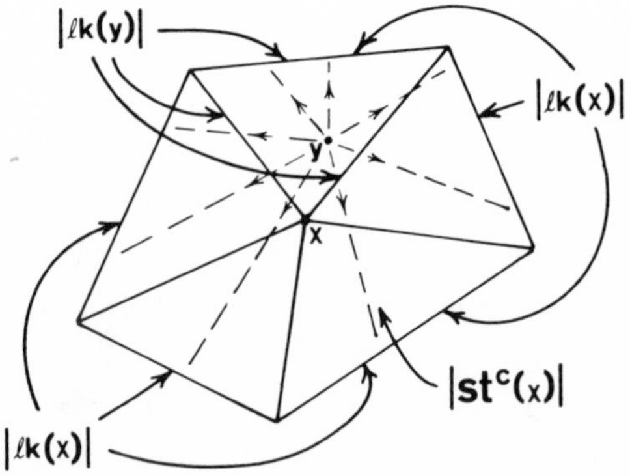

FIGURE 6.11

all the previous ones and is independent of the triangulation (K,φ).

(c) Discuss how one might extend the definition of orientability given in (b) to more general spaces such as n-dimensional manifolds.

6.6. EULER'S THEOREM REVISITED

Let us return to the theorem of Euler which we discussed in Sec. 1.1 and give a rigorous proof of that theorem. In fact, we are now in a position to study the Euler characteristic in a much more general situation. To begin with, note that the Euler characteristic in previous discussions had to do with two separate ideas. On the one hand, it was a purely combinatorial concept about abstract simplicial complexes whose definition simply involved counting simplices and taking an alternating sum of such numbers; and on the other, we wanted it to be a number that is associated to a space, rather than a particular triangulation.

DEFINITION. If K is a simplicial complex, let $n_q(K)$ denote the number of q-simplices in K and define the Euler characteristic of K, $\chi(K)$, by

$$\chi(K) = \sum_{q=0}^{\dim K} (-1)^q n_q(K)$$

In order to show that this combinatorially defined number $\chi(K)$ is actually an invariant associated with the underlying space $|K|$ we shall show that it is related to the Betti numbers, $\beta_q(K)$, of K.

THEOREM 9 (The Euler-Poincaré Formula). Let K be a simplicial complex. Then

$$\chi(K) = \sum_{q=0}^{\dim K} (-1)^q \beta_q(K)$$

Proof: Observe that $\partial_q : C_q(K) \to B_{q-1}(K)$ is onto and has kernel $Z_q(K)$. Similarly, the natural projection $Z_q(K) \to H_q(K)$ is onto and has kernel $B_q(K)$. It follows from this and Theorem 4 in Appendix B that $\operatorname{rk} C_q(K) = \operatorname{rk} B_{q-1}(K) + \operatorname{rk} Z_q(K)$ and $\operatorname{rk} Z_q(K) = \operatorname{rk} H_q(K) + \operatorname{rk} B_q(K)$. But $\operatorname{rk} C_q(K) = n_q(K)$, and so we get that

$$\begin{aligned}
\chi(K) &= \sum_{q=0}^{\dim K} (-1)^q n_q(K) = \sum_{q=0}^{\dim K} (-1)^q \operatorname{rk} C_q(K) \\
&= \sum_{q=0}^{\dim K} (-1)^q [\, \operatorname{rk} B_{q-1}(K) + \operatorname{rk} H_q(K) + \operatorname{rk} B_q(K)\,] \\
&= \sum_{q=0}^{\dim K} (-1)^q \operatorname{rk} H_q(K) + \sum_{q=0}^{\dim K} (-1)^q [\, \operatorname{rk} B_{q-1}(K) + \operatorname{rk} B_q(K)\,] \\
&= \sum_{q=0}^{\dim K} (-1)^q \operatorname{rk} H_q(K)
\end{aligned}$$

because the $\operatorname{rk} B_k(K)$'s cancel each other in the sum. Since $\beta_q(K) = \operatorname{rk} H_q(K)$, we are done.

Theorem 9 shows that $\chi(K)$ depends only on the homotopy type of $|K|$ because we already know this to be true for the Betti numbers. Furthermore, the formula in the theorem suggests a way to define the Euler characteristic of a space which is intrinsic and does not involve first choosing a triangulation.

DEFINITION. If X is a polyhedron, define the Euler characteristic of X, $\chi(X)$, by

$$\chi(X) = \sum_{q=0}^{\dim X} (-1)^q \beta_q(X)$$

The next corollary is merely a restatement of Theorem 9. It makes explicit the fact that the number $\chi(X)$, which is a homotopy-type invariant by definition, is actually easily computed by a simple counting. As a special case we obtain the topological invariance of the Euler characteristic of surfaces as defined in Secs. 1.2 and 3.4. (In Sec. 1.2 we may assume that the regions are triangular similar to the way it was done in the proof of Euler's theorem.)

COROLLARY. Let X be a polyhedron and assume that (K,φ) is a triangulation of X. Then

$$\chi(X) = \sum_{q=0}^{\dim K} (-1)^q n_q(K)$$

The arguments we have just given are also applicable to the connectivity numbers of a space. In particular, we have

THEOREM 10. Let K be a simplicial complex. Then

$$\chi(K) = \sum_{q=0}^{\dim K} (-1)^q \varkappa_q(K)$$

Proof: The proof of this theorem is the same as that of Theorem 9 except that one uses the mod 2 groups $C_q(K;\mathbb{Z}_2)$, $Z_q(K;\mathbb{Z}_2)$, $B_q(K;\mathbb{Z}_2)$, and $H_q(K;\mathbb{Z}_2)$ and works with the dimensions of these vector spaces rather than the ranks of the corresponding groups in the usual homology theory with integer coefficients. A crucial observation, however, is the fact that $\dim C_q(K;\mathbb{Z}_2) = n_q(K)$.

COROLLARY. If X is a polyhedron, then

$$\chi(X) = \sum_{q=0}^{\dim X} (-1)^q \varkappa_q(X)$$

In the special case of surfaces we can use this corollary to give a simple geometric interpretation to connectivity numbers. Let S be a surface. To begin with, we know from Problem 6.11(b) and Lemma 6(a) that $\varkappa_0(S) = 1 = \varkappa_2(S)$. Therefore, the corollary to Theorem 10 implies that $\varkappa_1(S) = 2 - \chi(S)$. This shows that the first (and only interesting) connectivity number $\varkappa_1(S)$ does not depend on the orientability of the surface S, which is quite different from what happens in the case of the first Betti number $\beta_1(S)$. Of course, this is only further evidence that homology theory with $\mathbb{Z}_2$ coefficients does not detect orientability properties of spaces. We saw this already in Sec. 6.5 with respect to the top dimensional homology groups of pseudomanifolds.

The main result we are after is:

THEOREM 11. The first connectivity number $\varkappa_1(S)$ of a surface S equals the maximum number of (distinct but not necessarily disjoint) simple closed curves in S along which one can cut and still have a connected set left over.

Outline of Proof: Define a number $\delta(S)$ by

$$\delta(S) = \max \{ k \mid S - (X_1 \cup X_2 \cup \cdots \cup X_k) \text{ is connected, where the } X_i \text{ are distinct simple closed curves in } S \}$$

We have to prove that $\varkappa_1(S) = \delta(S)$. The hypothesis that the X_i in the definition of $\delta(S)$ are "simple closed curves" will mean for us that there is a triangulation (K,φ) of S and subcomplexes L_i of K such that $\varphi(|L_i|) = X_i$ (and $X_i \approx S^1$). By considering the normal form for a surface which was given in Chapter 3, one can easily see that $\delta(S) \geq \varkappa_1(S) = 2 - \chi(S)$. Conversely, let Σ_i be the generator of $H_1(L_i;\mathbb{Z}_2) = Z_1(L_i;\mathbb{Z}_2) \subset Z_1(K;\mathbb{Z}_2)$, that is, Σ_i is the sum of the 1-simplices in L_i. Assume that $k > \varkappa_1(S)$. Since $\varkappa_1(S)$ is the dimension of the vector space $H_1(K;\mathbb{Z}_2)$, there is a 2-chain $c \in C_2(K;\mathbb{Z}_2)$ with $\partial_2(c) = a_1\Sigma_1 + a_2\Sigma_2 + \cdots + a_k\Sigma_k$, where $a_i \in \{0,1\}$ and not all the a_i are zero. The chain c cannot be the sum of all the 2-simplices of K, because it was shown in the proof of Lemma 6 that then $\partial_2(c) = 0$. Since we are assuming that $\partial_2(c) \neq 0$, at least one 2-simplex σ of K does not belong to c. Using this fact one shows that $S - (X_1 \cup X_2 \cup \cdots \cup X_k)$ is not connected. This contradicts our initial hypothesis and proves that we cannot have $k > \varkappa_1(S)$. In other words, $\delta(S) \leq \varkappa_1(S)$.

Theorem 11 answers a question raised earlier in Problem 3.16(b). It also justifies the terminology.

Finally, the fact that one can get a topological invariant for a simplicial complex K by taking an alternating sum of the numbers

$n_q(K)$ leads us naturally to the following question: What other functions of the numbers $n_q(K)$ are topological invariants? For example, the integer $7\left(\sum_{q=0}^{\dim K} (-1)^q n_q(K)\right)$ is a topological invariant since $\chi(K)$ is. More generally, if $f(x) = \sum_{i=0}^{t} a_i x^i$, $a_i \in \mathbb{Z}$, is a polynomial, define $\chi_f(K)$ by

$$\chi_f(K) = \sum_{i=0}^{t} a_i (\chi(K))^i$$

Clearly, $\chi_f(K)$, thought of as a function of the $n_q(K)$, is a topological invariant. Our question then becomes: Are there any functions of the numbers $n_q(K)$ which are topological invariants other than those, as in the case of $\chi_f(K)$, that are functions of $\chi(K)$? The surprising answer is that basically there are not! We shall only outline a proof of this result. The general idea behind the proof is easy enough, but filling in the details would involve a lengthy digression which is left to the interested reader.

First of all, one has to extend the notion of a simplicial complex and define a cell complex.

DEFINITION. Any space which is homeomorphic to D^k, $k \geq 0$, will be called a k-dimensional cell or k-cell.

The dimension of a cell is well defined by Theorem 4. Cells are going to be to cell complexes what simplices are to simplicial complexes. More precisely,

DEFINITION. A cell complex is a finite collection C of cells which is defined inductively as follows: Let k be the number of cells in C and set $|C| = \bigcup_{e \in C} e$. As usual, $|C|$ is called the underlying space of C. If $k = 0$, then $C = \emptyset = |C|$. Suppose that we have already defined what we mean by a cell complex with $k - 1$ cells, $k \geq 1$. Then any collection C of k cells will be called a cell complex provided that there are cell complexes C_1 and C_2 (containing k - 1 or fewer cells) such that $C_2 \subset C_1 \subset C$, $e = c\ell(|C| - |C_1|)$ is a cell in C, and $\partial e = |C_2|$.

DEFINITION. If C is a nonempty cell complex, then the dimension of C, dim C, is defined to be the maximum of all the dimensions of the cells of C. If $C = \emptyset$, set dim C = -1.

Clearly, every simplicial complex is a cell complex. The main difference between a cellular decomposition of a space X (which means a pair (C,h), where C is a cell complex and $h : |C| \to X$ is a homeomorphism) and a triangulation of X is that cells have a flexible number of faces. Figure 6.12 attempts to indicate the inductive aspect of the definition of a cell complex. The underlying space of an arbitrary cell complex is obtained from a finite collection of points by inductively attaching cells along their boundary to the part that has already been constructed in a compatible way. Figure 6.12a-d show the various stages of such a process leading to a cellular decomposition of D^2 in Fig. 6.12d which does not correspond to a triangulation. On the other hand, it is not hard to show (by induction) that even though a cell complex may not be a simplicial complex, its underlying space is nevertheless a polyhedron.

It is possible to define a homology theory for cell complexes merely by copying what was done in the case of simplicial complexes. In other words, one can define the notion of an oriented cell and then the group of q-chains, $C_q(C)$, of C is obtained by taking formal linear combinations of oriented q-cells in C. There is also a natural boundary map $\partial_q : C_q(C) \to C_{q-1}(C)$ and $H_q(C) = \text{Ker } \partial_q / \text{Im } \partial_{q+1}$ is called the q-th homology group of C. Given a continuous map $f : |C| \to |C'|$ between the underlying spaces of two cell complexes C and C', there is induced a natural homomorphism $f_{*q} : H_q(C) \to H_q(C')$. The groups $H_q(C)$ and homomorphisms f_{*q} satisfy all the properties their simplicial analogs did. Finally, one can prove the important theorem which asserts that if C is a cell complex and K is a simplicial complex such that $|C| \approx |K|$, then $H_q(C) \approx H_q(K)$. This means that one can obtain the homology groups of a polyhedron either from a simplicial or a cellular decomposition.

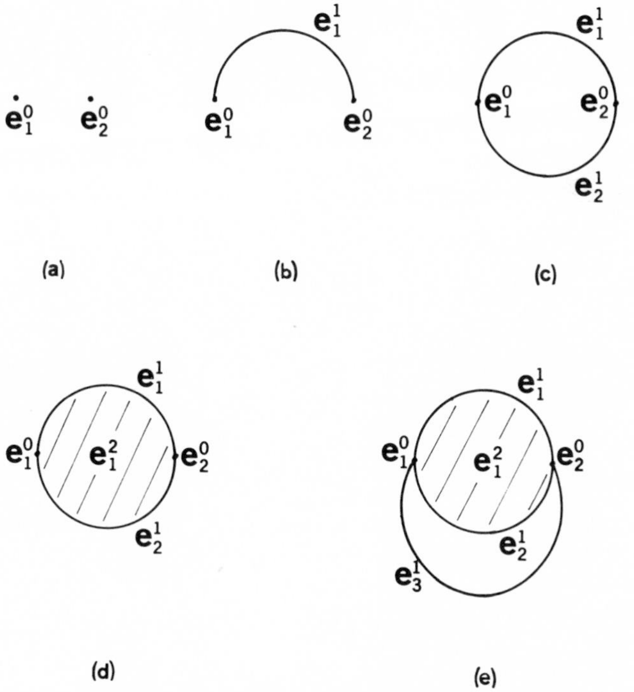

e_j^i REFERS TO THE j-TH i CELL

FIGURE 6.12

DEFINITION. If C is a cell complex, let $n_q(C)$ denote the number of q-cells in C and define the Euler characteristic of C, $\chi(C)$, by

$$\chi(C) = \sum_{q=0}^{\dim C} (-1)^q n_q(C)$$

DEFINITION. If $f(x_0,x_1,\cdots)$ is any function of the indeterminates $x_0,x_1,\cdots$, and if C is a cell complex, define $f(C) = f(n_0(C), n_1(C),\cdots)$. We shall say that f is a topologically invariant function if $f(C) = f(C')$ whenever $|C| \approx |C'|$.

For example, define $f(x_0, x_1, \cdots) = x_0 - x_1x_3 + x_7^2$ and $g(x_0, x_1, \cdots) = \sum_{q=0}^{\infty} (-1)^q x_q$. Then $f(C) = n_0(C) - n_1(C)n_3(C) + (n_7(C))^2$ and $g(C) = \chi(C)$.

THEOREM 12. The only topologically invariant functions $f(x_0, x_1, \cdots)$ on cell complexes are those which are functions of the Euler characteristic and the dimension, that is, if C and C' are cell complexes with $\dim C = \dim C'$ and $\chi(C) = \chi(C')$, then $f(C) = f(C')$.

Outline of Proof: First one has to show that $\chi(C)$ and $\dim C$ are in fact topological invariants. This is done in essentially the same way as it is done for simplicial complexes, but using the cellular homology groups instead. For example, one shows that $\chi(C) = \sum_{q=0}^{\dim |C|} (-1)^q \beta_q(|C|)$. The more interesting part of Theorem 12 is the converse.

Suppose that C and C' are cell complexes with $k = \dim C = \dim C'$ and $\chi(C) = \chi(C')$. We want to show that $f(C) = f(C')$. This will follow if we can find cell complexes C_1 and C_2 such that $|C_1| = |C|$, $|C_2| = |C'|$, and $n_q(C_1) = n_q(C_2)$ for all q. Let $m(C,C')$ be the largest integer m such that $m \leq k$ and $n_q(C) = n_q(C')$ for $0 \leq q \leq m$. [Note that $m(C,C') = -1$ if $n_0(C) \neq n_0(C')$.] The construction of C_1 and C_2 will proceed by induction on $m(C,C')$. If $m(C,C') = k$, then we are done if we set $C_1 = C$ and $C_2 = C'$. Suppose that we can construct C_1 and C_2 whenever $m(C,C') \geq m + 1$, where $-1 \leq m < k$, and let $m(C,C') = m$. Without loss of generality assume that $n_{m+1}(C) < n_{m+1}(C')$.

CASE 1. $m \leq k - 2$: Let e be an $(m + 2)$-cell in C, let e_1 be an $(m + 1)$-cell of C contained in ∂e, and let $x \in e - \partial e$ (see Fig. 6.13). If e_2 is the cone on ∂e_1 in e from x, we see that e_2 is an $(m + 1)$-cell with $\partial e_2 = \partial e_1$ which divides e into two $(m + 2)$-cells e' and e''. Let $C_1' = (C - \{e\}) \cup \{e_2, e', e''\}$. It is easy to see that C_1' is a cell complex such that $|C_1'| = |C|$,

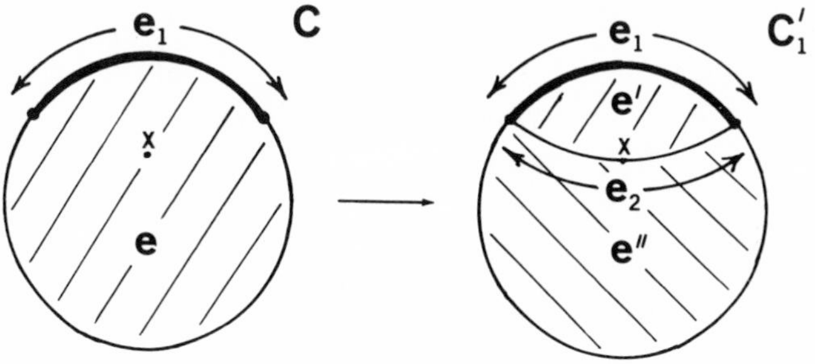

FIGURE 6.13

$n_{m+1}(C_1') = n_{m+1}(C) + 1$, $n_{m+2}(C_1') = n_{m+2}(C) + 1$, and $n_q(C_1') = n_q(C)$ for $0 \leq q \leq k$ and $q \neq m + 1$ or $m + 2$. In other words, we have increased the number of $(m + 1)$-cells in C by one without disturbing anything in lower dimensions. Therefore, if we repeat this procedure $n_q(C') - n_q(C)$ times we will get a cell complex C_1'' such that $|C_1''| = |C|$ and $n_q(C_1'') = n_q(C')$ for $0 \leq q \leq m + 1$. This means that $m(C_1'',C') \geq m + 1$ and our desired cell complexes C_1 and C_2 exist by the inductive hypothesis applied to C_1'' and C'.

CASE 2. $m = k - 1$: The construction of Case 1 would clearly not work because there are no $(k + 1)$-cells. Fortunately, the case $m(C,C') = k - 1$ cannot occur. This follows from the fact that

$$(-1)^k n_k(C) = \chi(C) - [n_0(C) - n_1(C) + \cdots + (-1)^{k-1} n_{k-1}(C)]$$

and

$$(-1)^k n_k(C') = \chi(C') - [n_0(C') - n_1(C') + \cdots + (-1)^{k-1} n_{k-1}(C')]$$

so that $n_k(C) = n_k(C')$ whenever $n_q(C) = n_q(C')$ for $0 \leq q \leq k - 1$ and $\chi(C) = \chi(C')$, that is, $m(C,C') \geq k - 1$ implies that $m(C,C') = k$.

Problems

6.19. Fill in the details which are missing from our proof of

(a) Theorem 11

(b) Theorem 12

6.20. Let S be a surface. We already know $H_0(S)$ and $H_2(S)$ from Problems 4.5(a) and 6.17, respectively. Compute $H_1(S)$ in terms of the orientability and genus.

6.7. RETRACTS AND THE BROUWER FIXED-POINT THEOREM

DEFINITION. Let $A \subset X$. Then A is said to be a retract of X if there exists a continuous map $r : X \to A$ such that $r|A = 1_A$. In that case r is called a retraction.

THEOREM 13. Let A and X be polyhedra. If A is a retract of X, then the homology groups of X "contain" the homology groups of A as a direct summand, that is, for each q there is an abelian group G_q and

$$H_q(X) \approx H_q(A) \oplus G_q$$

Proof: We may assume that $X = |K|$ and $A = |L|$, where K and L are simplicial complexes. Let $r : X \to A$ be a retraction. We shall show that

$$H_q(K) \approx H_q(L) \oplus \text{Ker } r_{*q}$$

This will clearly prove the theorem and at the same time identify the groups G_q.

Observe that the commutative diagram

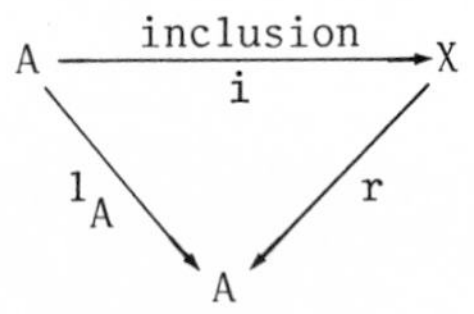

induces a commutative diagram

$$\begin{array}{ccc} H_q(L) & \xrightarrow{\ i_{*q}\ } & H_q(K) \\ & {\scriptstyle 1_{H_q(L)}} \searrow \quad \swarrow {\scriptstyle r_{*q}} & \\ & H_q(L) & \end{array}$$

that is, $1_{H_q(L)} = r_{*q} \circ i_{*q}$. Define

$$\psi_q : H_q(K) \to H_q(L) \oplus \operatorname{Ker} r_{*q}$$

by $\psi_q(u) = (\ r_{*q}(u),\ u - (i_{*q} \circ r_{*q})(u)\ )$. [To see that $u - (i_{*q} \circ r_{*q})(u)$ belongs to $\operatorname{Ker} r_{*q}$, note that

$$\begin{aligned} r_{*q}(u - (i_{*q} \circ r_{*q})(u)) &= r_{*q}(u) - (r_{*q} \circ i_{*q} \circ r_{*q})(u) \\ &= r_{*q}(u) - (1_{H_q(L)} \circ r_{*q})(u) \\ &= r_{*q}(u) - r_{*q}(u) \\ &= 0.] \end{aligned}$$

The map ψ_q is clearly a homomorphism. We shall show that it is bijective. Suppose that $u \in H_q(K)$ and $\psi_q(u) = 0$. Then $r_{*q}(u) = 0$ and $u - (i_{*q} \circ r_{*q})(u) = 0$. Therefore, $u = (i_{*q} \circ r_{*q})(u) = i_{*q}(0) = 0$, which shows that ψ_q is injective. Next, let $(a,v) \in H_q(L) \oplus \operatorname{Ker} r_{*q}$ and set $u = i_{*q}(a) + v$. Then

$$\psi_q(u) = (\ r_{*q}(i_{*q}(a) + v),\ i_{*q}(a) + v - (i_{*q} \circ r_{*q})(i_{*q}(a) + v)\ )$$

But $r_{*q}(i_{*q}(a) + v) = (r_{*q} \circ i_{*q})(a) + r_{*q}(v) = a + 0 = a$ and

$$\begin{aligned} (i_{*q} \circ r_{*q})(i_{*q}(a) + v) &= (i_{*q} \circ (r_{*q} \circ i_{*q}))(a) + (i_{*q} \circ r_{*q})(v) \\ &= i_{*q}(a) + i_{*q}(r_{*q}(v)) \\ &= i_{*q}(a) + i_{*q}(0) \\ &= i_{*q}(a) \end{aligned}$$

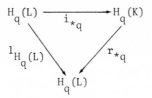

so that $i_{*q}(a) + v - (i_{*q}^{\cdot} \circ r_{*q})(i_{*q}(a) + v) = i_{*q}(a) + v - i_{*q}(a) = v$. In other words, $\psi_q(u) = (a,v)$ and ψ_q is also surjective. This proves that ψ_q is an isomorphism and finishes the proof of Theorem 13.

The next theorem is an easy corollary of Theorem 13. Intuitively, it asserts that it is impossible to "project" a disk onto its boundary without tearing or puncturing it.

THEOREM 14. S^{n-1} is not a retract of D^n for $n \geq 1$.

Proof: Assume that S^{n-1} were a retract of D^n. It would then follow from Theorem 13 that $\operatorname{rk} H_q(D^n) \geq \operatorname{rk} H_q(S^{n-1})$ for all q. This contradicts the computations of Corollary 2 in Sec. 6.1 and Lemma 2 in Sec. 6.2 and proves Theorem 14.

One of the standard applications of elementary homology theory is to the theory of fixed points of maps.

DEFINITION. Given a map $f : X \to X$, we shall say that $x \in X$ is a fixed point of f if $f(x) = x$.

Probably the most famous fixed-point theorem of all, which can be proved very easily using algebraic topology, is:

THEOREM 15 (Brouwer Fixed-point Theorem). Any continuous map $f : D^n \to D^n$, $n \geq 0$, has a fixed point.

If $n = 1$, it is easy to convince oneself on intuitive grounds of the validity of Theorem 15. From Fig. 6.14 one can see that the fixed points of a function f correspond to those points where the graph of f intersects the line $y = x$, and it is clear that the line which is the graph of f cannot possibly go from the line $x = -1$ to the line $x = 1$ without intersecting the line $y = x$ and at the same time stay in $D^1 \times D^1 \subset \mathbb{R}^2$.

Proof: If $n = 0$, then D^n is a point and the result is true trivially. Assume that $n \geq 1$ and suppose that the continuous map $f : D^n \to D^n$ has no fixed points. Define a map $r : D^n \to S^{n-1}$ as

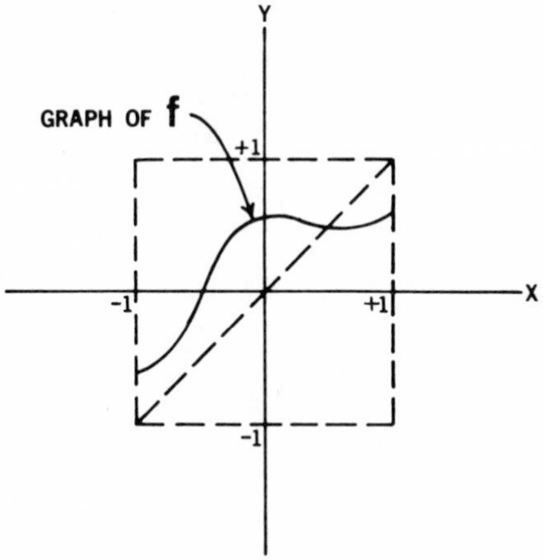

FIGURE 6.14

follows: Since $x \neq f(x)$, there is a unique line L_x in $\mathbb{R}^n$ which passes through x and $f(x)$ (see Fig. 6.15). Set

$$L_x^- = \{ x + t(x - f(x)) \mid t \in \mathbb{R} \text{ and } t \geq 0 \}$$

so that L_x^- is the part of the line L_x which is to the "left" of x. Clearly, L_x^- intersects S^{n-1} in a unique point [Problem 6.24 (a)]. Define $r(x) = L_x^- \cap S^{n-1}$. Then r is a continuous map [Problem 6.24(b)] which is a retraction of D^n onto S^{n-1}. Since this contradicts Theorem 14, f must have had a fixed point after all and Theorem 15 is proved.

REMARK. We just saw how Theorem 14 implies Theorem 15. In fact, the two theorems are equivalent because if $r : D^n \to S^{n-1}$ were a retraction, then the map $f : D^n \to D^n$ given by $f(x) = -r(x)$ would have no fixed points, so that Theorem 15 implies Theorem 14.

There is an important generalization of Brouwer's fixed-point theorem which was proved by Lefschetz [Lef 1] in 1926. It concerns the fixed points of a continuous map $f : X \to X$, where X is an arbitrary compact polyhedron. We have not included it here because it assumes familiarity with the "trace" of a homomorphism. One conse-

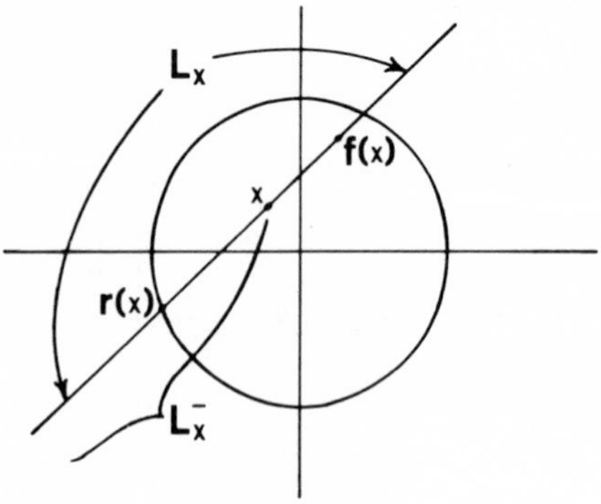

FIGURE 6.15

quence of Lefschetz's theorem is that if X has the homotopy type of a point, then f always has a fixed point. (Brouwer's theorem is the case $X = D^n$.)

Problems

6.21. Let σ be a k-simplex, $k \geq 1$. Prove that $(\sigma \times 0) \cup (\partial\sigma \times [0,1])$ is a retract of $\sigma \times [0,1]$.

6.22. Prove that S^{n-1}, $n \geq 1$, is a retract of $D^n - x$ for every $x \in D^n - S^{n-1}$.

6.23. Let $x_0 \in S^1$. Show that the two circles $(S^1 \times x_0) \cup (x_0 \times S^1)$ are not a retract of the torus $S^1 \times S^1$.

6.24. Using the notation in the proof of Theorem 15, finish the proof of that theorem by showing:

(a) L_x^- intersects S^{n-1} in a unique point

(b) r is continuous.

6.25. Suppose that $f : S^n \to S^n$, $n \geq 0$, and that $f(S^n) \neq S^n$. Using Brouwer's fixed-point theorem prove that f has a fixed point.

6.26. Let $f : [a,b] \to [c,d]$ be a continuous surjective map. Prove that every such f has a fixed point in case $c \leq a$ and $b \leq d$. Give counterexamples to this assertion in the other cases.

6.27. It follows from Theorem 15 that every continuous map $f : A \to A$ has a fixed point whenever A is equal to a closed interval $[a,b]$. Show that this is no longer the case when we replace A by a half-open interval $(a,b]$ or $[a,b)$.

6.28. Find a space X and a map $f : X \to X$ such that X has the homotopy type of a point but f has no fixed points. (*Hint:* By the remarks about Lefschetz's theorem at the end of the section, X cannot be a polyhedron. Actually, the crucial property needed in the theorem is the compactness. Therefore, look for some nice "noncompact polyhedron" X.)

6.29. Let M be an n-dimensional pseudomanifold.

(a) Assume $\partial M \neq \emptyset$. Generalize Theorem 14 by proving that ∂M is not a retract of M.

(b) Assume $\partial M = \emptyset$ and $n \geq 1$. Prove that the identity map of M is not homotopic to a constant map. This fact also implies Theorem 14 because if $r : D^{n+1} \to S^n$ were a retraction, then $h_t : S^n \to S^n$, $h_t(x) = r(tx)$, would be a homotopy between the identity map of S^n and a constant map.

(*Hint:* Use Lemma 6.)

6.8. HISTORICAL COMMENTS

The death of Poincaré in 1912 roughly marks the end of the first period in the history of topology. The years between about 1910 and the beginning of the second World War saw the subject develop into an important and independent field of mathematics. A better understanding of homology theory was gained and greater rigor introduced. It was also during this time that topology split into two separate parts: general topology (the study of arbitrary topological spaces), which we have completely ignored in this book, and combinatorial topology (the study of simplicial complexes via combinatorial invariants).

Section 6.1. We have already mentioned that the first proof of the topological invariance of the homology groups is basically due to Alexander [Alex 1] in 1915. The notion of the homotopy type of a space was introduced by W. Hurewicz (1904-1956) in 1935. Although there had been some papers on homotopy theory previous to that, notably some important contributions by H. Hopf (1894-) [Hop 1, 3-7], it is Hurewicz who is usually considered to be the founder of homotopy theory proper because of a series of four papers [Hure] in 1935-1936 in which he defined some of the basic concepts of the theory. Since most of the tools provided by algebraic topology are not fine enough to distinguish between different homeomorphism types--at best they distinguish between different homotopy types--the homeomorphism problem remains a very difficult one which can only be approached indirectly, as, for example, in Theorem 3(b). In fact, there is absolutely no hope of a general solution anyway because it was shown by A. A. Markov (1903-) [Mar] in 1958 that it is not possible, not even in principle, to solve some of these homeomorphism problems by any finite algorithm. In particular, he showed that there is no finite algorithm which would enable one to determine when nice spaces such as two 4-dimensional manifolds are homeomorphic. In this connection let us also mention that there exist two compact 3-dimensional manifolds (lens spaces) which are homotopy equivalent but not homeomorphic.

The axiomatic development of homology theory which we hinted at is due to Eilenberg and N. E. Steenrod (1910-1971) in [Eil-S 1,2]. This approach unified a number of different theories that had been defined by that time such as "singular" homology theory (see Chapter 8) and the homology theories of E. Čech (1893-1960) [Ce 2] and L. Vietoris (1891-) [Vi 1] for arbitrary spaces. It also gave an impetus to further investigations of the algebraic aspect of algebraic topology. In fact, some algebraic topologists nowadays are really more algebraists than topologists.

Section 6.2. Our treatment of local homology groups is taken from Seifert and Threlfall's book [Se-T 1]. In particular, Theorem 2 appears there for the first time. Although our definitions have used the local simplicial structure of a space which is not unique, one can give definitions that are topologically invariant from the start. This was done by E. R. van Kampen [Kam] in 1929 using "singular" homology groups together with a notion of relative cycles and later by Čech [Ce 3] in 1933 who studied local Betti numbers of arbitrary spaces.

Section 6.3. Brouwer is without doubt the next most important figure in the history of algebraic topology after Poincaré. Poincaré may have founded combinatorial topology, but it was Brouwer who first used it to solve significant geometric problems. As we shall see, many of the applications discussed in this chapter and the next are due to Brouwer originally. Theorem 3(b) and Theorems 4 and 6 in the case of manifolds were first proved by Brouwer [Brou 3], although his proofs were different. The invariance of the dimension of triangulated n-dimensional manifolds, $n \leq 3$, had already been proved by J. Lüroth (1844-1910) [Lu] in 1907. Theorem C was proved by Brouwer in [Brou 6]. The case $n = 1$ in Theorem C is easy. The case $n = 2$ was first proved by A. Schoenflies (1853-1928) [Scho 1] in 1899. Simpler proofs were given by W. F. Osgood (1864-1943) [O] and F. Bernstein (1878-1956) [Bern] in 1900. R. L. Baire (1874-1932) [Bai] and J. S. Hadamard (1865-1963) [Had] showed in 1907 and 1910, respectively, that the truth of the theorem for all n followed from the generalized Jordan curve theorem (see Sec. 7.6). Lebesgue [Leb 2] gave an alternate proof of Theorem C in 1911 (see also [Leb 1]).

Dimension theory is an interesting and important topic on its own and does not concern only polyhedra but arbitrary spaces. The idea that we are living in a three-dimensional world is an old one so that geometers and others certainly have been aware of the concept of dimension for a long time. We have already mentioned efforts by Cayley [Cay 1,4], Grassmann [Gra], and Riemann [Riem 2] to work with n dimensions. Nevertheless, the concept remained vague and intuitive and was relatively unexplored until the late nineteenth century. Before that time "n-dimensional" meant quantities that needed, in some unspecified way, a minimum of n real parameters to describe their points. Cantor's discovery in [Can] of a bijective function between $\mathbb{R}$ and $\mathbb{R}^2$ showed that "dimension" had nothing to do with the "number" of points involved. Finally, Peano's curve in [Pea] which was a continuous map from [0,1] onto the square $[0,1] \times [0,1] \subset \mathbb{R}^2$ showed that continuous maps could raise the dimension, so that "dimension" could not be defined in terms of the number of continuous parameters required to describe a space. Poincaré [Poin 10] was one of the first to propose a definition, which was soon thereafter made precise independently by Brouwer [Brou 9], then K. Menger (1902-) [Meng], and P. Urysohn (1898-1924) [U 1,3]. It was essentially an inductive definition that involved looking at the lower dimensional subsets of a space which disconnect it. The important point, though, is that a satisfactory theory could be built on this definition. Dimension theory had come of age. Since then many equivalent definitions of the dimension of a space have been given. The interested reader can find a wealth of material on this beautiful subject in [Meng] and [Hure-W].

Section 6.4. The mod 2 homology theory was defined by Alexander and O. Veblen (1880-1960) [Alex-V] in 1913 but it had been introduced even earlier by H. Tietze (1880-1964) [Ti 1] in 1908. The topological invariance of this theory was proved by Veblen [Veb 4] in 1922. A few years later in 1926 Alexander [Alex 6] published a precise definition of simplicial homology theory with coefficients in $\mathbb{Z}_n$ and proved its invariance. In 1928 Lefschetz [Lef 2] suggested $\mathbb{Q}$ for coefficients, and arbitrary coefficients were introduced by L. Pontrjagin (1908-1960) [Pont 2] in 1934. Actually, one can see that once the group theory approach to homology theory was adopted, it was an easy exercise to extend the necessary definitions to the case of arbitrary coefficients. These ideas were more than empty abstract generalizations though because in conjunction with "cohomology" groups which were defined around that time this theory has significant consequences. Unfortunately, we cannot go into any details because these topics are too advanced for this book (see Chapter 8).

Section 6.5. The discussion of pseudomanifolds and orientability is an extension of ideas that developed naturally out of the study of surfaces. The relation between the orientability of a surface and the existence of a compatible orientation of all of its simplices was observed by Klein [Kle 1] in 1874.

Section 6.6. Theorem 9 was proved by Poincaré [Poin 4] in 1895. The definition of a cell complex, in a much more general form than is presented in this text, was introduced by J. H. C. Whitehead (1904-1960) [Wh] in 1949. The usefulness and importance of such complexes derives from the fact that cell decompositions of a space are much simpler than simplicial ones. One usually has many fewer cells to worry about and yet one can do everything that one can do with simplicial complexes. Theorem 12 was proved by D. Sullivan (1941-   ) in 1964 and recently other interesting related results have been discovered.

Section 6.7. Retracts were defined by K. Borsuk (1905-   ) [Bor 1] in 1931. The study of fixed points of maps is important not only in topology but in other areas of mathematics such as in the theory of differential equations. For example, a solution $y(x)$ to the differential equation

$$\frac{dy}{dx} = f(x,y)$$

over the interval $0 \leq x \leq 1$ with initial condition $y(0) = 0$ will satisfy the equation

$$y(x) = \int_0^x f(x,y(x))dx$$

Thus, one is looking for a fixed point to the operator T defined on a space of functions φ, where

$$T(\varphi) = \int_0^x f(x,\varphi(x))dx$$

Brouwer proved his fixed point theorem, Theorem 15, in 1910 (see [Brou 5]). Poincaré's interest in dynamics and celestial mechanics also led him to study fixed points. In 1912 he showed in [Poin 11] that whether or not a certain three-body problem had periodic orbits depended on the fixed points of an associated map. The Poincaré-Birkhoff fixed-point theorem was proved by G. D. Birkhoff (1884-1944) [Bir 1,2] in 1913. It stated that an area-preserving homeomorphism of an annulus which moved all points on the two bounding circles had at least two fixed points. Generalizations to function spaces were made by Birkhoff and O. D. Kellogg (1878-1932) [Bir-K] in 1922. J. P. Schauder (1899-1940) [Scha 1,2] in 1927 and 1930 and Schauder and J. Leray (1906-) [Scha-L] jointly in 1934 are responsible for the first important applications of fixed-point theory to the solution of differential equations. Finally, note that the problem of finding roots of an equation $f(x) = 0$ is equivalent to finding fixed points of the map $g(z) = f(z) + z$. For more on fixed points see the historical comments on Sec. 7.1 in Sec. 7.7 and also the nice article [Bin].

Chapter 7

MAPS OF SPHERES AND MORE APPLICATIONS

7.1. THE DEGREE OF A MAP; VECTOR FIELDS

The set $[S^n,S^n]$ of homotopy classes of maps of the n-sphere to itself is one of the first important objects that one encounters in homotopy theory. The solutions to a number of interesting geometric problems depend on knowing some properties of this set. We begin our study of $[S^n,S^n]$ by defining a map $\deg : [S^n,S^n] \to \mathbb{Z}$. The fact is that when one is studying a mathematical object about which little is known one can often get some information about it by constructing a map between it and some known object. This general approach lies at the heart of algebraic topology and we cannot emphasize it enough. It is the basis for what we have done and will do in this book.

Let $f : S^n \to S^n$, $n \geq 1$, be a continuous map and let (K,φ) be a triangulation of S^n. From Lemma 2 in Sec. 6.2 we know that $H_n(K) \approx \mathbb{Z}$. Furthermore, if G is any abelian group which is isomorphic to $\mathbb{Z}$, then any homomorphism $\alpha : G \to G$ is nothing but multiplication by k for some unique $k \in \mathbb{Z}$ (see Lemma 9 in Appendix B).

DEFINITION. Define the degree of f, $\deg f$, by $\deg f = k$, where $(\varphi^{-1} \circ f \circ \varphi)_*(a) = ka$ for all $a \in H_n(K)$.

Intuitively, the integer $\deg f$ tells us how many times f wraps S^n around itself. This idea will become clearer as we go along and as we look at some specific examples--expecially in the case $n = 1$. We note also that $\deg f$ is already determined by the equation $(\varphi^{-1} \circ f \circ \varphi)_*(u) = (\deg f)u$, where u is a generator for $H_n(K)$.

LEMMA 1. (a) $\deg f$ is a well-defined integer.

(b) If $f,g : S^n \to S^n$ are homotopic continuous maps, then $\deg f = \deg g$.

Proof: (a) Assume that (L,ψ) is another triangulation of S^n and consider the commutative diagram

$$\begin{array}{ccccc} |K| & \xrightarrow{\varphi} & S^n & \xleftarrow{\psi} & |L| \\ {\scriptstyle \varphi^{-1} \circ f \circ \varphi}\downarrow & & \downarrow{\scriptstyle f} & & \downarrow{\scriptstyle \psi^{-1} \circ f \circ \psi} \\ |K| & \xrightarrow[\varphi]{} & S^n & \xleftarrow[\psi]{} & |L| \end{array}$$

Suppose that $(\psi^{-1} \circ f \circ \psi)_*(b) = k_1 b$ for all $b \in H_n(L)$. We must show that $k = k_1$. Let $b \in H_n(L)$. Then

$$\begin{aligned} k_1 b &= (\psi^{-1} \circ f \circ \psi)_*(b) \\ &= (\psi^{-1} \circ \varphi \circ (\varphi^{-1} \circ f \circ \varphi) \circ \varphi^{-1} \circ \psi)_*(b) \\ &= (\psi^{-1} \circ \varphi)_*((\varphi^{-1} \circ f \circ \varphi)_*((\varphi^{-1} \circ \psi)_*(b))) \\ &= (\psi^{-1} \circ \varphi)_*(k((\varphi^{-1} \circ \psi)_*(b))) \\ &= k((\psi^{-1} \circ \varphi)_*((\varphi^{-1} \circ \psi)_*(b))) \\ &= k((\psi^{-1} \circ \varphi \circ \varphi^{-1} \circ \psi)_*(b)) \\ &= kb \end{aligned}$$

It follows that $k = k_1$ since b was arbitrary and $H_n(L) \approx \mathbb{Z}$. This proves (a).

(b) Let h_t be a homotopy between f and g. Then $\varphi^{-1} \circ h_t \circ \varphi$ is a homotopy between $\varphi^{-1} \circ f \circ \varphi$ and $\varphi^{-1} \circ g \circ \varphi$. Therefore, $(\varphi^{-1} \circ f \circ \varphi)_{*q} = (\varphi^{-1} \circ g \circ \varphi)_{*q}$ for all q and (b) is proved, finishing the proof of the lemma.

We see from Lemma 1(b) that deg induces a map

$$\deg : [S^n, S^n] \to \mathbb{Z} \quad n \geq 1$$

defined by $\deg [f] = \deg f$. Clearly, $\deg [1_{S^n}] = 1$ and $\deg [c] = 0$, where $c : S^n \to S^n$ is any constant map. (To see that $\deg c = 0$, assume that $c(S^n) = x_0$. We may also assume that $\varphi^{-1}(x_0)$ is a vertex of K for some triangulation (K,φ) of S^n. It follows that $\varphi^{-1} \circ c \circ \varphi$ is a constant simplicial map and therefore induces zero homomorphisms on the homology groups in dimension $q > 0$.) This shows in particular that the identity map of S^n is not homotopic to the constant map, which is a fact that we alluded to in Sec. 5.2 but had not been able to prove before. Let us compute the degrees of maps in a few more special cases.

LEMMA 2. Let $f : S^n \to S^n$, $n \geq 1$, be the reflection of S^n given by $f(x_1, x_2, \cdots, x_{n+1}) = (-x_1, x_2, \cdots, x_{n+1})$. Then $\deg f = -1$.

Proof: In order to make our computation of $\deg f$ easy we describe a convenient triangulation for S^n. First, consider the n-simplex $\sigma_n = v_0(n)v_1(n)\cdots v_n(n)$, $n \geq 1$, whose vertices $v_i(n) \in S^{n-1} \subset \mathbb{R}^n$ are defined inductively as follows (see Fig. 7.1): Let $e_i = (0,0,\cdots,0,1) \in \mathbb{R}^i$ and set $v_0(1) = -e_1$ and $v_1(1) = e_1$. Next, suppose that $v_i(n-1)$, $i = 0,1,\cdots,n-1$, has already been defined for $n \geq 2$. Define $v_i(n) = (3/5)v_i(n-1) - (4/5)e_n$ for $0 \leq i \leq n-1$ and $v_n(n) = e_n$.

Now define simplicial complexes K_{n+1} and L_n, $n \geq 1$, by $K_{n+1} = \overline{\sigma_{n+1}}$ and $L_n = \partial K_{n+1}$ and define $\varphi_n : |L_n| \to S^n$ by $\varphi_n(x) = x/|x|$. It is easy to check that φ_n is a homeomorphism and hence gives a triangulation of S^n. Finally, in order to simplify the notation, let us set $v_i = v_i(n+1)$, and define a simplicial map $f_n : L_n \to L_n$, $n \geq 1$, by $f_n(v_0) = v_1$, $f_n(v_1) = v_0$, and $f_n(v_i) = v_i$ for $2 \leq i \leq n$. Note that f_n can also be considered as a simplicial map from K_{n+1} to K_{n+1}.

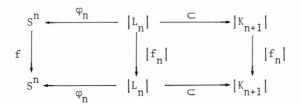

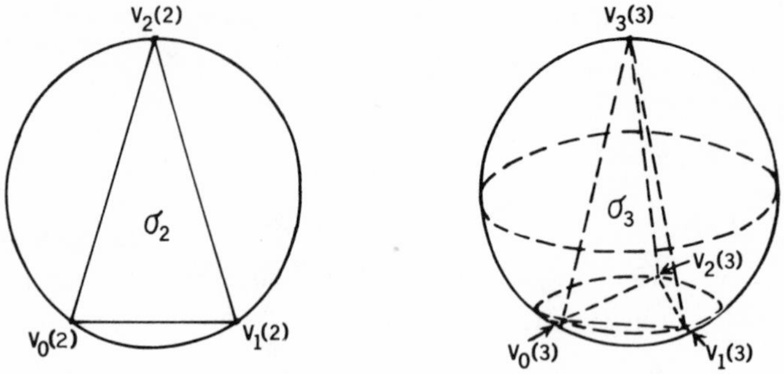

FIGURE 7.1

The point of all these definitions above is that we have a commutative diagram

$$\begin{array}{ccccc} S^n & \xleftarrow{\varphi_n} & |L_n| & \xrightarrow{\subset} & |K_{n+1}| \\ \big\downarrow f & & \big\downarrow |f_n| & & \big\downarrow |f_n| \\ S^n & \xleftarrow[\varphi_n]{} & |L_n| & \xrightarrow[\subset]{} & |K_{n+1}| \end{array}$$

The proof of commutativity is straightforward and left as an exercise to the reader. It should certainly be clear from the geometry of the situation by looking at Fig. 7.1. It follows that $\deg f$ is defined by the equation $(f_n)_*(z) = (\deg f)z$, where z is a generator of $H_n(L_n)$. But in the proof of Lemma 2 in Sec. 6.2 we showed that one generator of $Z_n(L_n) = H_n(L_n)$ is $\partial_{n+1}([v_0v_1\cdots v_{n+1}])$ and

$$\begin{aligned} (f_n)_*(\partial_{n+1}([v_0v_1\cdots v_{n+1}])) &= \partial_{n+1}((f_n)_*([v_0v_1\cdots v_{n+1}])) \\ &= \partial_{n+1}([v_1v_0v_2\cdots v_{n+1}]) \\ &= \partial_{n+1}(-[v_0v_1v_2\cdots v_{n+1}]) \\ &= -(\partial_{n+1}([v_0v_1\cdots v_{n+1}])) \end{aligned}$$

Therefore, $\deg f = -1$ and the lemma is proved.

LEMMA 3. If $f,g : S^n \to S^n$ are continuous maps, then $\deg (f \circ g) = (\deg f)(\deg g)$.

Proof: Let $s = \deg f$ and $t = \deg g$. Choose a triangulation (K,φ) for S^n and let $a \in H_n(K)$. Then

$$\begin{aligned}(\varphi^{-1} \circ (f \circ g) \circ \varphi)_*(a) &= ((\varphi^{-1} \circ f \circ \varphi) \circ (\varphi^{-1} \circ g \circ \varphi))_*(a)\\ &= (\varphi^{-1} \circ f \circ \varphi)_*((\varphi^{-1} \circ g \circ \varphi)_*(a))\\ &= (\varphi^{-1} \circ f \circ \varphi)_*(ta)\\ &= s(ta)\end{aligned}$$

Therefore, by definition, $\deg (f \circ g) = st = (\deg f)(\deg g)$ and Lemma 3 is proved.

One map of S^n to itself is of special interest and deserves a name.

DEFINITION. Let $n \geq 0$. The points x and $-x$ on S^n are called antipodal points and the map from S^n to itself which sends x to $-x$ is called the antipodal map.

LEMMA 4. If A is the antipodal map of S^n, then $\deg A = (-1)^{n+1}$.

Proof: Define $f_i : S^n \to S^n$, $1 \leq i \leq n + 1$, by

$$f_i(x_1,\cdots,x_{i-1},x_i,x_{i+1},\cdots,x_{n+1}) = (x_1,\cdots,x_{i-1},-x_i,x_{i+1},\cdots,x_{n+1})$$

Then $A = f_1 \circ f_2 \circ \cdots \circ f_{n+1}$, and so by Lemma 3 and induction $\deg A = (\deg f_1)(\deg f_2)\cdots(\deg f_{n+1})$. But Lemma 2 clearly implies that $\deg f_i = -1$ and therefore Lemma 4 is proved.

Our first application of the ideas developed in this section will be to the study of vector fields.

DEFINITION. Let $X \subset \mathbb{R}^n$. An $\mathbb{R}^n$-vector field on X is a continuous function $\sigma : X \to \mathbb{R}^n$. The $\mathbb{R}^n$-vector field σ is said to be nonzero if $\sigma(x) \neq 0$ for all $x \in X$.

By translating the vector $\sigma(x)$ to $x \in X$ we should identify the $\mathbb{R}^n$-vector field σ intuitively with a continuously varying collection of arrows, one at each point of X. If the space X is nice and one can talk about vectors being "tangent" to X, then an $\mathbb{R}^n$-vector field on X with the property that $\sigma(x)$ is "tangent" to X at x for each $x \in X$ is called simply a vector field on X. We make this precise only in the important case of $X = S^n$.

DEFINITION. A vector field on S^n, $n \geq 1$, is an $\mathbb{R}^{n+1}$-vector field σ on S^n such that $\sigma(x) \cdot x = 0$ for all $x \in S^n$.

The sphere S^n always has the zero vector field $\sigma_0 : S^n \to \mathbb{R}^{n+1}$, where $\sigma_0(x) = 0$ for all $x \in S^n$. This is not very interesting. What is interesting is to determine those n for which S^n admits a nonzero vector field. Note that S^n certainly has nonzero $\mathbb{R}^{n+1}$-vector fields σ. One such is the identity map of S^n. However, it is not always possible to find a σ whose vectors are tangent to S^n.

THEOREM 1. S^n admits a nonzero vector field if and only if n is odd.

Proof: First, assume that n is odd, say $n = 2k + 1$. Define a nonzero vector field $\sigma : S^n \to S^n \subset \mathbb{R}^{n+1}$ by

$$\sigma(x_1, x_2, \cdots, x_{2k+1}, x_{2k}) = (-x_2, x_1, \cdots, -x_{2k}, x_{2k+1})$$

Next, let $n = 2k$ be even and suppose that σ is a nonzero vector field on S^n. Without loss of generality we may assume that $|\sigma(x)| = 1$ for all $x \in S^n$, so that σ is actually a map from S^n into S^n. (If $|\sigma(x)| \neq 1$ for all $x \in S^n$, consider the vector field σ' given by $\sigma'(x) = \sigma(x)/|\sigma(x)|$. The map σ' is well defined because σ is never zero and clearly $|\sigma'(x)| = 1$.) Define $F : S^n \times [0,1] \to S^n$ by $F(x,t) = (\cos \pi t)x + (\sin \pi t)\sigma(x)$. Observe that $F(x,0) = x$ and $F(x,1) = -x$. In other words, F is a homotopy between the identity map of S^n and the antipodal map. The former has degree +1, and the latter, degree $(-1)^{2k+1} = -1$ (by Lemma 4). Since $+1 \neq -1$, we have a contradiction, and so no even-dimensional sphere can have a nonzero vector field.

COROLLARY. S^2 does not admit a nonzero vector field.

The corollary provides a solution to an amusing popular problem called "the hairy billiard ball problem." We assume that we are given a billiard ball that has hair on it which can be represented by little vectors that stick out from every point on the ball. In other words, the hairs correspond to a nonzero $\mathbb{R}^3$-vector field. The problem is to show that it is impossible to comb the ball in such a way so that all the hair is lying down smoothly. A little thought will convince one on intuitive grounds that no matter how the hair is combed there will always be a "cowlick" somewhere (see Fig. 7.2). The actual proof rests on the fact that a smooth combing would provide us with a nonzero vector field which is impossible by the corollary. Alternatively, we can say that every hairy billiard ball has at least one hair that is standing perpendicular to the ball, otherwise we could take the projection of the hair onto the tangent planes and again get a nonzero vector field.

There is an important generalization of Theorem 1.

DEFINITION. The vector fields $\sigma_1, \sigma_2, \cdots,$ and σ_k on S^n are said to be linearly independent if the vectors $\sigma_1(x), \sigma_2(x), \cdots,$ and $\sigma_k(x)$ (in $\mathbb{R}^{n+1}$) are linearly independent for every $x \in S^n$.

Instead of asking whether or not S^n admits one nonzero vector field one can ask the much preciser question: What is the maximal number of linearly independent vector fields on S^n? Write $n = (2a + 1)2^b$, where $b = c + 4d$ and $0 \leq c \leq 3$, and define $\rho(n) = 2^c + 8d$. Theorem 1 is a trivial corollary of the next theorem which provides a complete solution to the problem of vector fields.

THEOREM D. (a) *(Hurwitz-Radon-Eckmann)* There exist $\rho(n) - 1$ linearly independent vector fields on S^{n-1}.

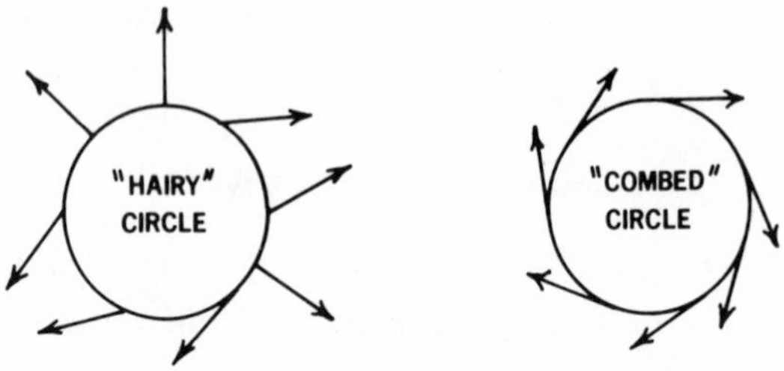

THE "HAIRY" CIRCLE CAN BE COMBED

(a)

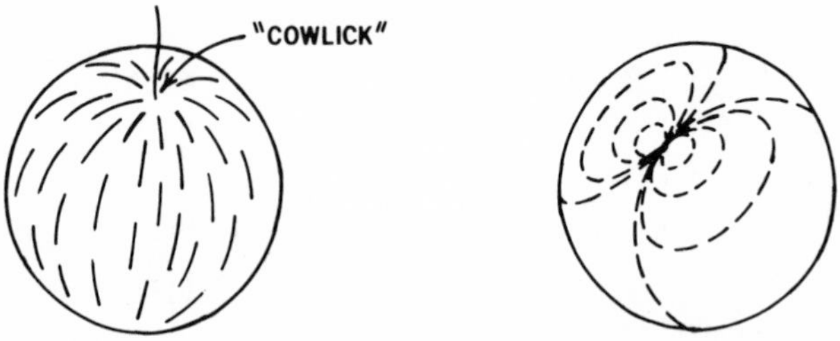

ATTEMPTS TO COMB THE "HAIRY" BILLIARD BALL

(b)

FIGURE 7.2

(b) *(Adams)* S^{n-1} does not admit $\rho(n)$ linearly independent vector fields.

The proof of Theorem D is far beyond the scope of this book.

Problems

7.1. Let $f, g : S^n \to S^n$, $n \geq 1$, be continuous maps.

(a) Prove that if $f(x) \neq g(x)$ for all $x \in S^n$, then $\deg f + (-1)^n (\deg g) = 0$.

(b) Prove that if $\deg f \neq (-1)^{n+1}$, then f has a fixed point.

(c) Prove that if $\deg f \neq 1$, then $f(x) = -x$ for some $x \in S^n$.

7.2. Prove that there is no nonzero $\mathbb{R}^2$-vector field σ on D^2 with the property that all the vectors of σ on S^1 either all point into D^2 or all out of D^2. (If $x \in S^1$ the vector $\sigma(x)$ is said to point into D^2, or out of D^2, if the angle θ between $\sigma(x)$ and x satisfies $\pi/2 < \theta < 3\pi/2$ or $-\pi/2 < \theta < \pi/2$, respectively.) Generalize this result to n dimensions.

7.3. Prove that given any nonzero $\mathbb{R}^2$-vector field on S^1 there will always be at least one pair of antipodal points on which the two vectors of the vector field are either in the same or opposite direction.

7.4. Show that every nonzero $\mathbb{R}^3$-vector field σ on S^2 will have at least one vector that is perpendicular to S^2, that is, $\sigma(x) = cx$ for some $x \in S^2$ and $0 \neq c \in \mathbb{R}$.

7.5. Show that an analog of the corollary to Theorem 1 holds for ellipsoids. More generally, consider an arbitrary "smooth" surface S in $\mathbb{R}^3$, where by "smooth" we essentially mean that it is possible to define a continuously varying tangent plane T_x to S at every point $x \in S$. In other words, S should not have any "corners." (Formally, T_x is defined to be the plane generated by all vectors at x which are tangent to a curve in S that passes through x.) Figure 7.3 shows some tangent planes to the torus. Define a vector field on S to be an $\mathbb{R}^3$-vector field σ with the property that $\sigma(x) + x \in T_x$ for all $x \in S$, that is, $\sigma(x)$ is tangent to S. Which of the smooth (orientable) surfaces S in $\mathbb{R}^3$ admit nonzero vector fields? Which do not? (A detailed analysis of these questions can be found in [Mi 3].)

7.6. (a) Let $f,g : S^{2n} \to S^{2n}$, $n \geq 0$, be continuous. Show that either f or g or $f \circ g$ has a fixed point. (*Hint:*

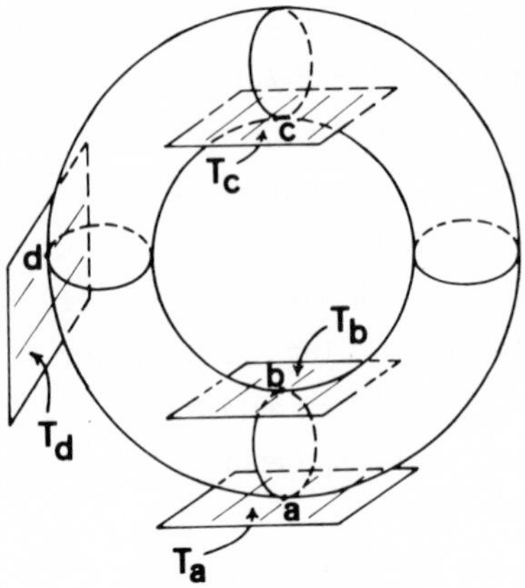

FIGURE 7.3

If the conclusion is false, then the points x, $g(x)$, and $(f \circ g)(x)$ are distinct for all $x \in S^{2n}$ and determine a unique circle C_x in S^{2n}. Now construct a nonzero vector field on S^{2n} by taking the unit tangent vector to C_x at x in the direction $g(x)$ for each $x \in S^{2n}$.)

(b) Prove that if $f : S^{2n} \to S^{2n}$, $n \geq 0$, is continuous, then $f \circ f$ has a fixed point.

(c) Prove that if $f : S^{2n} \to S^{2n}$, $n \geq 0$, is continuous, then $f(x) = x$ or $-x$ for some $x \in S^{2n}$.

7.2. THE BORSUK-ULAM THEOREM AND THE HAM SANDWICH PROBLEM

Let us begin by giving an alternate characterization of the degree of a continuous map $f : S^n \to S^n$. Let (K,μ) and (L,ν) be two triangulations of S^n. We know from Sec. 6.5 (Theorem 8(a) in particular) that the n-simplices of K and L can be appropriately oriented so that

$$\Sigma_K = \sum_{\text{n-simplices } \sigma \in K} [\sigma] \quad \text{and} \quad \Sigma_L = \sum_{\text{n-simplices } \tau \in L} [\tau]$$

are generators for $H_n(K)$ and $H_n(L)$, respectively. Assume that the map $g = \nu^{-1} \circ f \circ \mu : |K| \to |L|$ admits a simplicial approximation

$\lambda : K \to L$. If τ is an n-simplex of L, let $p(\tau,\lambda)$ be the number of n-simplices $\sigma \in K$ such that $\lambda_{\#}([\sigma]) = [\tau]$ and let $n(\tau,\lambda)$ be the number of n-simplices $\sigma \in K$ such that $\lambda_{\#}([\sigma]) = -[\tau]$. The numbers $p(\tau,\lambda)$ and $n(\tau,\lambda)$ are often easily computed from the geometry of the situation, and so the next lemma provides us with a useful method for determining the degree of a map.

LEMMA 5. (a) If $(\nu^{-1} \circ \mu)_*(\Sigma_K) = \Sigma_L$, then $\deg f = p(\tau,\lambda) - n(\tau,\lambda)$.

(b) If the total number, $p(\tau,\lambda) + n(\tau,\lambda)$, of n-simplices σ in K which get mapped onto τ by $|\lambda|$ is odd, then so is $\deg f$.

Proof: Let $k = p(\tau,\lambda) - n(\tau,\lambda)$ and observe that k is the coefficient of $[\tau]$ in $\lambda_{\#}(\Sigma_K)$. Therefore, $(\nu^{-1} \circ f \circ \mu)_*(\Sigma_K) = \lambda_*(\Sigma_K) = k\Sigma_L$ and the hypothesis in (a) implies that

$$(\deg f)\Sigma_K = (\mu^{-1} \circ f \circ \mu)_*(\Sigma_K) = (\mu^{-1} \circ \nu)_*((\nu^{-1} \circ f \circ \mu)_*(\Sigma_K))$$
$$= (\mu^{-1} \circ \nu)_*(k\Sigma_L) = k(\mu^{-1} \circ \nu)_*((\nu^{-1} \circ \mu)_*(\Sigma_K)) = k\Sigma_K$$

It follows that $\deg f = k$ and (a) is proved. A similar argument shows that $\deg f = -k$ if $(\nu^{-1} \circ \mu)_*(\Sigma_K) = -\Sigma_L$. But $(\nu^{-1} \circ \mu)_*(\Sigma_K)$ is either equal to Σ_L or $-\Sigma_L$ (because $\nu^{-1} \circ \mu$ is a homeomorphism) and $p(\tau,\lambda) + n(\tau,\lambda)$ has the same parity as $\pm(p(\tau,\lambda) - n(\tau,\lambda))$. This shows that $p(\tau,\lambda) + n(\tau,\lambda)$ and $\deg f$ are either both even or both odd and proves (b).

DEFINITION. Let $n,m \geq 0$. A map $f : S^n \to S^m$ is said to be antipodal preserving if $f(-x) = -f(x)$ for all $x \in S^n$.

The fundamental theorem of this section is

THEOREM 2. Every continuous antipodal-preserving map $f : S^n \to S^n$, $n \geq 0$, has odd degree.

NOTE. So far we have only defined the degree of f in case $n \geq 1$. Therefore, let us extend the definition to the case $n = 0$ as follows: If $f : S^0 \to S^0$, define the degree of f, $\deg f$, to be 1,-1, or 0 depending on whether or not f is the identity map, the antipodal map, or constant map, respectively. With this natural extension of the definition of degree, previous results about the degrees of maps on spheres could have been stated so as to include the case $n = 0$.

Proof of Theorem 2: The case $n = 0$ is trivial because an antipodal-preserving map from S^0 to S^0 cannot be constant. Now assume that the theorem has been proved for $n = m - 1$, $m \geq 1$, and let $n = m$. As usual, we have to translate everything in terms of simplicial complexes and simplicial maps, but since it will be important to preserve the symmetry of the situation this time, we are forced into having to make use of an especially large amount of notation. However, the definitions are formidable in appearance only and not in content.

The idea of the proof can already be seen in the case $n = 1$. Let us index the points of S^1 by the angle they make with the x-axis. As the point x moves from 0 to π, the point $f(x)$ will move from $f(0)$ to $f(\pi) = f(0) + \pi + 2k\pi$. Since the map f is completely determined by its action on the upper half of S^1, moving from π to 2π, moves $f(\pi)$ to $f(2\pi) = f(\pi) + \pi + 2k\pi$. In other words, as x moves from 0 to 2π, $f(x)$ will move from $f(0)$ to $f(2\pi) = f(\pi) + \pi + 2k\pi = f(0) + \pi + 2k\pi + \pi + 2k\pi = f(0) + 2\pi(2k + 1)$, so that S^1 will have been wrapped an odd number of times around itself, which means that $\deg f$ is odd. Now to return to the general case.

First of all, by letting $|K|$ be an appropriate polyhedron inscribed in S^m, it is easy to show that there is a triangulation (K,h) of S^m and a simplicial complex L with the following properties (see Fig. 7.4 for one possible (K,h) and L in case $m = 1$):

1. The vertices of K are points of S^m and $h : |K| \to S^m$ is given by $h(x) = x/|x|$.
2. If v is a vertex of K, then so is $-v$, and the map which sends v to $-v$ is a simplicial map from K to K. In particular, $x \in |K|$ if and only if $-x \in |K|$.

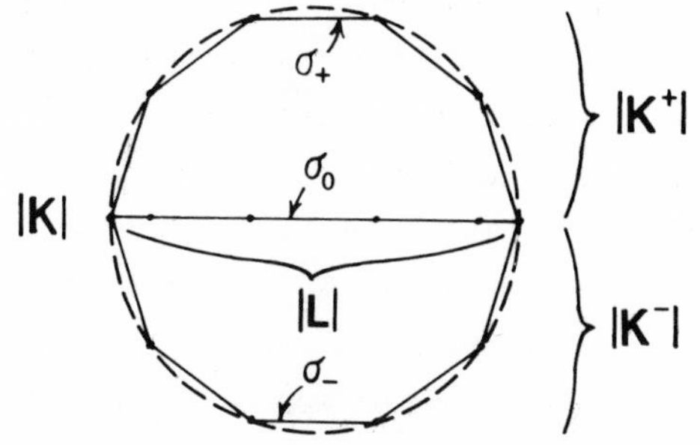

FIGURE 7.4

3. There are subcomplexes K_j and $K_j^{\pm}$ of K for $0 \leq j \leq m$ such that $K_0 = S^0 \subset K_1 \subset \cdots \subset K_m = K$, $h(|K_j|) = S^j \subset S^m$, $h(|K_j^{\pm}|) = S_{\pm}^j$, and $K_j^+ \cap K_j^- = K_{j-1}$. Set $K^{\pm} = K_m^{\pm}$.
4. If $\pi : |K| \to \mathbb{R}^m$ is the restriction of the natural projection from $\mathbb{R}^{m+1}$ to $\mathbb{R}^m$ [$(x_1,x_2,\cdots,x_{m+1}) \to (x_1,x_2,\cdots,x_m)$], then $|L| = \pi(|K|)$. Furthermore, $x \in |K|$ is a vertex of K if and only if $\pi(x)$ is a vertex of L, and the map which sends a vertex v of K to $\pi(v)$ is a simplicial map from K to L. Thus $\partial L = K_{m-1}$ and $|L|$ is the closed convex hull of $|\partial L| = |K_{m-1}|$.
5. The origin of $\mathbb{R}^m$ is an interior point of an m-simplex σ_0 in L. Let $\sigma_{\pm} = \pi^{-1}(\sigma_0) \cap |K^{\pm}|$ be the two m-simplices of K which get mapped onto σ_0 by π.

DEFINITION. We shall call a map $g : |K| \to |K|$ antipodal preserving if $g(-x) = -g(x)$ for all $x \in |K|$.

Now define $g : |K| \to |K|$ by $g = h^{-1} \circ f \circ h$. The map g is clearly antipodal preserving.

CLAIM 1. If k is sufficiently large, then g admits a simplicial approximation $\varphi : \text{sd}^k K \to K$ such that $|\varphi|$ is antipodal preserving.

To prove Claim 1 we first choose a $k \geq 0$ such that $g : |\text{sd}^k K| \to |K|$ has simplicial approximations. The integer k exists by the simplicial approximation theorem. It remains to be shown that there is at least one such approximation φ for which $|\varphi|$ is antipodal preserving. The important observation to make here is that if φ exists, then $|\varphi|$ is already determined on $|K_j|$ by its action on the "hemisphere" $|K_j^+|$. Therefore, we can construct φ inductively by assuming it is defined on $sd^k K_j$ and defining an extension to $sd^k K_{j+1}$ by first extending it to $sd^k K_{j+1}^+$ and then using the equation $\varphi(v) = -\varphi(-v)$ to extend it to $sd^k K_{j+1}^-$. To get the first extension from $sd^k K_j$ to $sd^k K_{j+1}^+$ we need a relative version of the simplicial approximation theorem, that is, we seek a simplicial approximation that agrees with the given one on $sd^k K_j$. Since g already admitted simplicial approximations on all of $sd^k K$, the proof of Theorem 2 in Sec. 5.5 carries over without any trouble to produce the desired extension. Furthermore, it is obvious that φ can be defined on K_0, so that we can get the induction started, and Claim 1 is proved. The details are left to the reader. Note that we need to have the original g antipodal preserving, because our construction of φ would not allow us to conclude that $\varphi|(sd^k K_{j+1}^-)$ is a simplicial approximation to $g|(|sd^k K_{j+1}^-|)$ without this hypothesis.

Next, let $\varphi : sd^k K \to K$ be a fixed simplicial approximation to g such that $|\varphi|$ is antipodal preserving. Set $\varphi_+ = \varphi|(sd^k K^+)$ and $\psi = \varphi|(sd^k K_{m-1})$. Define $r : |L| - \text{int}\ \sigma_0 = |L - \sigma_0| \to |\partial L|$ by $r(x) = h^{-1}(x/|x|)$. Since φ is a simplicial map and $\dim K_{m-1} = m - 1$, it follows from property (4) of L that $(\pi \circ |\psi|)(|K_{m-1}|) \subset |L - \sigma_0|$. Therefore, $g_1 = r \circ \pi \circ |\psi| : |K_{m-1}| \to |K_{m-1}|$ is defined and clearly $h \circ g_1 \circ h^{-1} : S^{m-1} \to S^{m-1}$ is an antipodal-preserving map. By our inductive hypothesis $h \circ g_1 \circ h^{-1}$ will have odd degree.

CLAIM 2. Let N be the number of m-simplices of $sd^k K^+$ which get mapped onto σ_0 by $\pi \circ |\varphi_+|$. Then N is odd.

Orient the m-simplices τ of K in such a way that

$$\Sigma = \sum_{\text{m-simplices } \tau \in K} [\tau]$$

is a generator of $Z_m(K) = H_m(K) \approx \mathbb{Z}$. It is easy to see by induction on k that the oriented simplices $[\tau]$ will induce an orientation on the m-simplices η of $\mathrm{sd}^k K$ so that

$$\mathrm{sd}_{\#}^k(\Sigma) = \sum_{\text{m-simplices } \eta \in \mathrm{sd}^k K} [\eta]$$

is a generator of $Z_m(\mathrm{sd}^k K) = H_m(\mathrm{sd}^k K) \approx \mathbb{Z}$. (Alternatively, note that by Problem 5.18 $\mathrm{sd}_{\#}^k = \mathrm{sd}_*^k : H_m(K) \to H_m(\mathrm{sd}^k K)$ is an isomorphism.) Let

$$\Sigma_+ = \sum_{\text{m-simplices } \tau \in K^+} [\tau]$$

$\Sigma_{m-1} = \partial_m(\Sigma_+) \in Z_{m-1}(K_{m-1}) = H_{m-1}(K_{m-1})$, and $\Sigma_{m-1}^k = \mathrm{sd}_{\#}^k(\Sigma_{m-1}) \in Z_{m-1}(\mathrm{sd}^k K_{m-1}) = H_{m-1}(\mathrm{sd}^k K_{m-1})$. Define

$$T = \{ \eta \mid \eta \text{ is an m-simplex of } \mathrm{sd}^k K^+ \text{ and } (\pi \circ |\varphi_+|)(\eta) = \sigma_0 \}$$

and define m-chains $A, B \in C_m(\mathrm{sd}^k K^+)$ by

$$A = \sum_{\text{m-simplices } \eta \in (\mathrm{sd}^k K^+ - T)} [\eta] \quad \text{and} \quad B = \sum_{\eta \in T} [\eta]$$

Finally, let ∂T denote the simplicial complex which consists of all the q-faces, $0 \leq q < m$, of the simplices in T.

Consider the commutative diagram

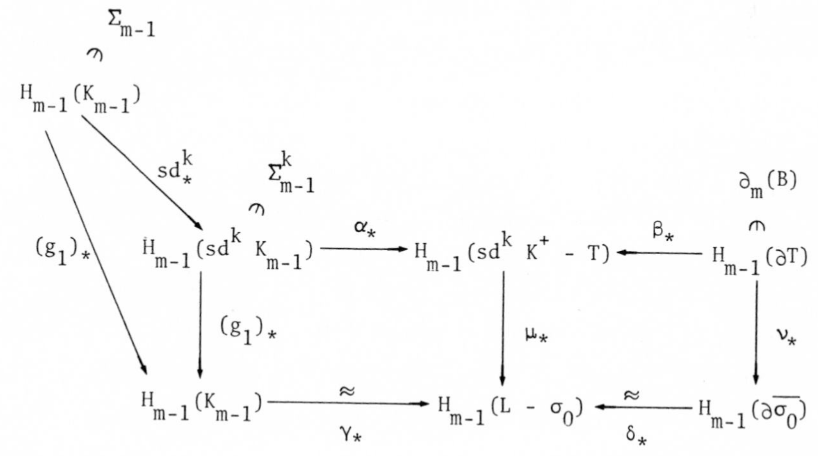

where μ_* and ν_* are induced by the appropriate restrictions of $\pi \circ |\varphi_+|$ and α_*, β_*, γ_*, and δ_* are induced by the natural inclusion maps. The maps on the bottom line of the diagram are isomorphisms because it is easy to see that the inclusion maps of $|K_{m-1}|$ and $\partial\sigma_0$ into $|L - \sigma_0|$ are homotopy equivalences. Since $h \circ g_1 \circ h^{-1}$ has odd degree, $(g_1)_*(\Sigma_{m-1}) = (\text{odd integer}) \cdot \Sigma_{m-1}$. (The case $m = 1$ has to be checked separately, but is easy. The point is that if $m \geq 2$, then Σ_{m-1} is a generator of $H_{m-1}(K_{m-1})$, and if $m = 1$, then Σ_{m-1} corresponds to $\pm(1,1) \in \mathbb{Z} \oplus \mathbb{Z}$ under the natural isomorphism of $H_{m-1}(K_{m-1})$ with $\mathbb{Z} \oplus \mathbb{Z}$.) Furthermore, $A \in C_m(\text{sd}^k\, K^+ - T)$ and $\partial_m(A) = \Sigma^k_{m-1} - \partial_m(B) \in Z_{m-1}(\text{sd}^k\, K^+)$, so that $\alpha_*(\Sigma^k_{m-1}) = \beta_*(\partial_m(B))$. Therefore, the commutativity of the diagram implies that

$$\begin{aligned}(\delta_* \circ \nu_*)(\partial_m(B)) &= (\mu_* \circ \beta_*)(\partial_m(B)) = \mu_*(\alpha_*(\Sigma^k_{m-1})) \\ &= (\gamma_* \circ (g_1)_*)(\text{sd}^k_*(\Sigma_{m-1})) = \gamma_*((g_1)_*(\Sigma_{m-1})) \\ &= (\text{odd integer}) \cdot \gamma_*(\Sigma_{m-1})\end{aligned}$$

In other words, if $u = (\delta_*^{-1} \circ \gamma_*)(\Sigma_{m-1}) \in H_{m-1}(\partial\overline{\sigma_0})$, then $\nu_*(\partial_m(B)) = (\text{odd integer}) \cdot u$. If $\eta \in T$, then clearly $\partial_m([\eta]) \in H_{m-1}(\partial T)$ gets mapped onto $\pm u$ in $H_{m-1}(\partial\overline{\sigma_0})$ by ν_*. Let a_+ and a_- denote the number of $\eta \in T$ such that $\partial_m([\eta])$ gets mapped onto $+u$ and $-u$, respectively. Then $\nu_*(\partial_m(B)) = \sum_{\eta \in T} \nu_*(\partial_m([\eta])) = (a_+ - a_-)u$, so that $a_+ - a_-$ is odd. On the other hand, $N = a_+ + a_-$. This proves Claim 2 because $a_+ - a_-$ and $a_+ + a_-$ have the same parity.

Next, observe that an m-simplex η in $\text{sd}^k\, K^+$ gets mapped onto σ_0 by $\pi \circ |\varphi_+|$ if and only if η gets mapped onto σ_+ or σ_- by $|\varphi_+|$. But $|\varphi|$ is antipodal preserving, so that the simplex η gets mapped onto σ_- by $|\varphi_+|$ if and only if the m-simplex $-\eta = \{ -x \in |\text{sd}^k\, K| : x \in \eta \}$ in $\text{sd}^k\, K$ gets mapped onto σ_+ by $|\varphi|$. This means that the integer N in Claim 2 is also the number of m-simplices of $\text{sd}^k\, K$ which get mapped onto σ_+ by $|\varphi|$. Since N is odd, Theorem 2 now follows from Lemma 5(b) and induction.

As an application of Theorem 2 we have

THEOREM 3 (Borsuk-Ulam). If $f : S^n \to \mathbb{R}^n$, $n \geq 0$, is a continuous map, then there is some $x \in S^n$ such that $f(x) = f(-x)$, that is, f takes on the same value at some pair of antipodal points of S^n.

Proof: Since the case $n = 0$ is trivial, assume that $n \geq 1$. If $f(x) \neq f(-x)$ for all $x \in S^n$, then we can define a continuous map $g : S^n \to S^{n-1}$ by $g(x) = (f(x) - f(-x))/|f(x) - f(-x)|$. The map $g_1 = g|S^{n-1} : S^{n-1} \to S^{n-1}$ is clearly an antipodal-preserving map and therefore has odd degree by Theorem 2. On the other hand, there is a homotopy $g_t : S^{n-1} \to S^{n-1}$ given by

$$g_t(y_1,y_2,\cdots,y_n) = g(ty_1,ty_2,\cdots,ty_n, 1 - t^2)$$

between g_1 and the constant map g_0, so that $\deg g_1 = 0$. Since 0 is not odd, we have a contradiction and Theorem 3 must be true.

Theorem 3 implies

THEOREM 4. If $f : S^n \to S^m$ is an antipodal-preserving continuous map, then $m \geq n$.

Proof: Suppose that $m < n$. Then $S^m \subset \mathbb{R}^{m+1} \subset \mathbb{R}^n$, so that f can be thought of as a map from S^n to $\mathbb{R}^n$. By Theorem 3, $f(x) = f(-x)$ for some $x \in S^n$. This clearly contradicts the fact that f is antipodal preserving and proves the theorem.

Actually, Theorems 3 and 4 are equivalent. To see how Theorem 4 implies Theorem 3, note that if $f : S^n \to \mathbb{R}^n$ is as in Theorem 3 and $f(x) \neq f(-x)$ for all $x \in S^n$, then the map $g : S^n \to S^{n-1}$ given by $g(x) = (f(x) - f(-x))/|f(x) - f(-x)|$ is an antipodal-preserving map, which is impossible by Theorem 4.

The Borsuk-Ulam theorem has some well-known amusing consequences. One is a solution to the so-called "ham sandwich problem." A ham sandwich is assumed to consist of three parts: a piece of ham between two slices of bread. The problem is to show that there is a plane (generated from the cut of a knife) which cuts in half, not only the whole sandwich, but also the three constituent pieces separately.

Anyone who has ever eaten a ham sandwich surely has discovered that this is always possible, but mathematicians, as usual, like to generalize. They have reformulated the problem in n dimensions without any restrictions on the location of each of the pieces of the "sandwich." In order not to get bogged down in technicalities about Lebesgue measurable sets in $\mathbb{R}^n$ and their volumes and lose the intuitive content of the next corollary, we have chosen not to make the hypotheses very precise.

COROLLARY 1 (The Generalized Ham Sandwich Problem). If $X_1, X_2, \cdots, X_n$ are n "nice" bounded subsets of $\mathbb{R}^n$, then there is a hyperplane in $\mathbb{R}^n$ which bisects each of the sets X_i simultaneously into sets of equal volumes.

REMARK. Open or convex sets in $\mathbb{R}^n$ are always "nice," so that the corollary applies to all those in particular. One can find weaker conditions on the X_i by studying the proof and determining those properties of the X_i that are needed to make it work.

Proof: If $a = (a_1, a_2, \cdots, a_n) \in S^{n-1}$, then all hyperplanes in $\mathbb{R}^n$ which are perpendicular to the vector a can be written in the form

$$P_d(a) = \{ (x_1, x_2, \cdots, x_n) \in \mathbb{R}^n \mid a_1x_1 + a_2x_2 + \cdots + a_nx_n = d \}$$

for some $d \in \mathbb{R}$. Each of these hyperplanes $P_d(a)$ divides $\mathbb{R}^n$ into two parts

$$A_d^+(a) = \{ (x_1, x_2, \cdots, x_n) \in \mathbb{R}^n \mid a_1x_1 + a_2x_2 + \cdots + a_nx_n \geq d \}$$

and

$$A_d^-(a) = \{ (x_1, x_2, \cdots, x_n) \in \mathbb{R}^n \mid a_1x_1 + a_2x_2 + \cdots + a_nx_n \leq d \}$$

with $A_d^+(a) \cap A_d^-(a) = P_d(a)$. The closed half-spaces $A_d^+(a)$, $A_d^-(a)$ correspond to the parts of $\mathbb{R}^n$ which are to the "right," respectively, "left" of $P_d(a)$, where the vector a is by definition said to point to the "right" (see Fig. 7.5a).

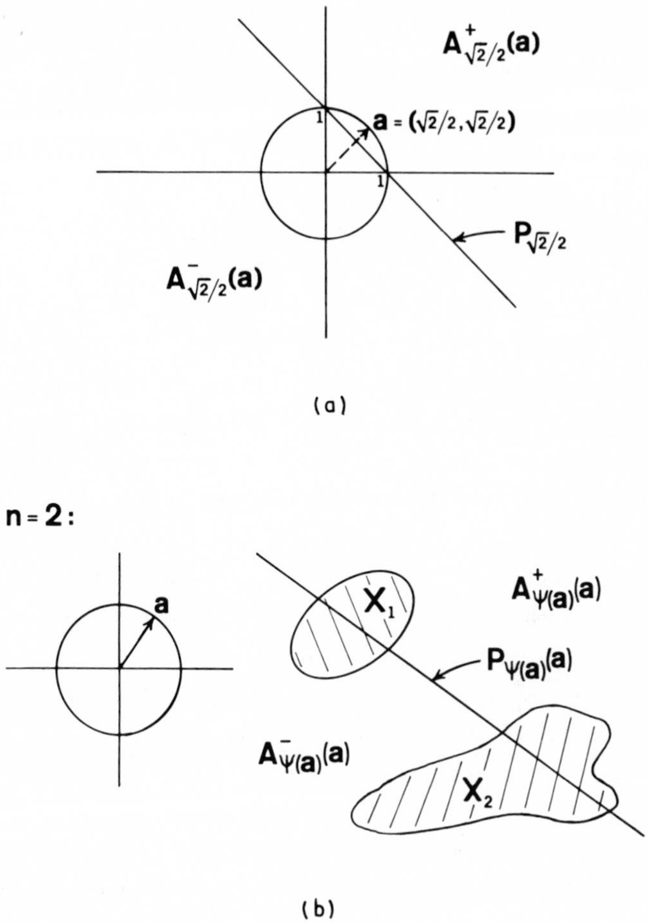

FIGURE 7.5

If $X \subset \mathbb{R}^n$, let vol(X) denote the n-dimensional volume of X. We shall assume that the X_i satisfy:

(*) The functions $\varphi_i^{\pm} : S^{n-1} \times \mathbb{R} \to \mathbb{R}$ given by $\varphi_i^{\pm}(a,d) = \text{vol}(A_d^{\pm}(a) \cap X_i)$ are continuous for all i.

(**) For each $a \in S^{n-1}$ there is a unique $d_a \in \mathbb{R}$ such that

$\mathrm{vol}(A^+_{d_a}(a) \cap X_1) = \mathrm{vol}(A^-_{d_a}(a) \cap X_1)$ and the function ψ : $S^{n-1} \to \mathbb{R}$ given by $\psi(a) = d_a$ is continuous.

Property (*) asserts that the volume of the part of X_i to either the right or left of some hyperplane $P_d(a)$ varies continuously with a and d. Property (**) says that for each $a \in S^{n-1}$ there is a unique hyperplane perpendicular to a which bisects X_1 by volume.

Note that $\varphi^+_i(a,d) = \varphi^-_i(-a,-d)$ because $A^+_d(a) = A^-_{-d}(-a)$. Intuitively, "right" and "left" get reversed when we pass from a to -a. Also, $\psi(-a) = -\psi(a)$.

Now define functions $f_i : S^{n-1} \to \mathbb{R}$, $i = 2,3,\cdots,n$ by $f_i(a) = \varphi^+_i(a,\psi(a))$. In other words, $f_i(a)$ is computed by first finding the appropriate hyperplane $P_{\psi(a)}(a)$ which is perpendicular to a and which bisects X_1 and then evaluating the volume of X_i to the right of $P_{\psi(a)}(a)$ [see Fig. 7.5b]. By (*) and (**) the functions f_i are continuous. Define $f : S^{n-1} \to \mathbb{R}^{n-1}$ by $f(a) = (f_2(a),f_3(a),\cdots, f_n(a))$. It then follows from the Borsuk-Ulam theorem that $f(a) = f(-a)$ for some $a \in S^{n-1}$. This means that $f_i(a) = f_i(-a)$ for $2 \le i \le n$, or $\varphi^+_i(a,\psi(a)) = f_i(a) = f_i(-a) = \varphi^+_i(-a,\psi(-a)) = \varphi^-_i(a,\psi(a))$. In other words, the hyperplane $P_{\psi(a)}(a)$ bisects all of the X_i and the corollary is proved.

COROLLARY 2. No matter how one may divide S^n into $n + 1$ sets, all but possibly one of which is closed, at least one of these sets will contain a pair of antipodal points.

Proof: Suppose that $S^n = A_1 \cup A_2 \cup \cdots \cup A_{n+1}$, where the A_i, $1 \le i \le n$, are closed subsets. Define functions $f_i : S^n \to \mathbb{R}$, $1 \le i \le n$, by $f_i(x) = \mathrm{dist}(x,A_i)$. It is easy to check that the f_i are continuous. Define $f : S^n \to \mathbb{R}^n$ by $f(x) = (f_1(x),f_2(x),\cdots,f_n(x))$. By the Borsuk-Ulam theorem there is a point $a \in S^n$ such that $f(a) = f(-a)$, that is, $f_i(a) = f_i(-a)$ for $1 \le i \le n$. If $f_i(a) = f_i(-a) = 0$ for some i, then both a and -a must belong to A_i since A_i

is closed and only the points of A_i are at distance 0 from A_i. On the other hand, if $f_i(a) = f_i(-a) > 0$ for $1 \leq i \leq n$, then neither a nor $-a$ belongs to A_i for $1 \leq i \leq n$. Therefore, a and $-a$ must belong to A_{n+1} because S^n is covered by the A_i, $1 \leq i \leq n + 1$. This proves the corollary.

The next corollary has to do with the problem of when one space is a subspace of another space. In particular, it should be intuitively clear on dimensional grounds that S^n can never be a subspace of $\mathbb{R}^m$ if $m < n$. One should also have a strong feeling that it is impossible to imbed S^n in $\mathbb{R}^n$, but for a different reason, namely, that any map of S^n into $\mathbb{R}^n$ would somehow cover its image at least twice. The Borsuk-Ulam theorem justifies both our intuitions because it asserts that any map $f : S^n \to \mathbb{R}^m$, $m \leq n$, maps at least one pair of antipodal points of S^n onto the same point in $\mathbb{R}^m$, that is, f can never be an inclusion map, which is an injection. We have therefore proved

COROLLARY 3. If $n \geq m$, then S^n is never homeomorphic to any subspace of $\mathbb{R}^m$.

There are two other theorems which are related to the Borsuk-Ulam antipode theorem:

THEOREM E (Kakutani-Yamabe-Yujobô). If $f : S^n \to \mathbb{R}$ is a continuous map, then there exist $x_1, x_2, \cdots, x_{n+1} \in S^n$ such that $x_i \cdot x_j = 0$ for $i \neq j$ and $f(x_1) = f(x_2) = \cdots = f(x_{n+1})$.

THEOREM F (Dyson-Yang). A continuous map $f : S^n \to \mathbb{R}$ maps the $2n$ end points of some n mutually orthogonal diameters into a single value.

We cannot prove these theorems here, but there is an interesting corollary to Theorem E which is worth mentioning. It states, for example, that one can always circumscribe a square around any compact subset in the plane $\mathbb{R}^2$ and a cube around any compact subset of $\mathbb{R}^3$.

COROLLARY. If X is a nonempty compact subset of $\mathbb{R}^n$, then there exists an n-cube C which contains X and such that X meets each face of C.

Proof: We shall use the same notation as in the proof of the ham sandwich problem (Corollary 1 to Theorem 3). Define a function $\alpha_+ : S^{n-1} \to \mathbb{R}$ by $\alpha_+(a) = d$, where $X \subset A_d^-(a)$ and $X \cap P_d(a) \neq \emptyset$. It is easy to see that α_+ is a well-defined continuous function. Now define $g : S^{n-1} \to \mathbb{R}$ by $g(a) = \alpha_+(a) + \alpha_+(-a)$. Intuitively, $g(a)$ measures the distance between the two (parallel) planes with the property that they are perpendicular to a, they both touch X, and the set X is entirely contained in the space between them (see Fig. 7.6). By Theorem E there exist points $x_1, x_2, \cdots, x_n \in S^{n-1}$ such

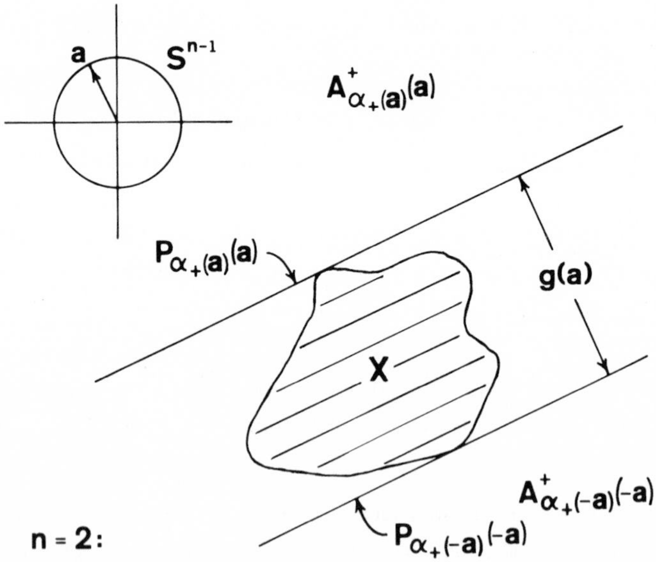

FIGURE 7.6

that $x_i \cdot x_j = 0$ for $i \neq j$ and $g(x_1) = g(x_2) = \cdots = g(x_n)$. If C is the closure of the bounded component of

$$\mathbb{R}^n - \bigcup_{i=1}^{n} (P_{\alpha_+(x_i)}(x_i) \cup P_{\alpha_+(-x_i)}(-x_i))$$

then C is the desired n-cube all of whose faces meet X, and the corollary is proved.

Problems

7.7. Prove that S^n is not homeomorphic to any proper subset of itself.

7.8. Prove that if $f : S^n \to S^n$, $n \geq 0$, has even degree then f maps a pair of antipodal points to the same point. (*Hint:* If the conclusion is false, then show that f is homotopic to an antipodal-preserving map.)

7.9. (a) Prove that if $f : S^n \to S^n$, $n \geq 0$, has the property that $f(x) = f(-x)$ for all $x \in S^n$, then f has even degree.

(b) Assume that $f : S^n \to S^n$, $n \geq 0$, has odd degree. Show that f maps at least one pair of antipodal points to antipodal points.

7.10. Let us change our definition of antipodal point on a sphere as follows: Fix a point $p \in D^{n+1} - S^n$. If $x \in S^n$, then the line which passes through x and p intersects S^n in x and another point x_p. Call x_p an antipodal point of x with respect to p (see Fig. 7.7). Note that if $p = 0$, then $x_p = -x$ is the usual antipodal point. Call a map $f : S^n \to S^m$ antipodal preserving with respect to p if $f(x_p) = (f(x))_p$ for all $x \in S^n$ and show that Theorems 2-4 still remain true in this context. In other words, the theorems in question involve an intrinsic property of pairs of points, of which being antipodal in the usual sense is only a special case. In fact, there are quite general theorems of the type: If $f : X \to Y$ is a continuous map, then at least one pair of "antipodal" points of X get mapped to the same point of Y (see [Ag]). The cor-

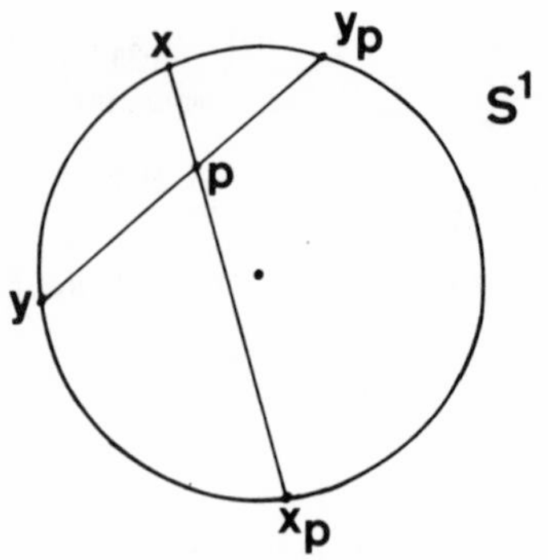

FIGURE 7.7

rect notion of two points x and y in a space X to be "antipodal" appears to be essentially that there are at least two distinct paths from x to y in X which have minimal length. The paths of minimal length from x to $-x$ in S^n are those moving along a great circle through x and $-x$.

7.11. Prove that for any nonzero $\mathbb{R}^n$-vector field on D^n there exists at least one pair of antipodal points of S^{n-1} for which the vectors are parallel.

7.12. Prove that if n closed sets cover D^n, then at least one of these sets contains a pair of antipodal points from S^{n-1}.

7.3. MORE ON DEGREES OF MAPS; ZEROS OF POLYNOMIALS

This section continues our study of the map $\deg : [S^n,S^n] \to \mathbb{Z}$ but specializes to the case $n = 1$.

For $m \in \mathbb{Z}$, define

$$\alpha_m : S^1 \to S^1$$

by $\alpha_m(z) = z^m$, where we are thinking of points of $\mathbb{R}^2$ as complex numbers. Since every point of S^1 is specified by an angle θ (see Fig. 7.8), an alternate definition of α_m is given by setting $\alpha_m(\theta)$

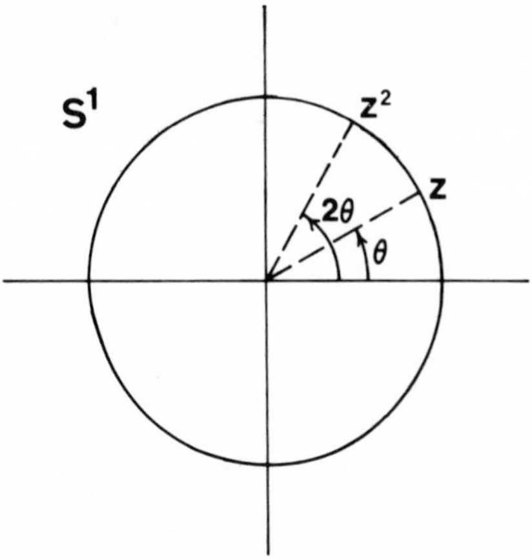

FIGURE 7.8

$= m\theta$. Note that α_0 is a constant map, α_1 is the identity, and α_m "wraps" S^1 m times around itself.

LEMMA 6. $\deg \alpha_m = m$.

Proof: Case 1. $m \geq 3$: Let Q_k, $w_j(k)$, and $e_j(k)$ be as in Sec. 3.2 and let K_k, $k \geq 3$, be the obvious simplicial complex determined by ∂Q_k, that is, $K_k = \{w_0(k),\cdots,w_{k-1}(k),e_1(k),\cdots,e_k(k)\}$. There is a natural triangulation (K_k,φ_k) of S^1, namely, $\varphi_k(z) = z/|z|$. Define $T_m : |K_m| \to |K_m|$ by $T_m = \varphi_m^{-1} \circ \alpha_m \circ \varphi_m$. In order to prove that $\deg \alpha_m = m$, we can either show that $(T_m)_*(a) = ma$ for every $a \in H_1(K_m)$ or try to use Lemma 5. We shall use the latter approach, although both involve pretty much the same calculations. Since T_m is not a simplicial map unless $m = 0$ or 1, define a homeomorphism $h_m : |K_{m^2}| \to |K_m|$ and a simplicial map $\psi_m : K_{m^2} \to K_m$ by $h_m = \varphi^{-1} \circ \varphi_{m^2}$ and $\psi_m(w_{km+s}(m^2)) = w_s(m)$ for $0 \leq s < m$. It is easy to check that $|\psi_m| = \varphi_m^{-1} \circ \alpha_m \circ \varphi_{m^2}$, so that $|\psi_m| = T_m \circ h_m$. Thus, we have a commutative diagram

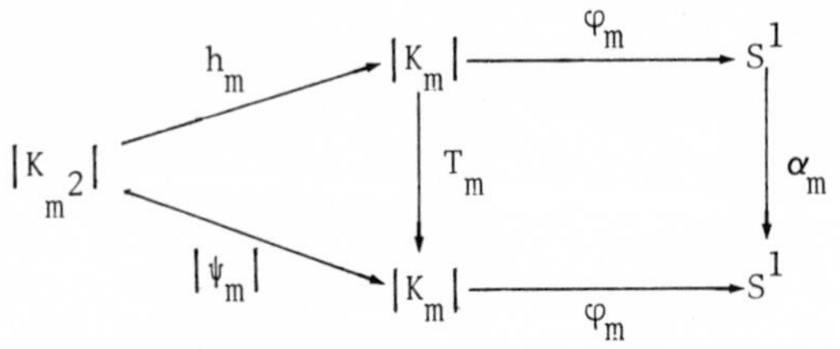

Now $H_1(K_k) \approx \mathbb{Z}$, and if we let $[e_i(k)]$ denote the oriented 1-simplex $[w_{i-1}(k)w_i(k)]$, then $\Sigma_k = \sum_{i=1}^{k} [e_i(k)] \in C_1(K_k)$ is actually a cycle which is a generator for $H_1(K_k)$. In Problem 7.13 you are asked to show that $(h_m)_*(\Sigma_{m^2}) = \Sigma_m$. [$(h_m)_*(\Sigma_{m^2})$ clearly equals $\pm\Sigma_m$ because h_m is a homeomorphism.] Also, $|\psi_m|$ maps m 1-simplices of K_{m^2} onto each 1-simplex of K_m. In fact, $p(e_i(m),\psi_m) = m$ and $n(e_i(m),\psi_m) = 0$, so that $\deg \alpha_m = m$ by Lemma 5(a) and Case 1 is proved.

Case 2. $m = 2$: The only reason that this case cannot be handled in Case 1 is that $\partial Q_2 = S^1$ is not the underlying space of a simplicial complex and therefore we must modify the proof slightly. All we have to do though is let K_4 and K_8 play the role of K_m and K_{m^2} in the proof of Case 1 and proceed as before.

Case 3. $m = 0$ or 1: Trivially, $\deg \alpha_0 = 0$ and $\deg \alpha_1 = 1$.

Case 4. $m < 0$: Observe that $\alpha_{-1} : S^1 \to S^1$ is also defined by $\alpha_{-1}(x_1,x_2) = (x_1,-x_2)$, where S^1 is considered as a subset of $\mathbb{R}^2$. It follows from Lemma 2 that $\deg \alpha_{-1} = -1$. Next, let $m < -1$. Since $\alpha_m = \alpha_{-1} \circ \alpha_{-m}$ and $-m > 0$, we get from Lemma 3 and Cases 1 and 2 that $\deg \alpha_m = \deg(\alpha_{-1} \circ \alpha_{-m}) = (\deg \alpha_{-1})(\deg \alpha_{-m}) = (-1)(-m) = m$. This finishes the proof of Lemma 6.

An immediate corollary of Lemma 6 is that $\deg : [S^1,S^1] \to \mathbb{Z}$ is onto. The lemma also enables us to prove an important theorem in algebra.

DEFINITION. Let $f : A \to B$ and let $w \in B$. Any $a \in A$ such that $f(a) = w$ is called a root of or solution to the equation $f(x) = w$, and if such an a exists, then the equation is said to be solvable. If $B = \mathbb{R}^n$, then a root or zero of f is defined to be a root of the equation $f(x) = 0$.

THEOREM 5 (The Fundamental Theorem of Algebra). Any complex polynomial of degree ≥ 1 has a complex root, that is, if $f : \mathbb{C} \to \mathbb{C}$ is a polynomial function given by

$$f(z) = a_n z^n + a_{n-1} z^{n-1} + \cdots + a_1 z + a_0$$

where $a_i \in \mathbb{C}$, $n \geq 1$, and $a_n \neq 0$, then f has a root.

Proof: Let us suppose that f does not have a root and show that this assumption leads to a contradiction. There is no loss of generality if we assume that $a_n = 1$. Choose a positive real constant M such that M is larger than any of the numbers $n|a_{n-1}|$, $n|a_{n-2}|$, $\ldots$, $n|a_0|$, or 1 and let $t \in [0,1]$. Then

$$\begin{aligned} |(a_{n-1}z^{n-1} + a_{n-2}z^{n-2} + \cdots + a_0)t| &\leq |a_{n-1}||z^{n-1}| + \cdots + |a_0| \\ &< |z^n| \end{aligned}$$

whenever $|z| \geq M$. It follows that if we set $h_t(z) = z^n + (a_{n-1}z^{n-1} + a_{n-2}z^{n-2} + \cdots + a_0)t$, then $h_t(z) \neq 0$ for $|z| \geq M$ and the map $\varphi_t : S^1 \to S^1$ given by $\varphi_t(z) = h_t(Mz)/|h_t(Mz)|$ is well defined. Next, define $\psi_t : S^1 \to S^1$ by $\psi_t(z) = f(Mtz)/|f(Mtz)|$. The map ψ_t is well defined because f is never zero. We have now shown that φ_t is a homotopy between $\alpha_n = \varphi_0$ and φ_1 and that ψ_t is a homotopy between $\varphi_1 = \psi_1$ and the constant map ψ_0. In other words, α_n is homotopic to a constant map ψ_0, which is a contradiction since $\deg \psi_0 = 0$ whereas by Lemma 6 $\deg \alpha_n = n \geq 1$. This proves Theorem 5.

Recall that a polynomial $f(z) = a_n z^n + a_{n-1} z^{n-1} + \cdots + a_0$ is said to have degree n if $a_n \neq 0$. The degree of the zero polynomial is defined to be 0.

COROLLARY. Every complex polynomial f of degree n splits into n linear factors, that is, $f(z) = a(z - c_1)\cdots(z - c_n)$, where a, $c_i \in \mathbb{C}$.

Proof: First let us prove

CLAIM 1. If $g(z)$ is a complex polynomial and $c \in \mathbb{C}$, then $g(z) = (z - c)h(z) + b$ for some complex polynomial $h(z)$ and some $b \in \mathbb{C}$.

The claim is proved by induction on the degree of $g(z) = b_m z^m + b_{m-1} z^{m-1} + \cdots + b_0$. It is trivial if the degree of $g(z)$ is zero and the inductive step follows from the observation that if the degree of $g(z)$ is $m \geq 1$, then $g(z) - b_m z^{m-1}(z - c)$ is a polynomial of degree less than m for which the result is already assumed to hold.

CLAIM 2. If $g(z)$ is a complex polynomial and if $g(z_0) = 0$ for some $z_0 \in \mathbb{C}$, then $g(z) = (z - z_0)h(z)$ for some polynomial $h(z)$.

It follows from Claim 1 that $g(z) = (z - z_0)h(z) + b$, where $b \in \mathbb{C}$. Therefore, $0 = g(z_0) = (z_0 - z_0)h(z_0) + b$, and $b = 0$. This proves Claim 2.

Finally, we prove the corollary. Let n be the degree of f. If $n = 0$ or 1, the result is obvious. Assume that it is true for $n - 1$, where $n \geq 2$. By Theorem 5 there is a $z_0 \in \mathbb{C}$ such that $f(z_0) = 0$. By Claim 2, $f(z) = (z - z_0)h(z)$. Since $h(z)$ has degree $n - 1$, it splits into linear factors by the inductive hypothesis and hence so does f. Therefore, the corollary is proved by induction.

Problem

7.13. Prove the claim made in the proof of Lemma 6 that $(h_m)_*(\Sigma_{m^2}) = \Sigma_m$.

7.4. LOCAL DEGREES; SOLVABILITY OF EQUATIONS AND SOME COMPLEX ANALYSIS

The intermediate value theorem is one of the first theorems that one learns in elementary calculus. It asserts that if $f : [a,b] \to \mathbb{R}$ is a continuous function and if $y \in [f(a),f(b)]$, then there is an $x_0 \in [a,b]$ such that $f(x_0) = y$. One can interpret this nice theorem as a criterion for when the equation $f(x) = y$ has a solution for a given y. It is natural to try and generalize this result and study the solvability of the equation $f(x) = y$, where $f : D^n \to \mathbb{R}^n$, $y \in \mathbb{R}^n$, and $n \geq 1$. We shall assume that $f(S^{n-1})$ is known and consider only points $y \notin f(S^{n-1})$. Since there is no natural linear ordering on the points of $\mathbb{R}^n$ for $n \geq 2$, it is clear that our answer in the general case cannot be phrased in terms of y being "between" points of $f(S^{n-1})$. Furthermore, we see from Fig. 7.9 that line segments between such points do not have to be contained in the image of f. The example also shows that a point can be completely surrounded by $f(D^n)$ and yet still not be in the image of f. It is therefore clear that the criterion for being able to solve $f(x) = y$ is going to be substantially more complicated than in the case $n = 1$.

Let $f : S^{n-1} \to \mathbb{R}^n$, $n \geq 1$, be a continuous map and let $y \in \mathbb{R}^n - f(S^{n-1})$.

DEFINITION. Define the local degree of f at y, $\deg(f;y)$, by $\deg(f;y) = \deg \varphi$, where $\varphi : S^{n-1} \to S^{n-1}$ is given by $\varphi(x) = (f(x) - y)/|f(x) - y|$.

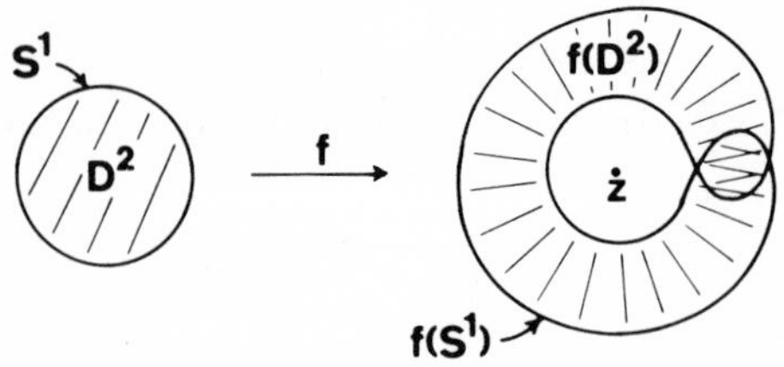

FIGURE 7.9

If $n = 2$, then $\deg(f;y)$ is sometimes called the winding number of f around y because it counts the number of times that the closed path $f(S^1)$ winds around y. There is an alternate way to define the winding number (see [Ch-S]) which we shall indicate briefly and which is helpful in getting a better intuitive understanding of our definition. First, it is convenient to describe what is meant by the angle swept out by a path $\varphi : [a,b] \to \mathbb{R}^2$ around a point y not in its image. Let P be a partition of $[a,b]$, that is, $P = \{t_0, t_1, \cdots, t_k\}$ where $a = t_0 < t_1 < \cdots < t_k = b$.

DEFINITION. Define the angle $A(\varphi,y,P)$ swept out by the path φ around y with respect to the partition P by

$$A(\varphi,y,P) = \sum_{j=1}^{k} [\arg(\varphi(t_i) - y) - \arg(\varphi(t_{i-1}) - y)]$$

It is obvious that $A(\varphi,y,P)$ depends on the partition P that is chosen. For example, if $P' = \{a,b\}$ then $A(\varphi,y,P')$ is somewhere between π and $3\pi/2$ for the path in Fig. 7.10a, whereas, if we choose a partition P'' as in Fig. 7.10b, then we see that $A(\varphi, y,P'') = A(\varphi,y,P') + 2\pi$. This example also points out the need to take a sufficiently fine partition P of $[a,b]$ if we want $A(\varphi,y,P)$ to correspond to the number of times that φ winds around y. If $P = \{t_0, t_1, \cdots, t_k\}$, then let $|P|$, called the norm of P, denote the largest of the numbers $|t_1 - t_0|$, $|t_2 - t_1|$, $\cdots$, and $|t_k - t_{k-1}|$. One can prove that

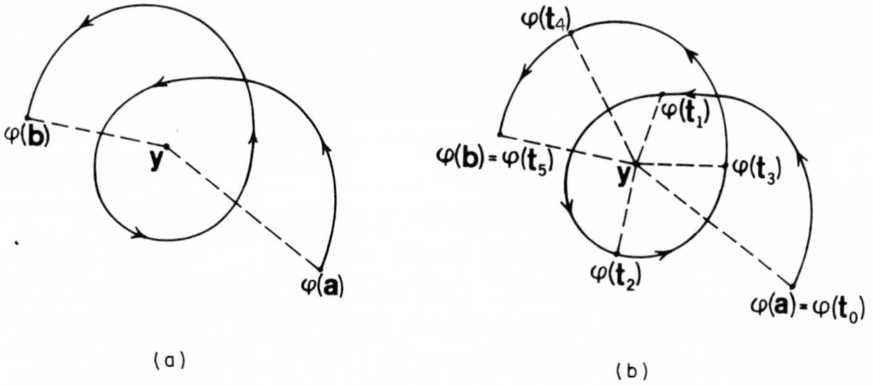

FIGURE 7.10

(1) There is an $\epsilon > 0$ such that if $|P'|$ and $|P''|$ are less than ϵ, then $A(\varphi,y,P') = A(\varphi,y,P'')$.

Because of (1) we are justified in defining the angle $A(\varphi,y)$ swept out by the path φ around y by $A(\varphi,y) = A(\varphi,y,P)$, where $|P|$ is sufficiently small. Next, one proves that if φ is a closed path, then it sweeps out an angle which is a multiple of 2π:

(2) If $\varphi(a) = \varphi(b)$, then $A(\varphi,y) = 2\pi d$, for some $d \in \mathbb{Z}$.

For convenience, one discards the factor of 2π.

DEFINITION. If $\varphi : [a,b] \to \mathbb{R}^2$ is a closed path and if y is not in the image of φ, then define the winding number $W(\varphi,y)$ of φ around y by

$$W(\varphi,y) = A(\varphi,y)/2\pi$$

Since a closed path is, of course, nothing but a map of the circle, the main fact that connects the notion of the winding number with that of the local degree is

(3) Let $\varphi : [0,2\pi] \to \mathbb{R}^2$ be a closed path and y a point not in the image of φ. Define $f_\varphi : S^1 \to \mathbb{R}^2$ by $f_\varphi(e^{i\theta}) = \varphi(\theta)$. Then $W(\varphi,y) = \deg(f_\varphi;y)$.

This concludes our short digression on winding numbers and we return to our discussion of local degrees, but it is a good exercise to fill in the details that were left out above. In Fig. 7.11 the numbers indicate the local degrees (or winding numbers) of the points in the various regions for the maps of S^1 which trace out the curves once in the indicated direction assuming that S^1 has been oriented in the usual counterclockwise fashion.

The fundamental properties of the local degree function are collected in the next lemma, which shows that the local degree is unchanged under certain deformations.

LEMMA 7. Let $n \geq 1$.

(a) Let $f : S^{n-1} \to \mathbb{R}^n$ be the constant map $f(S^{n-1}) = z_0$. If $y \in \mathbb{R}^n - z_0$, then $\deg(f;y) = 0$.

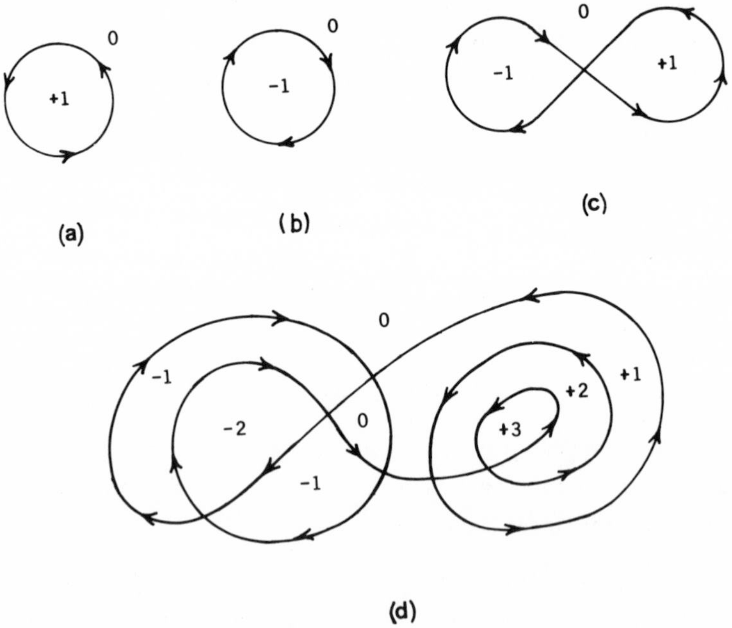

FIGURE 7.11

(b) *(Poincaré-Bohl)* Let $h_t : S^{n-1} \to \mathbb{R}^n$ be a homotopy between two maps $f = h_0$, $g = h_1 : S^{n-1} \to \mathbb{R}^n$ and suppose that y is not in the image of any of the maps h_t for $t \in [0,1]$. Then $\deg(f;y) = \deg(g;y)$.

(c) Let $f : S^{n-1} \to \mathbb{R}^n$ be a map and suppose that y_0 and y_1 belong to the same component of $\mathbb{R}^n - f(S^{n-1})$. Then $\deg(f;y_0) = \deg(f;y_1)$.

Proof: (a) This part of the lemma follows immediately from the fact that the map $x \to (f(x) - y)/|f(x) - y| = (z_0 - y)/|z_0 - y|$ from S^{n-1} to S^{n-1} is constant and that the degree of a constant map is zero.

(b) If $n = 1$, then the homotopy is constant (Problem 7.15) and $h_t = f = g$, so that the result is obvious. Assume that $n \geq 2$ and define $\varphi_t : S^{n-1} \to S^{n-1}$ by $\varphi_t(x) = (h_t(x) - y)/|h_t(x) - y|$. By definition, $\deg(f;y) = \deg \varphi_0$ and $\deg(g;y) = \deg \varphi_1$. But φ_t is a homotopy between φ_0 and φ_1 and so $\deg \varphi_0 = \deg \varphi_1$ by Lemma 1(b).

(c) Since y_0 and y_1 belong to the same component of $\mathbb{R}^n - f(S^{n-1})$, let $\gamma : [0,1] \to \mathbb{R}^n - f(S^{n-1})$ be a path from y_0 to y_1. Parallel translation of $f(S^{n-1})$ gives us a homotopy $h_t : S^{n-1} \to \mathbb{R}^n$, $t \in [0,1]$, which is defined by $h_t(x) = f(x) + (y_0 - \gamma(t))$ (see Fig. 7.12). Observe that $h_0 = f$ and y_0 does not belong to $h_t(S^{n-1})$ for any $t \in [0,1]$. Therefore, $\deg(f;y_0) = \deg(h_1;y_0)$ by part (a) of this lemma. Since $h_1(x) - y_0 = f(x) + (y_0 - \gamma(1)) - y_0 = f(x) - y_1$, so that $(h_1(x) - y_0)/|h_1(x) - y_0| = (f(x) - y_1)/|f(x) - y_1|$, it follows that $\deg(h_1;y_0) = \deg(f;y_1)$ and (c) is proved.

We can now prove

THEOREM 6 (The Generalized Intermediate Value Theorem). Let $f : D^n \to \mathbb{R}^n$, $n \geq 1$, be a continuous map and suppose that $y \in \mathbb{R}^n - f(S^{n-1})$. If $\deg(f|S^{n-1};y) \neq 0$, then the equation $f(x) = y$ has a solution in $\operatorname{int} D^n$.

Proof: Suppose that $y \notin f(D^n)$. Since $h_t : S^{n-1} \to \mathbb{R}^n - y$, $h_t(x) = f(tx)$, is a homotopy between $f|S^{n-1}$ and a constant map, it

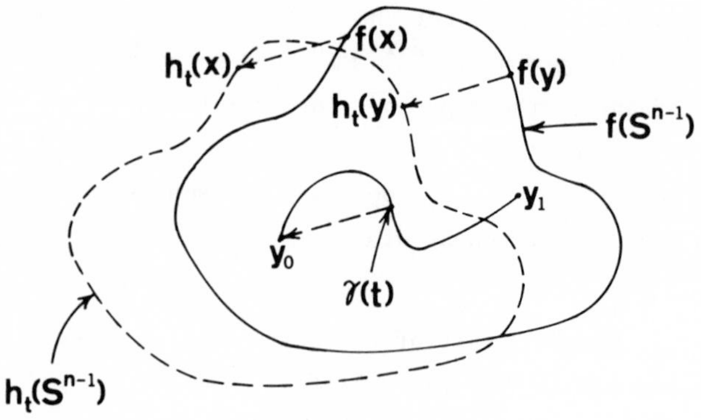

FIGURE 7.12

follows from Lemma 7(a) and (b) that $\deg(f|S^{n-1};y) = 0$, which contradicts the hypothesis of the theorem. Therefore, $y \in f(D^n)$ and Theorem 6 is proved.

Given an equation $f(x) = y$ there are two types of questions that can be asked about it. The first obviously is whether or not there are any solutions at all. Theorem 6 gave the best possible answer to this question, because we see from the example in Fig. 7.13 that the case $\deg(f|S^{n-1};y) = 0$ is indeterminate and no general conclusion can be drawn. The second question and the one we want to consider now is, given that there are solutions to $f(x) = y$, how many are there and how might one be able to locate them? We shall specialize to the case $n = 2$, so that we are really doing complex analysis but from a topological point of view. (A more detailed discussion of the connection between complex analysis and topology via the theory of Riemann surfaces can be found in [Bla, Chapters 3 and 4].)

DEFINITION. Let $\gamma : S^1 \to \mathbb{C}$ and let f be a complex-valued function defined on the image of γ. If $w \notin f(\gamma(S^1))$, then call $\deg(f \circ \gamma;w)$ the local degree of f along γ at w and denote it by $\deg(f|\gamma;w)$.

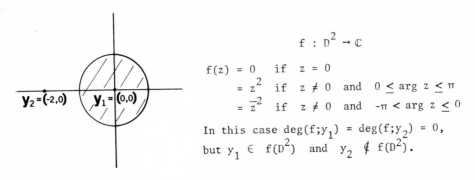

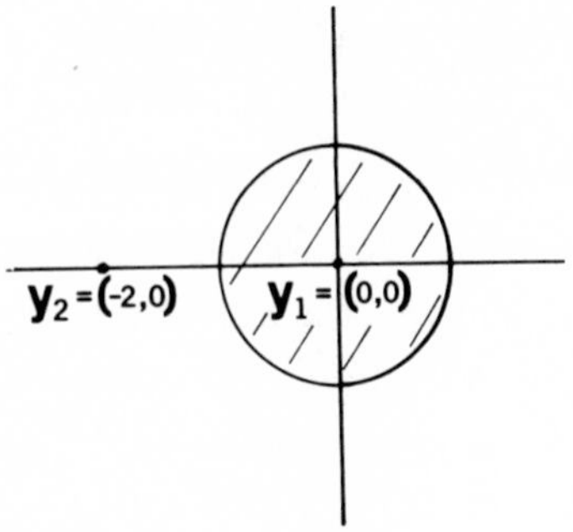

$$f : D^2 \to \mathbb{C}$$

$$f(z) = \begin{cases} 0 & \text{if } z = 0 \\ z^2 & \text{if } z \neq 0 \text{ and } 0 \leq \arg z \leq \pi \\ \bar{z}^2 & \text{if } z \neq 0 \text{ and } -\pi < \arg z \leq 0 \end{cases}$$

In this case $\deg(f;y_1) = \deg(f;y_2) = 0$, but $y_1 \in f(D^2)$ and $y_2 \notin f(D^2)$.

FIGURE 7.13

Let U be an open subset of $\mathbb{C}$ and $f : U \to \mathbb{C}$ a continuous map. Fix $w \in \mathbb{C}$ and assume that $f(z) = w$ has only a finite number of distinct roots $z_1, z_2, \cdots, z_k \in U$. Since U is open and the z_i's are distinct, there is a $\delta > 0$ such that $U_i = \{ z : |z - z_i| \leq \delta \} \subset U$ for each i and $U_i \cap U_j = \emptyset$ if $i \neq j$. Define $\alpha_i : S^1 \to \mathbb{C}$ by $\alpha_i(z) = z_i + \delta z$.

DEFINITION. The integer $\deg(f|\alpha_i;w)$ is called the multiplicity of the root z_i of the equation $f(z) = w$ and is denoted by $m_f(z_i,w)$.

It is easy to check, using Lemma 7(b), that the multiplicity $m_f(z_i,w)$ of z_i does not depend on the particular δ that is chosen in the definition (Problem 7.17). The standard example to keep in mind is the map $z \to z^n$ for $n \geq 1$. In this case, $z^n = w$ has n distinct solutions if $w \neq 0$. On the other hand, if $w = 0$, then $z = 0$ is the only solution and it has "multiplicity" n, so that by Lemma 6 our definition agrees with the one used in analysis. (In analysis one usually says that z_0 is a zero of $f(z)$ with multiplicity n if $f(z) = (z - z_0)^n h(z)$, where $h(z_0) \neq 0$.)

Next, let $\psi : D^2 \to \mathbb{C}$ be a map which is a homeomorphism onto its image.

DEFINITION. We shall call ψ orientation preserving if $\deg(\psi|S^1; \psi(0)) = 1$.

Intuitively, ψ is orientation preserving if the path $\psi|S^1$ traverses its image in a counterclockwise fashion. For example, the identity map 1_{D^2} is orientation preserving, whereas the map which sends z to its complex conjugate $\bar{z}$ is not.

We come now to the important

THEOREM 7. Let U be an open subset of $\mathbb{C}$, $f : U \to \mathbb{C}$, and $w \in \mathbb{C}$. Assume that the equation $f(z) = w$ has precisely k distinct roots $z_1, z_2, \cdots, z_k \in U$. Let K be a simplicial complex such that $|K| \subset U$ and $z_i \in |K| - |\partial K|$. Finally, assume that $\psi : D^2 \to |K|$ is a homeomorphism which is orientation preserving. Set $\psi_0 = \psi|S^1$. Then

$$\begin{aligned} \deg(f|\psi_0;w) &= \sum_{i=1}^{k} m_f(z_i,w) \quad \text{if } k \geq 1 \\ &= 0 \qquad\qquad\qquad\quad \text{if } k = 0 \end{aligned}$$

Proof: First assume that $k \geq 1$. Let δ, U_i, and α_i be as in the definition of $m_f(z_i,w)$. There is no loss of generality if we assume that the z_i belong to the interior of distinct 2-simplices σ_i in K and that U_i is contained in σ_i (see Fig. 7.14). Let $\mu_i : U_i \to \partial\sigma_i$ be the map which sends $x \in \partial U_i$ onto that unique point of $\partial\sigma_i$ which lies on the ray from z_i through x, equivalently, x belongs to the line segment from z_i to $\mu(x) \in \partial\sigma_i$. Define $\alpha_i' : S^1 \to \mathbb{C}$ by $\alpha_i'(z) = \mu_i(z_i + \delta z)$. It follows easily from Lemma 7(b) that $m_f(z_i,w) = \deg(f|\alpha_i;w) = \deg(f|\alpha_i';w)$.

Now let (K_3,φ_3) be the triangulation of S^1 and let Σ_3 be the generator of $H_1(K_3)$ which we defined in the proof of Lemma 6. If σ is a 2-simplex of K, then let $[\sigma]$ denote the simplex σ together with the orientation induced from the natural counterclockwise orientation of $\mathbb{R}^2$ and set

$$u_K = \sum_{\text{2-simplices } \sigma \in K} [\sigma] \in C_2(K)$$

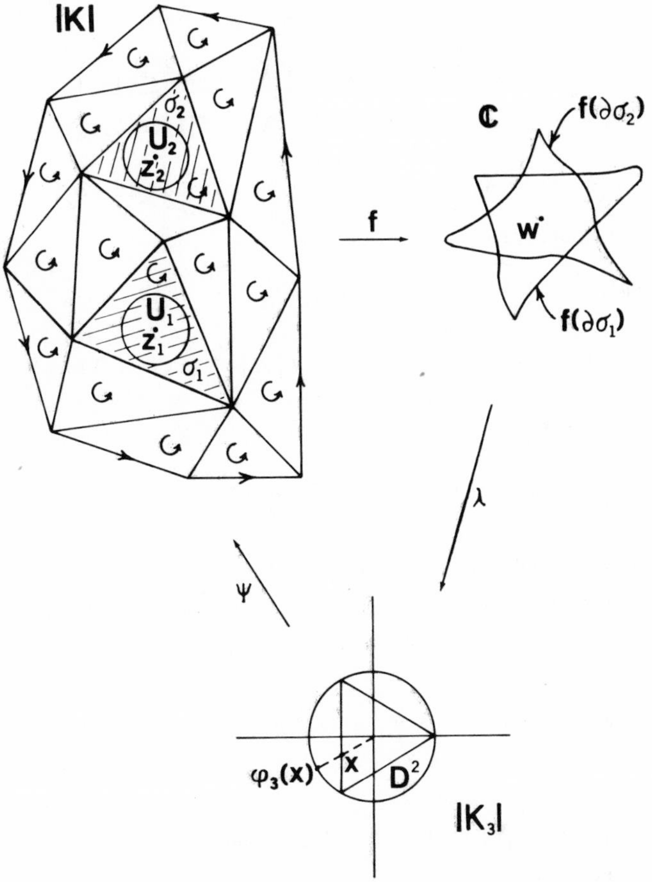

FIGURE 7.14

Alternatively, since ψ is orientation preserving, the class u_K, and hence the oriented simplices $[\sigma]$, can also be defined by the equation $(\psi \circ |\varphi_3|)_*(\Sigma_3) = \partial_2(u_k) \in Z_1(\partial K) = H_1(\partial K)$. Define λ : $\mathbb{C} - w \to S^1$ by $\lambda(z) = (z - w)/|z - w|$. It follows from the definitions that

$$(\deg(f|\alpha_i^!;w))\Sigma_3 = (|\varphi_3|^{-1} \circ \lambda \circ f \circ \alpha_i^! \circ |\varphi_3|)_*(\Sigma_3)$$
$$= (|\varphi_3|^{-1} \circ \lambda \circ (f|\partial\sigma_i))_*(\partial_2([\sigma_i]))$$

and

$$\begin{aligned}(\deg(f|\psi_0;w))\Sigma_3 &= (|\varphi_3|^{-1} \circ \lambda \circ f \circ \psi \circ |\varphi_3|)_*(\Sigma_3) \\ &= (|\varphi_3|^{-1} \circ \lambda \circ (f|(|\partial K|)))_*(\partial_2(u_K))\end{aligned}$$

But $\partial_2(u_K)$ and $\partial_2(\sum_{i=1}^{k} [\sigma_i])$ are homologous in $Z_2(K - \{\sigma_1,\sigma_2,\cdots,\sigma_k\})$ and therefore

$$(|\varphi_3|^{-1} \circ \lambda \circ (f|(|\partial K|)))_*(\partial_2(u_K)) = \sum_{i=1}^{k} (|\varphi_3|^{-1} \circ \lambda \circ (f|\partial\sigma_i))_*(\partial_2([\sigma_i]))$$

This shows that

$$\deg(f|\psi_0;w) = \sum_{i=1}^{k} \deg(f|\alpha_i';w) = \sum_{i=1}^{k} m_f(z_i,w)$$

and proves the theorem in case $k \geq 1$.

If $k = 0$, then $f \circ \psi$ is homotopic to a constant in $\mathbb{C} - w$, and Theorem 7 follows from Lemma 7(a) and (b).

DEFINITION. A map $f : X \to Y$ will be called finite-to-one if $f^{-1}(y)$ is finite for every $y \in Y$.

COROLLARY. Let $f : \mathbb{C} \to \mathbb{C}$ be a continuous finite-to-one map with the property that $\lim_{|z|\to\infty} f(z)$ exists and is equal to c (c may be infinite). Fix $w_0 \in \mathbb{C}$, $w_0 \neq c$, and set $n_0 = \sum_{i=1}^{k} m_f(z_i,w_0)$, where $z_1,z_2,\cdots,z_k$ are the distinct roots of the equation $f(z) = w_0$. (We let $n_0 = 0$ if $k = 0$.) Then $f(z) = w$ has at least $|n_0|$ roots for every $w \in \mathbb{C}$ provided that we count each root with its multiplicity.

Proof: Since there is nothing to prove if $k = 0$, assume that $k \geq 1$. Let $w \in \mathbb{C}$ and let $z_1',z_2',\cdots,z_t'$ be the distinct roots of the equation $f(z) = w$. Let K_n be a simplicial complex such that $|K_n| = [-n,n] \times [-n,n] \subset \mathbb{R}^2 = \mathbb{C}$ and let $\psi_n : D^2 \to |K_n|$ be the obvious orientation-preserving homeomorphism obtained by radial expansion. Set $\psi_{n,0} = \psi_n|S^1$. The proof of the corollary is divided into three cases.

Case 1. $\lim_{|z|\to\infty} f(z) = \infty$: Choose n so large that $\{z_1, z_2, \cdots, z_k, z_1', z_2', \cdots, z_t'\} \subset |K_n| - |\partial K_n|$ and $|f(x)| > |w|, |w_0|$ for all $x \in |\partial K_n|$. Then w and w_0 belong to the same component of $\mathbb{C} - f(|\partial K_n|)$ and by Lemma 7(c), $\deg(f|\psi_{n,0};w) = \deg(f|\psi_{n,0};w_0)$. Therefore, Theorem 7 implies that $\sum_{i=1}^{t} m_f(z_i',w) = \sum_{i=1}^{k} m_f(z_i,w_0) = n_0$, which proves the corollary in this case.

Case 2. $\lim_{|z|\to\infty} f(z) = c \neq \infty$ or w: We can proceed as in case 1, because if n is large enough then $f(|\partial K_n|)$ will be arbitrarily close to c and therefore w and w_0 will again belong to the same component of $\mathbb{C} - f(|\partial K_n|)$.

Case 3. $\lim_{|z|\to\infty} f(z) = c = w \neq \infty$: By choosing a sufficiently large n we may assume that $\{z_1, z_2, \cdots, z_k\} \subset |K| - |\partial K_n|$ and $|f(x) - c| \leq |c - w_0|$ for every $x \in |\partial K_n|$. Therefore, Theorem 7 and the definition of the local degree imply that $n_0 = \deg(f|\psi_{n,0};w_0) = \deg \nu$, where $\nu : S^1 \to S^1$ is given by $\nu(z) = (f(\psi_n(z)) - w_0)/|f(\psi_n(z)) - w_0|$. Furthermore, because of our hypothesis that $f(|\partial K_n|)$ is contained in the ball of radius $|c - w_0|$ around c, it is obvious that $-(c - w_0)/|c - w_0|$ does not belong to the image of ν (see Fig. 7.15). This clearly means that ν is homotopic to a constant map and $\deg \nu = 0$ (see Lemma 9 in Sec. 7.5). Thus $n_0 = 0$ and the corollary is trivially true.

Very loosely speaking, what the corollary shows is that if a function $f : \mathbb{C} \to \mathbb{C}$ is n-to-1 at one point, then it is n-to-1 at every point (if we count multiplicities). The condition that $\lim_{|z|\to\infty} f(z)$ should exist has a good explanation because that is precisely what is needed if we wanted to extend f to a continuous map $F : S^2 \to S^2$, where we are thinking of S^2 as $\mathbb{C} \cup \infty$ [define $F(\infty) =$

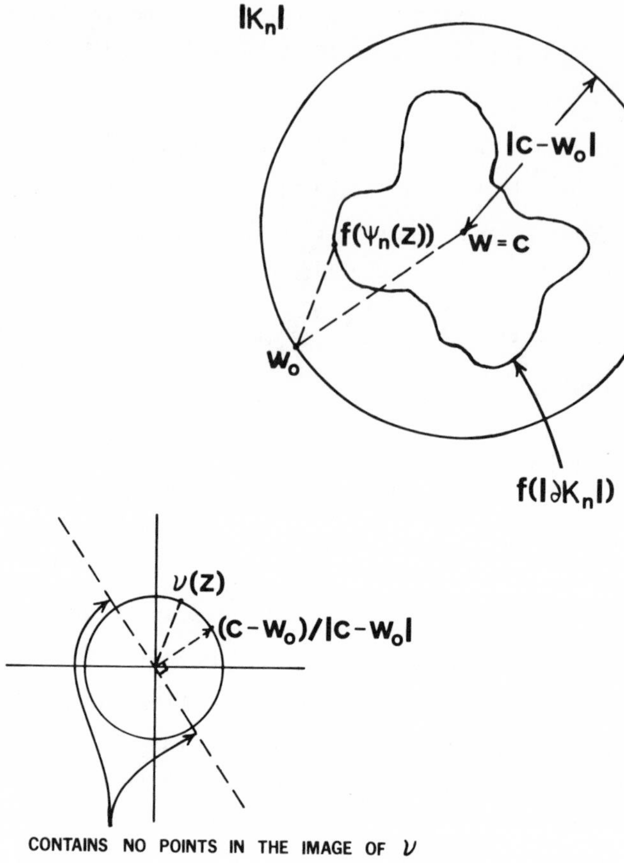

FIGURE 7.15

$\lim_{|z|\to\infty} f(z)]$. In fact, the corollary is fundamentally really about maps $F : S^2 \to S^2$ and their preimages. There is a natural generalization to the case of maps $F : S_1 \to S_2$ between orientable surfaces S_1 and S_2 which we shall discuss shortly. Before we go on to this topic, however, we shall stay with functions $f : \mathbb{C} \to \mathbb{C}$ a little longer and shall briefly consider the problem of how one might locate solutions to the equation $f(z) = w$. The reason is that there is an important theorem in complex analysis which is relevant to this question and which is easy to prove with the tools that we have at hand. For simplicity, we let $w = 0$.

THEOREM 8 (Rouché's Theorem). Let $f,g : \mathbb{C} \to \mathbb{C}$, $\gamma : S^1 \to \mathbb{C}$ be continuous maps and suppose that $|f(\gamma(z))| < |g(\gamma(z))| < \infty$ for all $z \in S^1$. Then $\deg((f + g)|\gamma;0) = \deg(g|\gamma;0)$.

Proof: Let $0 \le t \le 1$. Define $h_t : S^1 \to \mathbb{C}$ by $h_t(z) = tf(\gamma(z)) + g(\gamma(z))$. Since $|tf(\gamma(z))| \le |f(\gamma(z))| < |g(\gamma(z))|$ for all $z \in S^1$, we see that $h_t(z)$ can never be equal to zero. But h_t is a homotopy between $(f + g) \circ \gamma$ and $g \circ \gamma$. Therefore, the theorem now follows from Lemma 7(b).

In order to indicate how this toplogical version of Rouché's theorem can help locate roots of functions we shall work out an example. Let $f(z) = 2z^4 - 2z^3 + 6z^2 - 12z + 7$. If $|z| = 3$, then $|-2z^3 + 6z^2 - 12z + 7| \le 2 \cdot 27 + 6 \cdot 9 + 12 \cdot 3 + 7 = 151 < 162 = |2z^4|$. Define $\gamma : S^1 \to \mathbb{C}$ by $\gamma(z) = 3z$. It follows from Rouché's theorem that $\deg(f|\gamma;0) = \deg(2\gamma^4;0)$. Since $\deg(2\gamma^4;0) = 4$ (essentially by Lemma 6), we can apply Theorem 7 and conclude that f has at least 4 roots (counted with their multiplicities) in the disk of radius 3 around 0. Being a polynomial of degree 4, f can have no other roots. For more applications of this type and more precise results using the winding numbers $W(\varphi,y)$ and the angles $A(\varphi,y)$ swept out by paths see [Ch-S, Sec. 36], [Bla, Sec. 4.4], and [Alexf-H].

We can also use Rouché's theorem to prove a weakened version of the Brouwer fixed-point theorem: Suppose that $f : \mathbb{C} \to \mathbb{C}$ is a function with the property that $|f(z)| < 1$ for all $z \in S^1$. Then $\deg([f(z) - z]|S^1;0) = \deg(z|S^1;0) = 1$ by Rouché's theorem and Lemma 6, and hence $f(z) - z$ has a root in the interior of D^2 by Theorem 6, that is, $f(z_0) = z_0$ for some $z_0 \in D^2 - S^1$.

Now we come to the last topic of this section. The proper setting for the preceding results is in the context of maps between oriented surfaces. Recall that if M is a closed orientable n-dimensional pseudomanifold, then $H_n(M) \approx \mathbb{Z}$ (Theorem 8(a) in Sec. 6.5).

DEFINITION. A closed oriented pseudomanifold is a triple $(M,(K,\varphi),\mu)$, where M is a closed orientable pseudomanifold of some dimension n, (K,φ) is a triangulation of M, and μ is one of the two possible generators of $H_n(K)$ [$-\mu$ is then the other]. The pair $((K,\varphi),\mu)$ is called an orientation for M. If $(M_i,(K_i,\varphi_i),\mu_i)$, $i = 1,2$, are closed oriented pseudomanifolds of dimension n and $f : M_1 \to M_2$ is a continuous map, define the degree of f, deg f, by the equation $(\varphi_2^{-1} \circ f \circ \varphi_1)_*(\mu_1) = (\deg f)\mu_2$. If $M = M_1 = M_2$ and $f : M \to M$, then we shall always assume that $K_1 = K_2$, $\varphi_1 = \varphi_2$, and $\mu_1 = \mu_2$ in the definition for the degree of f.

Note that deg f does not really depend on the triangulations (K_i,φ_i) but rather on the choice of generators μ_i. By this we mean that we can replace the triangulations (K_i,φ_i) by any other triangulations (K_i',φ_i') and the definition of deg f with respect to the oriented pseudomanifolds $(M_i,(K_i,\varphi_i),\mu_i)$ and $(M_i,(K_i',\varphi_i'),\mu_i')$, where $\mu_i' = ((\varphi_i')^{-1} \circ \varphi_i)_*(\mu_i)$, will be the same. (If $H_n(M)$ were an intrinsically defined group, then we would have defined an oriented pseudomanifold to be simply a pair (M,μ), where μ would be a generator of $H_n(M)$.) In the future, therefore, whenever it is convenient to pass from one triangulation to another in problems about degrees of maps, we shall do so without any further justification--it being understood that the μ_i then changes to the specific μ_i' as above. Also, it is easy to check that the above definition of degree agrees with our earlier definition in the case of maps $f : S^n \to S^n$.

Now let us again specialize to the case of surfaces. We do this only in order to keep the discussion on a concrete level, although everything that we shall do is true for higher dimensional manifolds.

Suppose that $(S_i,(K_i,\varphi_i),\mu_i)$, $i = 1,2$, are oriented surfaces and $f : S_1 \to S_2$ is a continuous map. Let $w \in S_2$. As before, we want to study solutions to the equation $f(z) = w$. Let $g = \varphi_2^{-1} \circ f \circ \varphi_1 : |K_1| \to |K_2|$.

THEOREM 9. If $\deg f \neq 0$, then the equation $f(z) = w$ always has a solution.

Proof: Suppose not and that $w \notin f(S_1)$. Since S_1 is compact, $f(S_1)$ actually misses a little neighborhood around w in S_2. By initially replacing K_1 and K_2 by appropriate subdivisions, if necessary, there is no loss of generality if we assume that $v = \varphi_2^{-1}(w)$ is a vertex of K_2, that g admits a simplicial approximation $\psi : K_1 \to K_2$, and that $|\psi|(|K_1|) \cap st(v) = \emptyset$. By definition we have that $(\deg f)\mu_2 = g_*(\mu_1) = \psi_*(\mu_1)$. Since a generator for $H_2(K_2)$ is just the sum of the 2-simplices of K_2 oriented in a compatible fashion and since the image of $|\psi|$ misses $st(v)$ entirely, it follows that $\psi_*(\mu_1) = 0$. Therefore, $\deg f = 0$, which contradicts the hypothesis that $\deg f \neq 0$, and proves the theorem.

Theorem 9, like Theorem 6, is the best possible general result (see Problem 7.22). Next, we want to prove an analog of Theorem 7. Suppose that $z_1, z_2, \cdots, z_k \in S_1$ are the distinct roots of the equation $f(z) = w$. By subdividing K_1 and K_2 as much as necessary we may assume that

(i) the $v_i = \varphi_1^{-1}(z_i)$ are all vertices of K_1 and $v = \varphi_2^{-1}(w)$ is a vertex of K_2

(ii) $|st^c(v_i)| \cap |st^c(v_j)| = \emptyset$ for $i \neq j$

(iii) $g(|st^c(v_i)|) \subset |st^c(v)|$

(iv) g admits a simplicial approximation $\psi : K_1 \to K_2$ such that $|\psi|^{-1}(|st^c(v)|) = \bigcup_{i=1}^{k} |st^c(v_i)|$ and $\psi^{-1}(v) = \{v_1, v_2, \cdots, v_k\}$

The only assumption that deserves any comment is (iv). To get such a special ψ one first shows, using (ii) and (iii), that each $g|(|\ell k(v_i)|)$ has a simplicial approximation ψ_i which sends the vertices of $\ell k(v_i)$ to the vertices of $\ell k(v)$. Next, one extends ψ_i to $st^c(v_i)$ by sending v_i to v, and finally these partial maps are extended to the desired map ψ essentially via the simplicial approximation theorem (or rather its proof). The main consequences of the conditions on ψ in (iv) are that $|\psi|(|\ell k(v_i)|) \subset |\ell k(v)|$ [in other words, $\psi|\ell k(v_i)$ is a simplicial map from $\ell k(v_i)$ to $\ell k(v)$] and that only simplices from the $st^c(v_i)$ get mapped onto simplices of $st^c(v)$ by $|\psi|$. These properties will play crucial roles in subsequent proofs, for which the diagram below may be helpful:

$$\begin{array}{ccccccc}
z_i & \in & S_1 & \xrightarrow{\quad f \quad} & S_2 & \ni & w \\
\uparrow & & \uparrow \varphi_1 & & \uparrow \varphi_2 & & \uparrow \\
v_i & \in & |K_1| & \xrightarrow[|\psi|]{g} & |K_2| & \ni & v \\
 & & \cup & & \cup & & \\
 & & |st^c(v_i)| & \xrightarrow[|\psi|]{g} & |st^c(v)| & & \\
 & & \cup & & \cup & & \\
 & & |\ell k(v_i)| & \xrightarrow[|\psi|]{} & |\ell k(v)| & &
\end{array}$$

Now orient the 2-simplices of K_1 and K_2 in such a way that

$$\mu_1 = \sum_{\text{2-simplices } \sigma \in K_1} [\sigma] \quad \text{and} \quad \mu_2 = \sum_{\text{2-simplices } \tau \in K_2} [\tau]$$

Let

$$\Gamma_i = \sum_{\text{2-simplices } \sigma \in st^c(v_i)} [\sigma] \in C_2(st^c(v_i))$$

and

$$\Gamma = \sum_{\text{2-simplices } \tau \in st^c(v)} [\tau] \in C_2(st^c(v))$$

Finally, set $\nu_i = \partial_2(\Gamma_i) \in Z_1(\ell k(v_i))$ and $\nu = \partial_2(\Gamma) \in Z_1(\ell k(v))$. Since we are dealing with surfaces it is easy to see that $|\ell k(v_i)| \approx S^1 \approx |\ell k(v)|$. Therefore, $Z_1(\ell k(v_i)) = H_1(\ell k(v_i)) \approx \mathbb{Z} \approx Z_1(\ell k(v)) = H_1(\ell k(v))$ and the ν_i and ν are clearly generators of the respective groups. This means that $\psi_\#(\nu_i) = k_i \nu$ for some $k_i \in \mathbb{Z}$.

DEFINITION. The integer k_i is defined to be the multiplicity of the root z_i of the equation $f(z) = w$ with respect to the orientations $((K_i, \varphi_i), \mu_i)$ and is denoted by $m_f(z_i, w)$.

The proof of the next lemma is easy and is left as an exercise.

LEMMA 8. The multiplicity $m_f(z_i,w)$ is well defined and depends only on the "orientations" μ_i.

NOTE. We should really write $m_f(z_i,w,\mu_1,\mu_2)$, but because the orientations are usually assumed to be fixed we omit any reference to them for simplicity. It follows easily from the definitions though that $m_f(z_i,w,\mu_1,\mu_2) = -m_f(z_i,w,-\mu_1,\mu_2) = -m_f(z_i,w,\mu_1,-\mu_2) = m_f(z_i,w,-\mu_1,-\mu_2)$.

The main theorem is

THEOREM 10. Suppose that $(S_i,(K_i,\varphi_i),\mu_i)$, $i = 1,2$, are oriented surfaces and $f : S_1 \to S_2$ is a continuous map. Let $w \in S_2$ and suppose that the equation $f(z) = w$ has only k distinct roots $z_1, z_2,\cdots,z_k \in S_1$. Then

$$\begin{aligned} \deg f &= \sum_{i=1}^{k} m_f(z_i,w) \qquad \text{if } k \geq 1 \\ &= 0 \qquad\qquad\qquad\quad \text{if } k = 0 \end{aligned}$$

Proof: We shall use the notation leading up to the definition of $m_f(z_i,w)$.

Case 1. $k \geq 1$:

CLAIM. $\psi_\#(\Gamma_i) = m_f(z_i,w)\Gamma$.

The claim follows from the fact that $\partial_2(\psi_\#(\Gamma_i)) = \psi_\#(\partial_2(\Gamma_i)) = \psi_\#(\nu_i) = m_f(z_i,w)\nu = m_f(z_i,w)\partial_2(\Gamma) = \partial_2(m_f(z_i,w)\Gamma)$ and that ∂_2 is injective since $H_2(st^c(v_i)) = 0$.

Using the claim we get that

$$\begin{aligned} (\deg f)\mu_2 &= g_*(\mu_1) = \psi_*(\mu_1) = \psi_\#(\mu_1) \\ &= \psi_\#[(\sum_{i=1}^{k} \Gamma_i) + (\mu_1 - \sum_{i=1}^{k} \Gamma_i)] \end{aligned}$$

$$= \sum_{i=1}^{k} \psi_\#(\Gamma_i) + \psi_\#(\mu_1 - \sum_{i=1}^{k} \Gamma_i)$$

$$= [\sum_{i=1}^{k} m_f(z_i,w)]\Gamma + \psi_\#(\mu_1 - \sum_{i=1}^{k} \Gamma_i)$$

Since the 2-chains Γ and $\psi_\#(\mu_1 - \sum_{i=1}^{k} \Gamma_i)$ have no oriented 2-simplices in common, it follows that $\deg f = \sum_{i=1}^{k} m_f(z_i,w)$. This proves the theorem in case 1.

Case 2. $k = 0$: In this case, $|\psi|(|K_1|) \cap |st^c(v)| = \emptyset$, so that no oriented 2-simplex from $st^c(v)$ appears in $\psi_\#(\mu_1)$, although they certainly are present in μ_2. Therefore, $\psi_\#(\mu_1)$ must be the zero cycle, which means that $\deg f = 0$, finishing the proof of Theorem 10.

It is instructive to compare the proof of Theorem 10 with that of Theorem 7, whose corollary we can now also generalize.

COROLLARY 1. Suppose that $(S_i,(K_i,\varphi_i),\mu_i)$, $i = 1,2$, are oriented surfaces and $f : S_1 \to S_2$ is a continuous finite-to-one map. Fix $w_0 \in S_2$ and set $n_0 = \sum_{i=1}^{k} m_f(z_i,w)$, where $z_1,z_2,\cdots,z_k \in S_1$ are the distinct roots of the equation $f(z) = w_0$. (Again let $n_0 = 0$ if $k = 0$.) Then $f(z) = w$ has at least $|n_0|$ roots for every $w \in S_2$ provided that we count each root with its multiplicity.

Proof: Let $w \in S_2$ and let $z_1',z_2',\cdots,z_t'$ be the distinct roots to the equation $f(z) = w$. The corollary now follows immediately from Theorem 10 which asserts that $\sum_{i=1}^{t} m_f(z_i',w) = \deg f = \sum_{i=1}^{k} m_f(z_i,$ $w_0)$.

Note that Corollary 1, in contrast to the corollary to Theorem 7, does not have any strange conditions on f. The proof is also completely trivial (modulo Theorem 10).

We shall finish this section with an application of Theorem 10 to rational complex-valued functions. Recall that a complex function f is said to be rational if $f = p/q$, where p and q are complex polynomials. Assume that p and q have no common factors.

DEFINITION. The degree of f, $\deg f$, is defined by

$$\deg f = \max \{\deg p, \deg q\}$$

One can show that $\deg f$ is well defined and extends the definition of the degree of a polynomial. Our object now is to show that $\deg f$ has a topological definition. In particular, this will provide another proof of the fact that $\deg f$ is well defined. Therefore, extend f to a function $F : S^2 = \mathbb{C} \cup \infty \to S^2 = \mathbb{C} \cup \infty$ by defining $F(z) = \infty$ for every z such that $q(z) = 0$ and letting $F(\infty) = \lim_{|z|\to\infty} f(z)$. The map F is well defined and continuous.

COROLLARY 2. $\deg f = |\deg F|$.

Proof: Let $n = \deg p$, $m = \deg q$, and fix a particular orientation of S^2. If $f = 0$, then both f and F are constant and the result is trivial. The rest of the proof is divided into two cases.

Case 1. $n \geq m \geq 0$: We know from the corollary to Theorem 5 that p has n roots. Furthermore, $n \geq m$ implies that $F(\infty) \neq 0$. Therefore, $F^{-1}(0)$ consists of k points $z_1, z_2, \cdots, z_k$ and $f(z) = (z - z_i)^{n_i} h_i(z)$, where the h_i are rational functions, $a_i = h_i(z_i) \neq 0$ or ∞, $n_i > 0$, and $n_1 + n_2 + \cdots + n_k = n$. Define $\alpha_i : \mathbb{C} \cup \infty \to \mathbb{C} \cup \infty$ by $\alpha_i(z) = (z - z_i)^{n_i}$ for $z \in \mathbb{C}$ and $\alpha_i(\infty) = \infty$. An easy argument similar to the one in Rouché's theorem shows that $m_f(z_i,0) = m_{a_i\alpha_i}(z_i,0) = m_{\alpha_i}(z_i,0)$. Since $m_{\alpha_i}(z_i,0) = n_i$ by Lemma 6, it follows from Theorem 10 that $\deg F = \sum_{i=1}^{k} m_f(z_i,0) = \sum_{i=1}^{k} n_i = n$, which proves Corollary 2 in this case.

Case 2. $m > n \geq 0$: Define $G : \mathbb{C} \cup \infty \to \mathbb{C} \cup \infty$ by $G(z) = 1/z$ for $z \in \mathbb{C} - 0$, $G(0) = \infty$, and $G(\infty) = 0$. Another easy calculation shows that $\deg G = -1$. Since $F = G \circ (G \circ F)$ and since $\deg (G \circ F) = m$ by case 1 applied to $1/f$, it follows from Lemma 3 that $\deg F = (\deg G)(\deg (G \circ F)) = (-1)(m) = -m$, finishing the proof of Corollary 2.

Problems

7.14. (a) Fill in the details which were left out in our discussion of $A(\varphi,y)$ and the winding number $W(\varphi,y)$. In particular, prove the assertions (1)-(3). Also prove the following additivity property of $A(\varphi,y)$:

(4) Let $\varphi : [a,c] \to \mathbb{R}^2$ be a path and y a point not in the image of φ. Suppose that $a \leq b \leq c$. If $\varphi_1 : [a,b] \to \mathbb{R}^2$ and $\varphi_2 : [b,c] \to \mathbb{R}^2$ are the restrictions of φ, then $A(\varphi,y) = A(\varphi_1,y) + A(\varphi_2,y)$.

(b) Determine the winding numbers of the points from the various regions into which the plane is divided by the curve in Fig. 7.16.

7.15. In the proof of Lemma 7(b) it was claimed that the homotopy h_t is constant when $n = 1$. Prove this.

7.16. Show that when $n = 1$ then Theorem 6 is equivalent to the

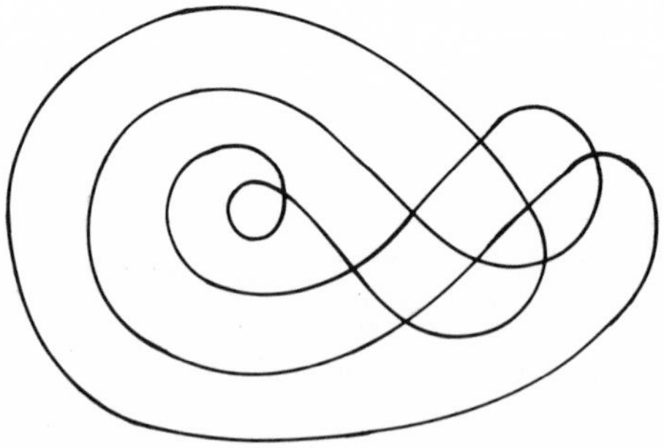

FIGURE 7.16

usual intermediate value theorem.

7.17. Let U be an open subset of $\mathbb{C}$ and $f : U \to \mathbb{C}$ a continuous map. Suppose that the equation $f(z) = w$ has only a finite number of distinct roots $z_1, z_2, \cdots, z_k \in U$. Show that $m_f(z_i, w)$ is well defined. In fact, prove that if $\alpha, \beta : S^1 \to U - \{z_1, z_2, \cdots, z_k\}$ are homotopic maps, then $\deg(f|\alpha; w) = \deg(f|\beta; w)$.

7.18. Define $f : \mathbb{R}^2 \to \mathbb{R}^2$ by $f(x,y) = (x,y)$ if $y \geq 0$ and $f(x,y) = (x,-y)$ if $y \leq 0$ and observe that f maps two points onto every point above the x axis, one point onto every point on the x axis, and no point onto a point below the x axis. Explain why this f does not contradict the corollary to Theorem 7. This example shows that our comment following the proof of the corollary, that if a function is n-to-1 at one point it is so at every point, is an oversimplification of the situation. If we had restricted ourselves to holomorphic functions $f : \mathbb{C} \to \mathbb{C}$, however, then it would have been correct.

7.19. Isolate the zeros of $f(z) = 2z^3 - 2z^2 - 16z + 23$. (*Hint:* Expand $f(z)$ about $z = 2$ and compare with the $(z - 2)^2$ part via Rouché's theorem.)

7.20. Use Theorem 7 and Rouché's theorem to give another proof of the Fundamental Theorem of Algebra.

7.21. If $f : S^n \to S^n$, then prove that the integer, $\deg f$, as defined in this section with respect to any orientation of S^n is the same as the integer, $\deg f$, as defined in Sec. 7.1. Generalize this to show that if M is a closed orientable pseudomanifold and $f : M \to M$, then the definition of $\deg f$ is independent of the orientation that is chosen for M.

7.22. Define $f : S^2 = \mathbb{C} \cup \infty \to S^2 = \mathbb{C} \cup \infty$ by $f(0) = 0$, $f(\infty) = \infty$, $f(z) = z^2$ if $z \neq 0$ or ∞ and $0 \leq \arg z \leq \pi$, and $f(z) = \overline{z^2}$ if $z \neq 0$ or ∞ and $-\pi < \arg z \leq 0$. Show that $\deg f = 0$. On the other hand, a constant map from S^2 to S^2 also has degree 0. Therefore, since f is surjective, we see that it is impossible to say anything about what points are in the image of a map if it has degree 0. (Compare this situation with that in Theorem 9.)

7.23. Prove Lemma 8.

7.24. Theorem 7 (and generalizations such as Theorem 10) can be restated in terms of vector fields and there is an interesting application to geography. Let U be an open subset of $\mathbb{R}^n$, $n \geq 1$, and let σ be an $\mathbb{R}^n$-vector field on U. Suppose that $x_0 \in U$ is an isolated zero of σ, that is, $\sigma(x_0) = 0$ and, for some $\epsilon > 0$, $\sigma(x) \neq 0$ for all x with $0 < |x - x_0| \leq \epsilon$. Define the index of σ at x_0, $I(\sigma,x_0)$, to be the degree of the map from S^{n-1} to S^{n-1} which sends x to $\sigma(x_0 + \epsilon x)/|\sigma(x_0 + \epsilon x)|$. Check that the integer $I(\sigma,x_0)$ does not depend on the ϵ which is chosen. Now let us specialize to the case $n = 2$. Suppose that X is an island in some ocean. Mathematically, the ocean is represented by $\mathbb{R}^2$ and X is any subset of $\mathbb{R}^2$ which is homeomorphic to D^2. Let $f : X \to \mathbb{R}$ be the function which associates to each point of X its elevation above sea level. The gradient of f, $\nabla f = (\partial f/\partial x, \partial f/\partial y)$, is an $\mathbb{R}^2$-vector field on X. Here we assume that the partial derivatives of f exist and are continuous. Assume further that ∇f has only isolated zeros. What this means is that our island X has only peaks, pits, and passes which correspond to local maxima, minima, and saddle points of f, respectively (see Fig. 7.17). We are not allowing ridges of constant elevation. Let $h : D^2 \to X$ be an orientation-preserving homeomorphism and let $h_0 = h|S^1$. Assume that ∇f is nonzero and points into X at every point of ∂X, that is, no ocean water is running into X.

(a) Show that $\deg(\nabla f|h_0;0) = 1 = I(\nabla f,p)$ for each peak or pit $p \in X$ but that $I(\nabla f,p) < 0$ for each pass $p \in X$.

(b) If $p \in X$ is a pass, call $|I(\nabla f,p)|$ the multiplicity of the pass p. Using Theorem 7 prove that the number of peaks plus the number of pits minus the number of passes (counted with their multiplicities) equals 1. For

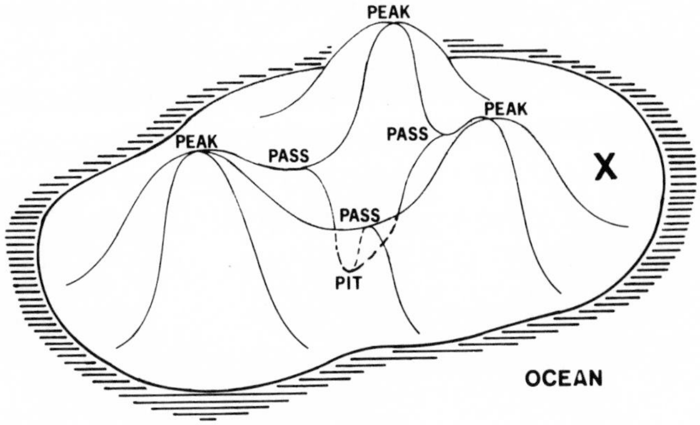

FIGURE 7.17

more on this topic see [Bla] and [Bal-C]. In [Bla] there is also a similar application to a problem in hydrodynamics.

7.5. EXTENDING MAPS

A very important problem in topology is the so-called "extension problem." One form of it can be represented by the diagram below:

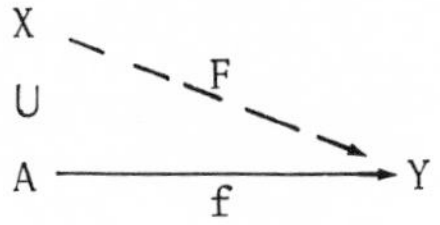

We are given a map f defined on some subspace A of a space X and we want to know if f extends to a map F which is defined on all of X. This is not always possible. For example, let $A = Y = S^n$, $X = D^{n+1}$, and $f = 1_{S^n}$. If the extension F existed, then it would be a retraction of D^{n+1} onto S^n which is impossible by Theorem 14 in Chapter 6. Quite a bit of algebraic topology is concerned with determining the obstructions to extensions of one sort or another. In this section we shall show that in certain situations extensions do exist; however, our goal is very limited and we are only interested in obtaining some results that are needed to prove the theorems in the next section.

DEFINITION. A map $f : X \to Y$ is said to be inessential if f is homotopic to a constant map. Otherwise, f is said to be essential.

Special cases of the next two lemmas have already been used implicitly on several occasions.

LEMMA 9. If $f : X \to S^n$ is a continuous map and $f(X) \neq S^n$, then f is inessential.

Proof: Since f is not surjective, we may assume (by rotating the sphere if necessary) that $e_{n+1} = (0,0,\cdots,0,1) \in S^n - f(X)$. Define $F : X \times [0,1] \to S^n$ by $F(x,t) = P_n^{-1}(tP_n(f(x)))$, where P_n : $S^n - e_{n+1} \to \mathbb{R}^n$ is the stereographic projection. The map F is a homotopy between f and the constant map $c : X \to S^n$, $c(X) = -e_{n+1}$, proving Lemma 9.

LEMMA 10. A map $f : S^n \to X$ is inessential if and only if f extends to a map $F : D^{n+1} \to X$.

Proof: Suppose that $f : S^n \to X$ is inessential and let h_t : $S^n \to X$ be a homotopy between $f = h_0$ and a constant map $c = h_1$. Note that every point $y \in D^{n+1} - 0$ can be written uniquely as tx, where $t \in (0,1]$ and $x \in S^n$. Define $F : D^{n+1} \to X$ by $F(tx) = h_{1-t}(x)$ for $x \in S^n$, $t \in (0,1]$, and $F(0) = h_1(S^1)$. It is easy to check that F is continuous.

Conversely, suppose that $f : S^n \to X$ extends to a map $F : D^{n+1} \to X$. Define $f_t : S^n \to X$, $t \in [0,1]$, by $f_t(x) = F(tx)$. Then f_t is a homotopy between f and a constant map and Lemma 10 is proved.

LEMMA 11. If X is a polyhedron and $\dim X < n$, then every continuous map $f : X \to S^n$ is inessential.

Proof: Let (K,φ) and (L,ψ) be triangulations of X and S^n, respectively. It suffices to show that every continuous map $g : |K| \to |L|$ is inessential. Therefore, let $\alpha : K \to L$ be any simplicial approximation to g. Since α is a simplicial map and since by the invariance of dimension theorem $\dim K = \dim X < n = \dim L$, $|\alpha|$ is not surjective. It follows from Lemma 9 that $|\alpha|$ is inessential. But $|\alpha| \simeq g$ (Lemma 6(a) in Sec. 5.3), so that g is inessential and Lemma 11 is proved.

COROLLARY. If $m < n$, then any map $f : S^m \to S^n$ extends to a map $F : D^{m+1} \to S^n$.

Proof: This corollary is an immediate consequence of Lemmas 10 and 11.

LEMMA 12. Let (X,Y) be a polyhedral pair. If $\dim (X,Y) \leq n$, then every map $f : Y \to S^n$ extends to a map $F : X \to S^n$.

Proof: Let $((K,L),\varphi)$ be a triangulation of (X,Y). Then $\dim (K,L) = \dim (X,Y) \leq n$, and it suffices to show that every map $f : |L| \to S^n$ extends to a map $F : |K| \to S^n$. The lemma will be proved by induction on the number $n_{(K,L)}$ of simplices in $K - L$. If $n_{(K,L)} = 0$, then $K = L$ and there is nothing to prove. Assume that we have proved the existence of F whenever $n_{(K,L)} < k$, for $k \geq 1$, and let $n_{(K,L)} = k$. If $\dim (K,L) = 0$, then $|K| - |L|$ consists of k points and F can be defined arbitrarily on these points. Therefore, assume that $\dim (K,L) \geq 1$ and let σ be a simplex of $K - L$ of maximal dimension. Let $K' = K - \sigma$. Clearly, K' is a simplicial complex which contains L, $\dim (K',L) \leq n$, and $n_{(K',L)} = n_{(K,L)} - 1$. By our inductive hypothesis f admits an extension $F' : |K'| \to S^n$. Furthermore, the corollary to Lemma 11 implies that $F'|\partial\sigma$ extends to a map $F'' : \sigma \to S^n$. Hence we can define an extension of f, $F : |K| \to S^n$, by $F|(|K'|) = F'$ and $F|\sigma = F''$. This completes the inductive step and Lemma 12 is proved.

We want to extend Lemma 12 in an important special case.

THEOREM 11. Let (S^n,X) be a polyhedral pair and let A be a set of points in $S^n - X$ with the property that A meets every component of $S^n - X$ in precisely one point. Then every map $f : X \to S^{n-1}$ extends to a map $F : S^n - A \to S^{n-1}$.

Proof: The proof consists of two steps.

The First Step: To show that f extends to a map $G : S^n - B \to S^{n-1}$, where B is a finite set of points in $S^n - X$.

Let $((K,L),\varphi)$ be a triangulation of (S^n,X) and let $f_1 = (f \circ \varphi)|(|L|) : |L| \to S^{n-1}$. It clearly suffices to show that f_1 extends to a map $F_1 : |K| - B_1 \to S^{n-1}$, where B_1 is a finite set of points in $|K| - |L|$.

Consider the simplicial complex $T = \{ \sigma \in K \mid \sigma \in L$ or $\dim \sigma \leq n - 1 \}$ which contains L. By Lemma 12 we know that f_1 extends to a map $F_2 : |T| \to S^{n-1}$. Next, let $\sigma \in K - T$. Ideally, we would like to extend $F_2|\partial\sigma$ to σ. Unfortunately, $\partial\sigma$ is homeomorphic to S^{n-1} and we know that not every map from S^{n-1} to S^{n-1} is inessential. That is the reason that we cannot expect to extend f_1 to all of $|K|$ in general. On the other hand, observe that $\partial\sigma$ is a retract of $\sigma - \hat{\sigma}$, where $\hat{\sigma}$ is the barycenter of σ (see Problem 6.22). Let $r_\sigma : \sigma - \hat{\sigma} \to \partial\sigma$ be a retraction and set $B_1 = \{ \hat{\sigma} \mid \sigma \in K - T \}$. Define the map $F_1 : |K| - B_1 \to S^{n-1}$ by $F_1|(|T|) = F_2$ and $F_1|(\sigma - \hat{\sigma}) = F_2 \circ r_\sigma$ for each $\sigma \in K - T$. The map F_1 is the desired extension of f_1, which completes the first step of the proof of Theorem 11.

The Second Step: Given that we have an extension $G : S^n - B \to S^{n-1}$ of f, where B is a finite set of points in $S^n - X$, to construct F.

Let $C_1,C_2,\cdots,C_k$ be the components of $S^n - X$ and let $A = \{a_1,a_2,\cdots,a_k\}$ with $a_i \in C_i$. Let $\overline{C}_i = C_i \cup X$. Set $B(i) = B \cap C_i$. Order the components C_i in such a way that $B(i) \neq \{a_i\}$ for $1 \leq i \leq t_B$ and $B(i) = \{a_i\}$ for $t_B + 1 \leq i \leq k$, where $0 \leq t_B \leq k$. The proof proceeds by induction on t_B. If $t_B = 0$, then $B = A$ and we are done because we can let $F = G$. Suppose that F can be constructed for $t_B = \ell - 1$, $1 \leq \ell \leq k$, and let $t_B = \ell$. For $y \in S^n$, define $D_r(y)$ to be the disk in S^n of radius r around y, that is, $D_r(y) = \{ z \in S^n : |y - z| \leq r \}$. The boundary $\partial D_r(y)$ of this disk is the set $\{ z \in S^n : |y - z| = r \}$.

CLAIM 1. We can find points $y_1, y_2, \cdots, y_s \in C_\ell$ and numbers r_i, $0 < r_i < 1$, such that $y_s = a_\ell$, $D_{r_i}(y_i) \subset C_\ell$, $D_{r_i}(y_i) \cap D_{r_{i-1}}(y_{i-1}) \neq \emptyset$, $U_i = D_{r_i}(y_i) \cup D_{r_{i+1}}(y_{i+1}) \cup \cdots \cup D_{r_s}(y_s)$ is connected, and $B(\ell) \cup \{a_\ell\} \subset U_1$ (see Fig. 7.18).

Let us assume Claim 1 for the moment. It says essentially that we can cover the finite set $B(\ell) \cup \{a_\ell\}$ by a finite chain of disks in C_ℓ each of which meets the next one in the chain. Since $\partial D_{r_i}(y_i)$ is a retract of $D_{r_i}(y_i) - u$ for all $u \in D_{r_i}(y_i) - \partial D_{r_i}(y_i)$ (see Problem 6.22), let $\rho_{i,u} : D_{r_i}(y_i) - u \to \partial D_{r_i}(y_i)$ be a retraction. If $s = 1$ in Claim 1, then extend $G|(\overline{C}_\ell - U_1)$ to a map $G' : C_\ell - y_1 \to S^{n-1}$ by defining $G'|(U_1 - y_1) = G \circ \rho_{1,y_1}$. Assume inductively that, given a map $G|(C_\ell - B(\ell))$, we are able to extend $G|X$ to $G' : \overline{C}_\ell - a_\ell \to S^{n-1}$ whenever $s = m - 1$, $m \geq 2$, and let $s = m$. Choose $z \in D_{r_2}(y_2) \cap D_{r_1}(y_1)$. We may assume that $[B(\ell) \cup$

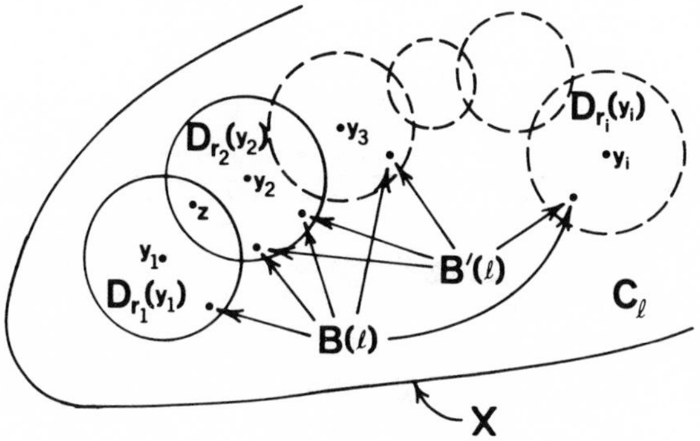

FIGURE 7.18

$\{a_\ell, z\}] \cap \partial D_{r_i}(y_i) = \emptyset$ for all i. (This is possible because X is a closed set, so that $S^n - X$ is open in S^n and we can replace the $D_{r_i}(y_i)$ by $D_{r_i+\epsilon}(y_i)$ for sufficiently small $\epsilon > 0$ if necessary.) Set $B'(\ell) = [B(\ell) \cup \{z\}] - [B(\ell) \cap D_{r_1}(y_1)]$ (see Fig. 7.18) and define $G'' : \overline{C}_\ell - B'(\ell) \to S^{n-1}$ by $G''|[\overline{C}_\ell - B(\ell) - D_{r_1}(y_1)] = G|[\overline{C}_\ell - B(\ell) - D_{r_1}(y_1)]$ and $G''|(D_{r_1}(y_1) - z) = G \circ \rho_{1,z}$. Since $B'(\ell) \subset U_2$ and since U_2 is connected and covered by $m - 1$ disks, our induction hypothesis applies to G'' and we can find a map $G' : \overline{C}_\ell - a_\ell \to S^{n-1}$ which extends $G''|X = G|X$. This means that we have established the following:

CLAIM 2. The map $G|X$ extends to a map $G' : \overline{C}_\ell - a_\ell \to S^{n-1}$.

Now let $B' = [B - B(\ell)] \cup \{a_\ell\}$ and let $G_1|[\overline{C}_i - B(i)] = G|[\overline{C}_i - B(i)]$ if $i \neq \ell$ and $G_1|[\overline{C}_\ell - \{a_\ell\}] = G'$. If $t_{B'}$ is defined similarly to t_B, then $t_{B'} = t_B - 1 = \ell - 1$, and so our inductive hypothesis implies that there is a map $F : S^n - A \to S^{n-1}$ with $F|X = G'|X = f$. Therefore, we have shown by induction that f always extends to a map F and Theorem 11 is proved modulo Claim 1.

Finally, we prove Claim 1. We show first that if Q is a finite set in C_ℓ, then there is a path $\gamma : [0,a] \to C_\ell$ such that $Q \subset \gamma([0,a])$. The path γ clearly exists if Q has no more than two points since C_ℓ is connected. Suppose that γ exists whenever Q has $n - 1$ points, $n \geq 3$, and assume that Q has n points. Choose $z \in Q$. By our inductive hypothesis there is a path $\gamma' : [0,a'] \to C_\ell$ such that $Q - \{z\} \subset \gamma'([0,a'])$. Since C_ℓ is connected there is a path $\gamma'' : [a',a' + 1] \to C_\ell$ from $\gamma'(a')$ to z. Let $a = a' + 1$ and define $\gamma : [0,a] \to C_\ell$ by $\gamma|[0,a'] = \gamma'$ and $\gamma|[a',a] = \gamma''$. Then $Q \subset \gamma([0,a])$ and we have shown that γ always exists by induction. Furthermore, it follows from the proof that we may let $\gamma(a)$ be any given point of Q.

In particular, there is a path $\gamma : [0,a] \to C_\ell$ with $[B(\ell) \cup \{a_\ell\}] \subset \gamma([0,a])$ and $\gamma(a) = a_\ell$. If $x \in [0,a]$, choose any r_x, $0 < r_x < 1$, so that $D_{r_x}(\gamma(x)) \subset C_\ell$. Let $I_x(\epsilon) = [x - \epsilon, x + \epsilon] \cap [0,a]$. For each $x \in [0,a]$ we can find an $\epsilon_x > 0$ so that $\gamma(I_x(\epsilon_x)) \subset D_{r_x}(\gamma(x))$. Since $[0,a]$ is compact, there is a finite set of points x_i, $0 = x_1 < x_2 < \cdots < x_s = a$ with the property that $[0,a] \subset [I_{x_1}(\epsilon_{x_1}) \cup I_{x_2}(\epsilon_{x_2}) \cup \cdots \cup I_{x_s}(\epsilon_{x_s})]$ and $I_{x_i}(\epsilon_{x_i}) \cap I_{x_{i-1}}(\epsilon_{x_{i-1}}) \neq \emptyset$. Let $r_i = r_{x_i}$ and $y_i = \gamma(x_i)$. Then r_i and $D_{r_i}(y_i)$ have the desired properties and Claim 1 is proved. This finishes the proof of Theorem 11.

The next theorem, which plays a very minor role in this book, is actually very important to modern algebraic topology; however, its real significance would only become apparent in a more advanced treatment of the subject than we are presenting here.

THEOREM 12 (The Homotopy Extension Theorem). If (X,Y) is a polyhedral pair, then every continuous map $f : (X \times 0) \cup (Y \times [0,1]) \to Z$ extends to a continuous map $F : X \times [0,1] \to Z$.

REMARK. The theorem asserts that we can fill in the dotted line in the diagram below

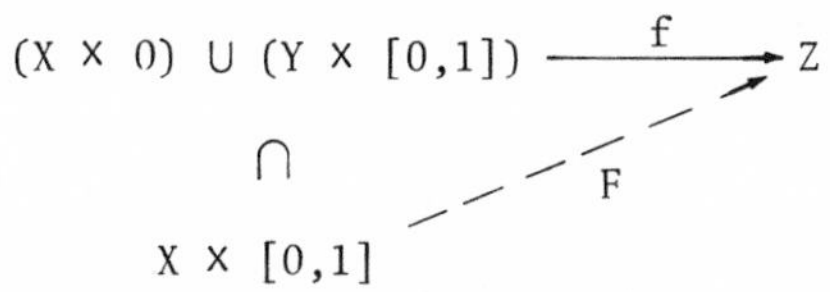

with a map F which makes the diagram commutative. Another way to state the result is that if we are given a map $g : X \to Z$ and a homotopy $(g|Y)_t$ of the partial map $g|Y$, then we can extend this homotopy to a homotopy g_t of the entire map g. This explains the name given the theorem.

Proof: Let $((K,L),\varphi)$ be a triangulation of (X,Y) and define $\varphi_1 : |K| \times [0,1] \to X \times [0,1]$ by $\varphi_1(x,t) = (\varphi(x),t)$. Let $f_1 = f \circ \varphi_1 : (|K| \times 0) \cup (|L| \times [0,1]) \to Z$. Clearly, f extends to the desired map F if and only if f_1 extends to a map $F_1 : |K| \times [0,1] \to Z$. We shall prove the existence of F_1 by induction on the number n of simplices in $K - L$. If $n = 0$, then $K = L$ and there is nothing to prove. Hence assume that $n > 0$ and let σ be a top-dimensional simplex of $K - L$. By the usual inductive hypothesis f_1 extends to a map $F_1' : (|K'| \times [0,1]) \cup (\sigma \times 0) \to Z$, where $K' = K - \sigma$. Let $r : \sigma \times [0,1] \to (\sigma \times 0) \cup (\partial\sigma \times [0,1])$ be a retraction (see Problem 6.21). Define $F_1 : |K| \times [0,1] \to Z$ by $F_1|(|K'| \times [0,1]) = F_1'$ and $F_1|(\sigma \times [0,1]) = F_1' \circ r$. The map F_1 obviously extends f_1, and so Theorem 12 is proved by induction.

COROLLARY. If (X,Y) is a polyhedral pair, then every inessential map $f : Y \to Z$ extends to an inessential map $F : X \to Z$.

Proof: Let $g_1 : Y \times [0,1] \to Z$ be a homotopy between f and a constant map, that is, $g_1(y,1) = f(y)$ and $g_1(y,0) = z_0 \in Z$. Define $g : (X \times 0) \cup (Y \times [0,1]) \to Z$ by $g(x,0) = z_0$ for all $x \in X$ and $g|(Y \times [0,1]) = g_1$. By Theorem 12, g extends to a map $G : X \times [0,1] \to Z$. If we define $F : X \to Z$ by $F(x) = G(x,1)$, then F extends f and F is inessential because G is a homotopy between F and a constant map. This proves the corollary.

Problems

7.25. Prove that if $f : S^n \to S^n$ is a continuous map with $f(-x) \neq f(x)$ for all $x \in S^n$, then f is surjective.

7.26. The results of this section can be used to give an alternate proof of the topological invariance of the dimension of a polyhedron. Prove the following without appealing to Theorem 4 in Sec. 6.3:

(*) Let K be a simplicial complex. A necessary and sufficient condition that $\dim K \leq n$ is that every map f :

$|L| \to S^n$, where L is a subcomplex of K, admits an extension $F : |K| \to S^n$.

(*Hint:* To prove necessity use Lemmas 11 and 12 but rephrased in terms of simplicial complexes so that one does not need Sec. 6.3. To prove sufficiency suppose that $\dim K > n$ and obtain a contradiction by extending a homeomorphism $h : \partial\sigma \to S^n$ to $|K|$, where σ is an $(n + 1)$-simplex of K. This would imply that $\partial\sigma$ is a retract of σ.) It is obvious that (*) is equivalent to Theorem 4 in Sec. 6.3 and shows that the dimension of a polyhedron is well defined.

7.27. Let X be a connected polyhedron and let $x_0 \in X$. Use the homotopy extension theorem to prove that any continuous map $f : S^n \to X$, $n \geq 0$, is homotopic to a map $g : S^n \to X$ such that $g(S^n_-) = x_0$.

7.6. THE JORDAN CURVE THEOREM AND OTHER SEPARATION THEOREMS

In this section we want to study the number of components of $\mathbb{R}^n - X$ or $S^n - X$ for subspaces X in terms of properties of X. The main goal is to prove the Jordan curve theorem. Unfortunately, we shall have to restrict ourselves not only to polyhedra X but also to the case where X is nicely (that is, "polyhedrally") imbedded. The latter assumption is somewhat restrictive; nevertheless, it is forced upon us because we shall have to extend maps defined on X to all of S^n and we are limited by the hypotheses in the results from the last section. With better theorems on extending maps we could weaken the restrictions on X.

DEFINITION. A polyhedron $X \subset \mathbb{R}^n$ is said to be polyhedrally imbedded if $(S^n, P_n^{-1}(X))$ is a polyhedral pair, where $P_n : S^n - (0,0,\cdots,0,1) \to \mathbb{R}^n$ is the stereographic projection.

THEOREM 13 (Borsuk's Separation Criterion). Let $X \subset \mathbb{R}^n$, $n \geq 1$, be a polyhedron which is polyhedrally imbedded. Let $x_0 \in \mathbb{R}^n - X$. Then

x_0 lies in the unbounded component of $\mathbb{R}^n - X$ if and only if the map $\varphi : X \to S^{n-1}$ given by $\varphi(x) = (x - x_0)/|x - x_0|$ is inessential.

Proof: Since X is compact it is bounded and therefore contained in a ball around the origin. Without loss of generality assume that X is contained in the interior of D^n and $x_0 = 0$.

If x_0 lies in the unbounded component, choose $x_1 \in \mathbb{R}^n - D^n$ and let $\gamma : [0,1] \to \mathbb{R}^n - X$ be a path connecting x_0 and x_1 (see Fig. 7.19). Define $F : X \times [0,1] \to S^{n-1}$ by $F(x,t) = (x - \gamma(t))/|x - \gamma(t)|$, and observe that $F(x,0) = x/|x| = \varphi(x)$ and $F(x,1) = (x - x_1)/|x - x_1|$. If we define $\psi : X \to S^{n-1}$ by $\psi(x) = F(x,1)$, then $x_1/|x_1| \in S^{n-1} - \psi(X)$. To see this, assume that $\psi(x) = (x - x_1)/|x - x_1| = x_1/|x_1|$. Then $x = (1 + |x - x_1|/|x_1|)x_1$, which is impossible because $|x| < 1$ and $|(1 + |x - x_1|/|x_1|)x_1| > |x_1| > 1$. Therefore, $S^{n-1} - \psi(X) \neq \emptyset$ and by Lemma 9 ψ is inessential. But F is a homotopy between φ and ψ, and so φ is inessential.

Conversely, assume that φ is inessential and let C be the component of $\mathbb{R}^n - X$ which contains x_0. If C is a bounded component, then consider the subspaces $X_1 = P_n^{-1}(X)$ and $C_1 = P_n^{-1}(C)$ of S^n. It follows from the hypothesis that X is polyhedrally imbedded that $(C_1 \cup X_1, X_1)$ is a polyhedral pair. Define $\varphi_1 : X_1 \to S^{n-1}$ by $\varphi_1 = \varphi \circ (P_n|X_1)$. Since φ is inessential, so is φ_1, and therefore by the corollary to Theorem 12 φ_1 extends to an inessential map $\Psi_1 : C_1 \cup X_1 \to S^{n-1}$. If we now define $\Psi : C \cup X \to S^{n-1}$ by $\Psi = \Psi_1 \circ (P_n^{-1}|X)$, then it is easy to see that Ψ is an extension of φ. By our initial hypothesis S^{n-1} is contained in the unbounded component of $\mathbb{R}^n - X$, so that $(C \cup X) \subset D^n$, and we can define an extension $r : D^n \to S^{n-1}$ of Ψ by $r(x) = \Psi(x)$ if $x \in (C \cup X)$ and $r(x) = x/|x|$ if $x \in (D^n - C)$. The map r is clearly a retraction of D^n onto S^{n-1}. This contradiction to Theorem 14 in Sec. 6.7 shows that x_0 must belong to the unbounded component of $\mathbb{R}^n - X$ and Theorem 13 is proved.

Theorem 13 allows us to prove the following useful criterion:

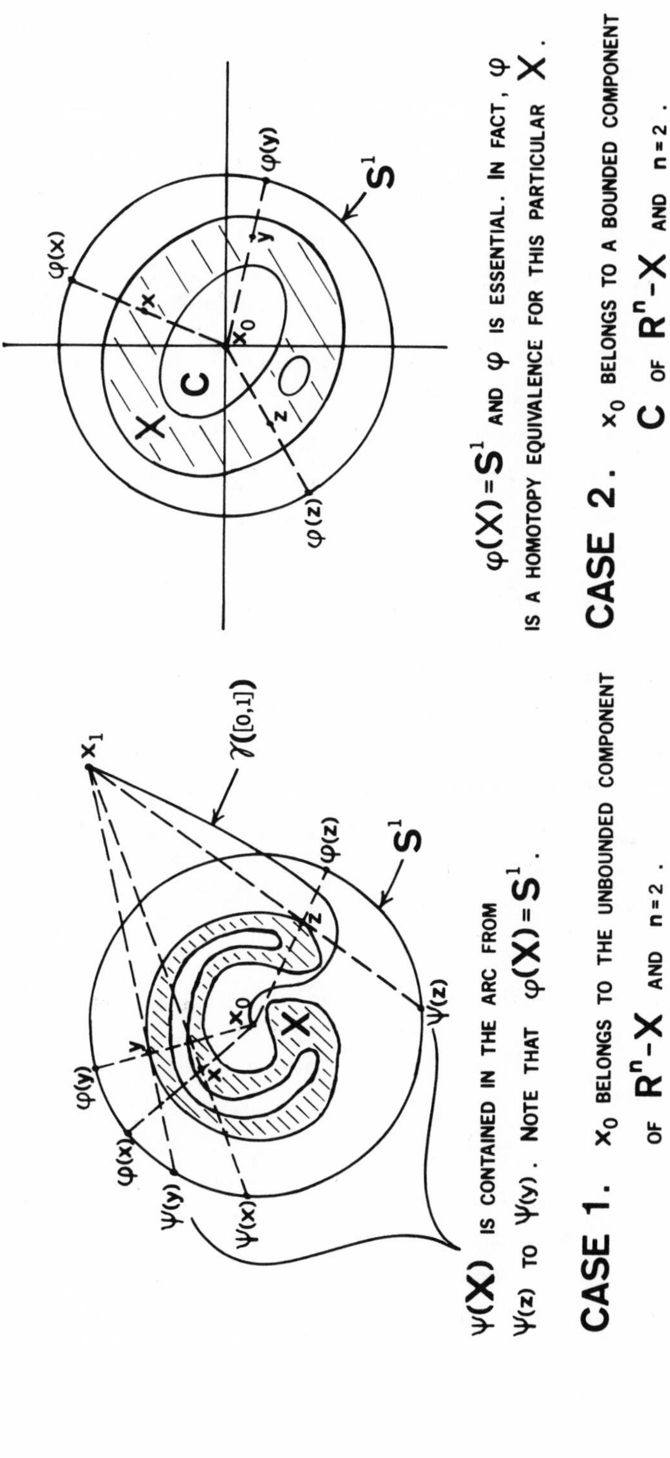
$\Psi(X)$ IS CONTAINED IN THE ARC FROM
$\Psi(z)$ TO $\Psi(y)$. NOTE THAT $\varphi(X)=S^1$.
CASE 1. x_0 BELONGS TO THE UNBOUNDED COMPONENT
OF R^n-X AND $n=2$.
(a)
$\varphi(X)=S^1$ AND φ IS ESSENTIAL. IN FACT, φ
IS A HOMOTOPY EQUIVALENCE FOR THIS PARTICULAR X.
CASE 2. x_0 BELONGS TO A BOUNDED COMPONENT
C OF R^n-X AND $n=2$.
(b)

FIGURE 7.19

THEOREM 14 (Borsuk). Let (S^n,X), $n \geq 1$, be a polyhedral pair with $X \neq S^n$. Then $S^n - X$ is connected if and only if every map $f : X \to S^{n-1}$ is inessential.

Proof: Suppose that $S^n - X$ is connected and let $f : X \to S^{n-1}$ be any map. Fix $z \in S^n - X$. By Theorem 11 f extends to a map $F : S^n - z \to S^{n-1}$. But $S^n - z$ is homeomorphic to $\mathbb{R}^n$ and so F is inessential. This shows that f is also inessential.

Conversely, assume that every map $f : X \to S^{n-1}$ is inessential. If $S^n - X$ is not connected, then let $x_1, x_2 \in S^n - X$ lie in distinct components and identify $S^n - x_1$ with $\mathbb{R}^n$. It follows that x_2 lies in a bounded component of $\mathbb{R}^n - X$. By Theorem 13, the map $\varphi : X \to S^{n-1}$ given by $\varphi(x) = (x - x_2)/|x - x_2|$ is essential, which contradicts our initial hypothesis. Therefore, $S^n - X$ must be connected and Theorem 14 is proved.

Finally, we come to the main result of this section. It is easy to see that the $(n - 1)$-sphere S^{n-1}, thought of as imbedded in S^n via the standard inclusion map, divides S^n into two pieces--the upper and lower hemispheres. Therefore, it is natural to ask whether any homeomorphic image of S^{n-1} in S^n divides S^n in the same way.

THEOREM G (The Generalized Jordan Curve Theorem). Let X be a subset of S^n which is homeomorphic to S^{n-1}. Then $S^n - X$ has two components whose closures intersect in X.

We are unable to prove Theorem G in its full generality here, but we can prove the following polyhedral version.

THEOREM 15. If (S^n,X), $n \geq 1$, is a polyhedral pair with X homeomorphic to S^{n-1}, then $S^n - X$ has two components.

Proof: Since we know that not every map from X to S^{n-1} is inessential, it follows from Theorem 14 that $S^n - X$ is not connected. We must still show that $S^n - X$ has precisely two components. As usual, we shall work with a triangulation $((K,L),\varphi)$ for (S^n,X) and show that $|K| - |L|$ has precisely two components.

It follows from the work in Sec. 6.3 and Sec. 6.5 that $\dim L = n - 1$, $\dim K = n$, and if we choose an $(n - 1)$-simplex σ in L, then σ is the face of exactly two n-simplices τ_1 and τ_2 of K. Let C_1 and C_2 be the components of $|K| - |L|$ which contain $\operatorname{int} \tau_1$ and $\operatorname{int} \tau_2$, respectively. Suppose that y is any point of $|K| - |L|$. If $y \in (\tau_1 \cup \tau_2)$, then y belongs to C_1 or C_2. Assume that $y \notin (\tau_1 \cup \tau_2)$.

CLAIM. $|K| - |L - \sigma|$ is connected.

Assuming the claim for the moment, there is therefore a path $\gamma : [a,b] \to (|K| - |L - \sigma|)$ from y to q, where q is the barycenter of τ_1. Let $t_0 = \text{g}\ell\text{b} \{ t \mid \gamma(t) \in (\tau_1 \cup \tau_2) \}$ (see Fig. 7.20). It is easy to see that $a < t_0 < b$, $\sigma \cap \gamma([a,t_0]) = \emptyset$, and $\gamma(t_0) \in (\partial\tau_1 \cup \partial\tau_2) - |L|$. If $\gamma(t_0) \in \partial\tau_j$, $j = 1$ or 2, then $\gamma|[a,t_0]$ is a path in $|K| - |L|$ from y to $\gamma(t_0) \in C_j$, which would show that $y \in C_j$. In other words, every point of $|K| - |L|$ belongs to C_1 or C_2 and so $|K| - |L|$ has at most two components. Since we already know that $|K| - |L|$ has at least two components, C_1 and C_2 must be distinct and Theorem 15 is proved modulo the claim.

To prove the claim first observe that $|L - \sigma|$ is a retract of $|L| - p$, where p is the barycenter of σ. To see this, let r_σ :

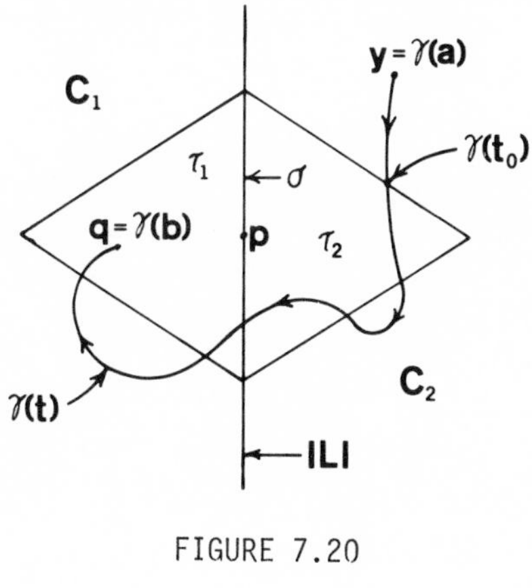

FIGURE 7.20

$\sigma - p \to \partial\sigma$ be a retraction and define a retraction $r : |L| - p \to |L - \sigma|$ by $r(x) = x$ if $x \notin \sigma$ and $r(x) = r_\sigma(x)$ if $x \in \sigma - p$. Now let $f : |L - \sigma| \to S^{n-1}$ be any continuous map. Then $f \circ r : |L| - p \to S^{n-1}$ is an extension of f. Since $(|L| - p) \approx [S^{n-1} - (\text{point})] \approx \mathbb{R}^{n-1}$, $f \circ r$ is homotopic to a constant map and restricting this homotopy to $|L - \sigma|$ shows that the arbitrary map f is inessential. The claim now follows from Theorem 14 applied to $|L - \sigma|$ and the proof of Theorem 15 is finished.

It is easy to see that the closures of the two components of $S^n - X$ in Theorem 15, call them A and B, intersect in X. If X were the standard S^{n-1} in S^n, then clearly A and B, being the upper and lower hemispheres of S^n, would in fact be homeomorphic to the n-disk D^n. Is this always the case? The answer to this question is "yes" provided that X possesses a "collar." We state the result without proof.

THEOREM H (The Generalized Schoenflies Theorem). Let h be an imbedding of $S^{n-1} \times D^1$ in S^n. Then the closure of each component of $S^n - h(S^{n-1} \times 0)$ is homeomorphic to D^n. [The set $h(S^{n-1} \times D^1)$ is called a collar of $h(S^{n-1} \times 0)$.]

Finally, note that an easy consequence of Theorem 15 is that any polyhedrally imbedded S^{n-1} in $\mathbb{R}^n$ divides $\mathbb{R}^n$ into two pieces, one bounded and the other unbounded. We also have the following theorem which in effect says that no homeomorph of D^k, separates S^n.

THEOREM 16. If (S^n, X), $n \geq 1$, is a polyhedral pair with X homeomorphic to D^k, $0 \leq k \leq n$, then $S^n - X$ is connected.

Proof: Since every map $f : X \to S^{n-1}$ is inessential, Theorem 16 follows from Theorem 14.

Problem

7.28. Give a proof of Theorem C (Invariance of Domain) using Theorem G (The Generalized Jordan Curve Theorem).

7.7. HISTORICAL COMMENTS

Section 7.1. The notion of the degree of a map is due to Brouwer. He introduced the concept for a very special case in his 1911 paper [Brou 3]. A year later in [Brou 5] he defined the degree and developed some of its properties in the general case of maps between oriented manifolds. Brouwer also considered vector fields and proved Theorem 1 at that time. The corollary was already known to Poincaré. There is a close connection between vector fields and differential equations. Suppose that $X \subset \mathbb{R}^n$ and let $\sigma : X \to \mathbb{R}^n$ be any $\mathbb{R}^n$-vector field on X. Define an integral curve of σ to be any differentiable map $\varphi : [a,b] \to \mathbb{R}^n$, $\varphi(t) = (\varphi_1(t),\varphi_2(t),\cdots,\varphi_n(t))$, such that $\varphi'(t) = (\varphi_1'(t),\varphi_2'(t),\cdots,\varphi_n'(t)) = \sigma(\varphi(t))$, that is, the tangent vectors to the curve φ are precisely the vectors from the vector field σ. (Here $\varphi_i'(t)$ denotes the usual derivative of the function $\varphi_i : [a,b] \to \mathbb{R}$ at t.) It is then clear that the solutions to a differential equation of the type

$$\frac{dy}{dx} = f(x,y)$$

are just the integral curves of the $\mathbb{R}^2$-vector field $(x,y) \to (1,f(x,y))$. See Fig. 7.21.

The study of differential equations in terms of integral curves to an appropriate vector field originated with Poincaré. This approach, which opens the way for the application of topological methods, leads to many important qualitative results and is today part of a large and active modern field called global analysis. Poincaré was motivated by problems in dynamics and celestial mechanics. In particular, singular points of a vector field, that is, points where the vector field is zero, often correspond to points of equilibrium in dynamical problems. One can define a notion of the index of a vector field at an isolated singular point (similar to the definition in Problem 7.24 which treated a special case). In 1885 Poincaré [Poin 1] showed that the sum of the indices of a tangent vector field on a surface S is related to the Euler characteristic of S. Var-

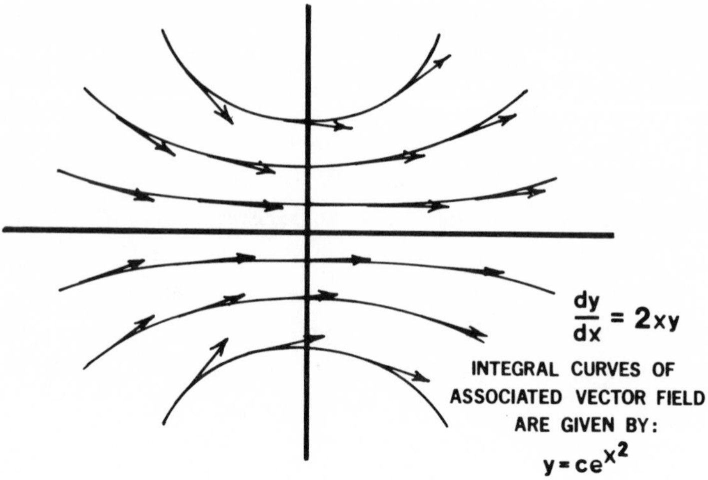

$\frac{dy}{dx} = 2xy$
INTEGRAL CURVES OF
ASSOCIATED VECTOR FIELD
ARE GIVEN BY:
$y = ce^{x^2}$

FIGURE 7.21

ious generalizations of this result by Brouwer and J. S. Hadamard (1865-1963) finally culminated in the very general Poincaré-Hopf theorem which was proved by Hopf [Hop 2] in 1926. An excellent reference for this topic is [Mi 3]. A related problem is the question of the existence and properties of closed integral curves for vector fields, which correspond to periodic solutions to differential equations. Poincaré considered this problem in various papers [Poin 1-3] published between 1881 and 1899 and we now have the well-known Poincaré-Bendixson theorem for which Poincaré had laid the groundwork but which was considerably extended by I. Bendixson (1861-1935) [Ben] in 1901. Finally, we should mention the important work of A. M. Liapunov (1857-1918) [Lia] in 1892 on the stability of solutions of differential equations which also pioneered the use of vector fields.

There is a connection between vector fields and fixed points of maps. For example, if we are given a tangent vector field on a surface S, then we can use that vector field to define a deformation of S to itself which moves a point p in the direction in which the vector at p is pointing. The fixed points of the map are just the singular points of the vector field.

Part (a) of Theorem D was proved independently by A. Hurwitz (1859-1919) [Hurw 2] and J. Radon (1887-1956) [Rad] in 1923. A special case is already due to Hurwitz in 1898 (see [Hurw 1]). B. Eckmann (1917-) [Ec] gave an elegant proof of the result in 1942 using group representation theory. Part (b) was the harder part and, although it was conjectured for a long time, it was not proved until 1962 by J. F. Adams (1930-) [Ad]. Adams' proof uses facts from many different fields in mathematics. It is a beautiful example of how the best mathematics has always been done by mathematicians with broad backgrounds and capabilities!

Section 7.2. Theorem 3 was first conjectured by S. M. Ulam (1909-) and then proved in 1933 by Borsuk [Bor 3], who also proved Theorem 2. Borsuk's proofs were "elementary" but long. Hopf later gave three briefer proofs. Corollary 1 was apparently first proved by H. D. Steinhaus (1887-1972) who also invented the name "ham sandwich theorem." Corollary 2 was proved by L. A. Liusternik (1899-) and L. G. Schnirelmann (1905-1938) in 1930.

S. Kakutani (1911-) [Kak] proved Theorem E and its corollary in the case $n = 2$ in 1942 and conjectured it for higher dimensions. This conjecture was proved by H. Yamabe (1923-1960) and Z. Yujobô in 1950 (see [Y-Y]). Theorem F for $n = 2$ was proved by F. J. Dyson (1923-) [Dys] in 1951. Dyson's conjecture for arbitrary n was proved by C. T. Yang (1923-) [Ya] in 1954.

Section 7.3. It is easy to generalize the proof of Lemma 6 and show that $\deg : [S^n,S^n] \to \mathbb{Z}$ is surjective for all $n \geq 1$ by explicitly constructing a map $\alpha_m^n : S^n \to S^n$ which "wraps" S^n m times around itself. Such a map α_m^n will have degree m. The surjectivity of $\deg$ was known to Brouwer in 1912 (see [Brou 5, p. 105]). It is also true that the map $\deg$ is injective, but this is harder. The following general result was proved by Hopf [Hop 1] in 1927:

THEOREM I (Hopf). If M^n is a closed oriented n-dimensional manifold, then the map

$$\deg : [M^n, S^n] \to \mathbb{Z}$$

given by deg ([f]) = deg f, is a bijection.

In other words, the homotopy class of a map $f : M^n \to S^n$ is completely determined by the degree of f (as defined in Sec. 7.4). The case $n = 2$ had already been proved by Brouwer [Brou 10] in 1921. Hopf initially assumed that M^n was triangulable, but in [Hop 3] in 1928 he pointed out that the notion of the degree of a map could be extended to maps between arbitrary topological manifolds. Theorem I remains valid in this more general situation. There is a corresponding result in the nonorientable case by replacing the degree of a map by the "mod 2" degree.

The first rigorous proof of Theorem 5 was given by Gauss in 1797 in his dissertation. Gauss attached a great importance to this theorem and gave a total of four different proofs (see [G 1]). It is not clear who stated it first. That m-th degree equations could have m roots was realized by P. Rothe [Rot] in 1608 and A. Girard (-1633) [Gi] commented on the number of roots to an equation in 1629. The term "fundamental theorem of algebra" appears to have been introduced by Gauss, although Euler seems to have stated the theorem in its present form in a letter in 1742. Predecessors of Gauss who attempted to prove the theorem were J. D'Alembert (1717-1783) [Da] in 1746, Euler [Eu 2] in 1749, F. Daviet de Foncenex (1734-1799) [Fon] in 1759, J. L. Lagrange (1736-1813) [Lag] in 1772, and P. S. Laplace (1749-1827) [Lap] in 1795. Of these, D'Alembert came the closest to giving a correct proof, and his name is also sometimes associated with the theorem.

Section 7.4. An analytic version of Theorem 6 is due to L. Kronecker (1823-1891). In 1869, 1873, and 1878 Kronecker [Kr] considered the question of common zeros of real-valued functions and developed the theory of the "Kronecker characteristic." Elements of this theory are even present already in Gauss' first and fourth proof of the fundamental theorem of algebra. Some early work of J. K. F. Sturm (1803-1855), J. J. Sylvester (1814-1897), C. Hermite

(1822-1901), C. G. J. Jacobi (1804-1851), and F. Brioschi (1824-1897) on the number of real roots of functions is also relevant (see [Dyc 1, p. 460-461]). What Kronecker did was to define a certain integral which he proved to be integer-valued and which is in fact equal to the local degree, as we have defined it, of the function in question. These ideas were developed further by Dyck [Dyc 1,2,3] in 1888, 1890, and 1895, respectively, and finally by Hadamard [Had] in 1910, where numerous applications can be found. The theory is also discussed in [Alexf-H]. The theory of local degrees as we have developed it is present in large part in Hopf's 1928 paper [Hop 3]. Lemma 7(b) is often referred to as the Poincaré-Bohl theorem and was proved independently by Poincaré [Poin 1] in 1886 and P. Bohl (1865-1921) [Bo] in 1904. The homology definition of the degree of a map between pseudomanifolds was first given by Hopf [Hop 7] in 1933. Relations between the number of solutions to equations and either the degree or the local degree of a map, such as were considered in Theorems 9 and 10, were studied by Kneser [Kn 1,3] in 1928 and 1930, respectively, and by Hopf [Hop 4] in 1930. Corollary 2 to Theorem 10 is an observation due to Brouwer [Brou 5] in 1912, which he used to justify the surjectivity of $\deg : [S^n, S^n] \to \mathbb{Z}$.

Relationships as in Problem 7.24(b) between the number of peaks, pits, and passes were discussed in papers by F. Reech (1805-) [Ree] in 1858, Cayley [Cay 2] in 1859, and J. C. Maxwell (1831-1879) [Max] in 1870.

Section 7.5. All of the results of this section can be found in [Alexf-H, Chapter 13]. Lemmas 9-12 and Theorem 11 appear either entirely or implicitly in papers by Hopf [Hop 5,7], by Borsuk [Bor 2], and by Alexandroff [Alexf 5]. Theorem 12 (The Homotopy Extension Theorem) was first proved by Hurewicz [Hure].

Section 7.6. Theorems 13 and 14 were proved by Borsuk [Bor 2] in 1932. In fact, Borsuk proved the theorems in the more general case where X is an arbitrary compact subset. With such a generalization we would have been able to prove Theorem G rather than just

Theorem 15. The crucial fact which is needed for this generalization is the following:

THEOREM J (Tietze-Urysohn). If A is a closed subset of $\mathbb{R}^n$, then every continuous function $f : A \to [0,1]$ or $\mathbb{R}$ extends to a continuous function on $\mathbb{R}^n$.

Theorem J was proved by Urysohn [U 2] in 1925. A special case had been proved by Tietze [Ti 2] in 1915. The Tietze-Urysohn extension theorem is one of the important theorems in topology.

Theorem G originated with Jordan [J 3] in 1893. Jordan considered simple closed curves in the plane (that is, the case $n = 2$ in Theorem G); however, his proof was incomplete because he assumed its validity in the case of polygonal curves and also did not give all the details in his argument. There is a long history of incorrect proofs of Jordan's theorem. N. J. Lennes (1874-1951) [Len] proved the theorem in the case of polygonal curves in 1903, as did Veblen [Veb 1] in 1904. In 1905 Veblen [Veb 2] proved the theorem for arbitrary curves. Later, Brouwer [Brou 2] and Alexander [Alex 3] gave simpler proofs in 1910 and 1920, respectively.

In 1902 Schoenflies [Scho 2] announced a converse to the Jordan theorem. Very roughly speaking, he asserted that a set of points which separates the plane into two parts and which is the common boundary of these two parts is a curve. Schoenflies developed his theory in a series of papers [Scho 3] from 1904 to 1906. In his 1906 paper he also "proved" that every imbedding of S^1 in $\mathbb{R}^2$ extends to an imbedding of D^2 in $\mathbb{R}^2$. This is clearly the case with the standard inclusion of S^1 in $\mathbb{R}^2$. The polygonal version of this theorem, something that Schoenflies had assumed, was proved by L. D. Ames (1869-1965) [Am] and G. A. Bliss (1876-1951) [Bli] in 1904. (The papers of Ames and Bliss also contained a proof of the polygonal Jordan curve theorem.) Schoenflies' original proofs contained errors which were corrected by Brouwer [Brou 1,8] in 1910 and 1912, respectively. An interesting example constructed by Brouwer in this connection shows that it is possible to find infinitely many regions (that is, open and connected sets) in $\mathbb{R}^2$ which have the same boundary. One may even assume that the regions are simply-connected (that is, every continuous map of S^1 into the region is homotopic, within the region, to a constant map). Schoenflies did not completely settle the question as to what characterizes a curve. This was finally accomplished by the famous Hahn-Mazurkiewicz theorem (see [Ha] and [Mazu]).

The extension of Jordan's theorem to n dimensions, namely, Theorem G, was proved in its full generality by Brouwer [Brou 7] in 1912. Brouwer also showed that a similar result held for closed $(n - 1)$-dimensional manifolds in S^n. Alexander [Alex 4] extended the theorem even further in 1922 by relating the Betti numbers of a closed set A in S^n with those of $S^n - A$. Actually the germ of his ideas was already present in [Leb 2]. Pontrjagin [Pont 1] finished the problem in 1931. The Alexander-Pontrjagin duality theorems are the source of many separation theorems.

The obvious extension of the Schoenflies theorem led to the following conjecture:

THE SCHOENFLIES CONJECTURE. Let X be a subset of S^n which is homeomorphic to S^{n-1}. Then the closure of each component of $S^n - X$ is homeomorphic to D^n.

An example given by M. L. A. Antoine (1888-) [Ant] in 1921 shows that this conjecture is false without some conditions on the flatness of the imbedding of X in S^n. In 1924 Alexander also gave a counterexample. His is the better known and usually referred to as the "Alexander horned sphere" (see [Alex 5, p. 176]). In both Antoine's and Alexander's example X is a wildly imbedded S^2 in S^3 with the property that not all circles in $S^3 - X$ can be shrunk to a point in $S^3 - X$, something that is clearly impossible if the conjecture above were true. The correct version of the generalized Schoenflies theorem, Theorem H, was finally proved by M. Brown (1931-) [Brw] in 1960, after B. Mazur (1937-) [Maz] had proven the result in 1959 assuming a simple "niceness" condition.

By the way, the Schoenflies theorem, that an imbedding $f : S^1 \to \mathbb{R}^2$ extends to an imbedding $F : D^2 \to \mathbb{R}^2$, is used to prove the triangulability of surfaces (see [Ahl-S]). A startling result of Milnor [Mi 1] in 1957 showed that the n-dimensional analog of this result is not valid in the "differentiable category," that is, there are "differentiable" imbeddings $f : S^6 \to \mathbb{R}^7$ which do not extend to "differentiable" imbeddings of D^7 in $\mathbb{R}^7$. Thus, there is a discrepancy between topological theorems and differentiable ones. Milnor's discovery is largely responsible for the tremendous energy that was devoted to the field of differentiable topology in the 1960s and 1970s.

Chapter 8

CONCLUDING REMARKS

In the previous chapters we have seen how isolated and seemingly insignificant observations about spaces gradually fell into place and led to a theory which enables us to see certain intrinsic properties of spaces that we would not have been aware of otherwise. Furthermore, this knowledge was shown to have many applications. In conclusion we would like to indicate some directions of further study to those students who have become as fascinated with algebraic topology as the author had when he originally encountered the subject. This book has just barely scratched the surface. As a matter of fact, except for some of the mathematical language and proofs, essentially everything that we have done was known by the 1920s and 1930s.

One concept which we did not discuss and one which is important when it comes to computing homology groups is that of relative homology groups. If L is a subcomplex of a simplicial complex K, then one can define relative homology groups $H_q(K,L)$. These groups were first defined by Lefschetz in 1927 and are gotten by looking at the groups $C_q(K)/C_q(L)$ and the induced boundary maps $\partial_q^r : C_q(K)/C_q(L) \to C_{q-1}(K)/C_{q-1}(L)$. The group $H_q(K,L)$ is defined to be $(\text{Ker } \partial_q^r)/(\text{Im } \partial_{q+1}^r)$. There is what is called an "exact sequence" which relates the homology groups of K, L, and (K,L), so that if one knows any two of them the third is fairly well determined. One can show that, for $q > 0$, $H_q(K,L)$ is isomorphic to $H_q(M)$, where M is a simplicial complex which triangulates the space obtained from $|K|$ by identifying all of the points of $|L|$ to a single point.

A big problem with simplicial homology groups is that they are defined for simplicial complexes, whereas we really want a group that is an intrinsic invariant of a polyhedron and whose definition does not depend on choosing a particular triangulation. The point is that problems of topology have to do with spaces and continuous maps not with certain subdivisions. We have experienced the awkwardness in our theory every time we had to replace a space by a triangulation of it. All this is avoided if one defines the "singular homology groups." The singular homology theory is due to Lefschetz [Lef 4] in 1933 and Eilenberg [Eil] in 1944, but aspects of this theory can already be found in [Alex 1], [Veb 3], and [Lef 3]. If $e_0 = (1,0,\cdots,0)$, $e_1 = (0,1,0,\cdots,0)$, $\cdots$, and $e_q = (0,\cdots,0,1)$ denote the standard unit vectors in $\mathbb{R}^{q+1}$, then $\Delta^q = e_0e_1\cdots e_q$ is called the standard q-simplex. Let $X \subset \mathbb{R}^n$.

DEFINITION. A continuous map $T : \Delta^q \to X$ is called a singular q-simplex of X. Let S_q be the set of singular q-simplices of X. Define the group of singular q-chains of X, $C_q^s(X)$, by

$$C_q^s(X) = \{ f : S_q \to \mathbb{Z} \mid f(T) = 0 \text{ for all but a finite number of } T \in S_q \}$$

[The group operation is the obvious one, namely, for $f,g \in C_q^s(X)$ and $T \in S_q$, $(f + g)(T) = f(T) + g(T)$.]

By identifying T with the map $f_T : S_q \to \mathbb{Z}$, where $f_T(T) = 1$ and $f_T(T') = 0$ is $T' \neq T$, we see (as in the case of the chain groups for a simplicial complex) that $C_q^s(X)$ is just the group of all finite linear combinations $n_1T_1 + n_2T_2 + \cdots + n_kT_k$, $n_i \in \mathbb{Z}$, of singular simplices T_i of X.

DEFINITION. Given a singular q-simplex $T : \Delta^q \to X$, define the i-th face of T, $\delta^iT : \Delta^{q-1} \to X$, by

$$(\delta^iT)(t_0,t_1,\cdots,t_{q-1}) = T(t_0,\cdots,t_{i-1},0,t_i,\cdots,t_{q-1})$$

There is a boundary homomorphism $\partial_q^s : C_q^s(X) \to C_{q-1}^s(X)$ which on a singular q-simplex $T : \Delta^q \to X$, is given by

$$\partial_q^s(T) = \sum_{i=0}^{q} (-1)^i(\delta^i T) \in C_{q-1}^s(X)$$

One can show, as in the simplicial theory, that $\partial_{q-1}^s \circ \partial_q^s = 0$.

DEFINITION. The q-th singular homology group, $H_q^s(X)$, is now defined by

$$H_q^s(X) = (\text{Ker } \partial_q^s)/(\text{Im } \partial_{q+1}^s)$$

DEFINITION. Given a continuous map $f : X \to Y$ between spaces X and Y, define a homomorphism

$$f_{\#q}^s : C_q^s(X) \to C_q^s(Y)$$

by the condition that $f_{\#q}^s(T) = f \circ T$ for every singular q-simplex $T : \Delta^q \to X$.

Again one can check that $\partial_q^s \circ f_{\#q}^s = f_{\#q-1}^s \circ \partial_q^s$, so that $f_{\#q}^s$ induces a homomorphism

$$f_{*q}^s : H_q^s(X) \to H_q^s(Y)$$

With the group $H_q^s(X)$ and maps f_{*q}^s we never have to worry about triangulations. Other intrinsically defined homology theories were defined by Vietoris [Vi 1] in 1927 and Alexandroff [Alexf 2-4] in 1928 and 1929 (see also [Vi 2]). These were variants of the so-called Čech homology theory defined by Čech [Ce 2] in 1932. Later it was proved that all three theories, the simplicial, the singular, and the Čech, give rise to isomorphic homology groups on polyhedra. Thus, one is free to use whichever theory one wants to in any particular problem; however, the singular homology theory is the one most widely used.

In 1930 Lefschetz introduced the notion of "pseudocycles" in his book [Lef 3], which was investigated further by Alexander [Alex 9] in 1935, H. Whitney (1907-) [Whi] in 1938, and Lefschetz [Lef 6] in 1942. These papers mark the beginning of cohomology theories. The word "cohomology" itself is due to Whitney.

Let K be a simplicial complex.

DEFINITION. The q-th (simplicial) cochain group of K, $C^q(K)$, is defined by

$$C^q(K) = \mathrm{Hom}(C_q(K), \mathbb{Z})$$

Define coboundary maps

$$\delta^q : C^q(K) \to C^{q+1}(K)$$

by the formula $[\delta^q(f)](c) = f(\partial_q(c))$ for every $f \in C^q(K)$ and $c \in C_{q+1}(K)$.

The fact that $\partial_{q-1} \circ \partial_q = 0$ implies immediately that $\delta^{q+1} \circ \delta^q = 0$, so that $\mathrm{Im}\ \delta^{q-1} \subset \mathrm{Ker}\ \delta^q$.

DEFINITION. The q-th (simplicial) cohomology group of K, $H^q(K)$, is defined by

$$H^q(K) = (\mathrm{Ker}\ \delta^q)/(\mathrm{Im}\ \delta^{q-1})$$

Suppose that K and L are simplicial complexes and $f : |K| \to |L|$ is a continuous map.

DEFINITION. Define a homomorphism

$$f^{\#q} : C^q(L) \to C^q(K)$$

by $[f^{\#q}(g)](c) = g(f_{\#q}(c))$ for $g \in C^q(L)$ and $c \in C_q(K)$.

Since $\delta^q \circ f^{\#q} = f^{\#q+1} \circ \delta^q$, $f^{\#q}$ induces a well-defined homomorphism

$$f^{*q} : H^q(L) \to H^q(K)$$

Given a space X, it is possible to define singular cohomology groups $H^q(X)$ by using the singular chain groups in the definitions above instead of the simplicial ones. At any rate, we see that the cohomology groups are obtained by a simple algebraic trick which consists of dualizing the constructions for the homology groups. Therefore, it may seem as if this would not lead to anything very new, but this is a mistaken first impression. It turns out that cohomology provides us with a much richer algebraic structure than does homology. The main reason for this is that one can multiply a class $u \in H^p(X)$ with a class $v \in H^q(X)$ to get a class $u \cup v \in H^{p+q}(X)$. This multiplication (called the "cup product") makes the cohomology groups of X into a "graded ring." In the simplicial case one first orders all the vertices of the simplicial complex K. Denote this ordering by "$<$." The cup product is now induced from the following product on the cochain level: if $f \in C^p(K)$ and $g \in C^q(K)$, then define $f \bullet g \in C^{p+q}(K)$ by the condition that $(f \bullet g)([v_0v_1\cdots v_{p+q}]) = f([v_0v_1\cdots v_p])g([v_pv_{p+1}\cdots v_{p+q}])$ for all oriented $(p + q)$-simplices $[v_0v_1\cdots v_{p+q}]$ of K with $v_0 < v_1 < \cdots < v_{p+q}$. Two distinct orderings of the vertices of K will induce isomorphic product structures on the cohomology groups. No such multiplication exists in general for homology. The ring structure in cohomology gives us therefore an additional invariant for spaces. For example, since the space X in Problem 6.8 has the same homology groups as the torus $S^1 \times S^1$, the homology theory does not distinguish between the two. On the other hand, X and $S^1 \times S^1$ have distinct cohomology rings (although the cohomology groups are the same), so that cohomology does tell these two spaces apart.

A very important fact about orientable n-dimensional closed manifolds M^n is that they satisfy Poincaré duality. More precisely, it can be shown that the groups $H_i(M^n)$ and $H_{n-i}(M^n)$ have the same rank and that the torsion subgroup of $H_i(M^n)$ is isomorphic to the torsion subgroup of $H_{n-i-1}(M^n)$. This is true for all i. The part about Betti numbers was proved by Poincaré in 1895 and the torsion part in 1900 (see [Poin 4,6]). The most natural statement of Poincaré duality makes use of cohomology groups and is the following: there is a natural homomorphism from $H^i(M^n)$ to $H_{n-i}(M^n)$ which is an isomorphism for all i. This implies the Poincaré duality theorem above because a general theorem, which is a purely algebraic fact and which is called "the universal coefficient theorem for cohomology," states that the cohomology groups are determined by the homology groups (and vice versa). In particular, it is true that for any polyhedron or compact space X, $H^i(X)$ and $H_i(X)$ have the same rank and the torsion subgroup of $H^i(X)$ is isomorphic to the torsion subgroup of $H_{i-1}(X)$. Beginning in the 1950s there was a great surge of interest in the theory of manifolds which resulted in a much better understanding of the structure of these nice spaces, and the key property which separates manifolds from other spaces is Poincaré duality.

The final remarks in this chapter will be on homotopy theory, a topic of major importance in algebraic topology, but one which we barely mentioned. In a sense, algebraic topology can be divided into two parts--homology theory and homotopy theory--although the two are intimately related. In homotopy theory the basic concept is that of a homotopy group. Let $x_0 \in X$ and $y_0 \in Y$. The points x_0 and y_0 will be called base points and are assumed fixed.

DEFINITION. A continuous map $f : X \to Y$ is base point preserving if $f(x_0) = y_0$. A base point preserving homotopy between base point preserving maps $f,g : X \to Y$ is a continuous map $h : X \times [0,1] \to Y$ such that $h(x,0) = f(x)$, $h(x,1) = g(x)$, and $h(x_0,t) = y_0$ for all $x \in X$ and $t \in [0,1]$. (As in the case of an ordinary homotopy the existence of h is equivalent to a one-parameter family of maps $h_t : X \to Y$ such that $h_0 = f$, $h_1 = g$, and $h_t(x_0) = y_0$ for all t.) Let $[f]_*$ denote the base point preserving homotopy class of a base point preserving map $f : X \to Y$ and let

$$[X,Y]_* = \{ [f]_* \mid f : X \to Y \}$$

There is a natural map from $[X,Y]_*$ to $[X,Y]$ which sends $[f]_*$ to $[f]$. If Y is a polyhedron, this map is actually surjective, but it is not always injective. Consider the case $X = S^n$, $n \geq 1$. One can define a natural operation in $[S^n,Y]_*$ which is indicated in Fig. 8.1 when $n = 1$ and 2. Let $[f]_*,[g]_* \in [S^n,Y]_*$ and

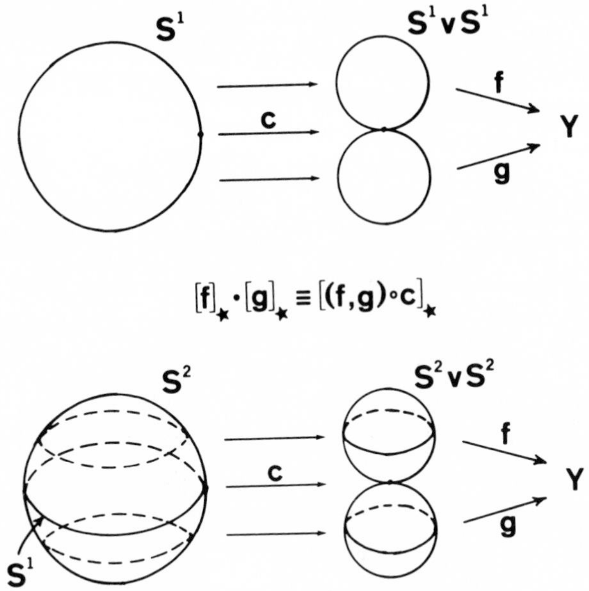

FIGURE 8.1

define $(f,g) : S^n \vee S^n \to Y$ to be the map which is f on the first factor of $S^n \vee S^n$ and g on the second. (The wedge $S^n \vee S^n$ was defined in Problem 6.5 and means that two copies of S^n are pasted together along their base points.) Let $c : S^n \to S^n \vee S^n$ be the collapsing map which sends S^{n-1} to the base point of $S^n \vee S^n$ and wraps the upper and lower hemisphere of S^n around the first and second factor of $S^n \vee S^n$, respectively. Define $[f]_* \cdot [g]_*$ to be the class $[(f,g) \circ c]_*$. This operation is well defined and makes $[S^n,Y]_*$ into a group, which is abelian if $n \geq 2$ but possibly non-abelian if $n = 1$. If we think of f and g as closed paths in Y in the case $n = 1$, then $[f]_* \cdot [g]_*$ is the homotopy class of the closed path where one first walks along f and then along g.

DEFINITION. The n-th homotopy group of Y, $\pi_n(Y,y_0)$, is defined to be the set $[S^n,Y]_*$ together with the operation just defined. The group $\pi_1(Y,y_0)$ is called the fundamental group of Y. Given a continuous base point preserving map $\varphi : Y \to Z$, define the map

$$\varphi_{*n} : \pi_n(Y,y_0) \to \pi_n(Z,z_0)$$

by $\varphi_{*n}([f]_*) = [\varphi \circ f]_*$.

It is easy to check that φ_{*n} is well defined [compare with Problem 5.7(a)] and a homomorphism. Elements of homotopy groups really correspond much better to "holes" in spaces than do elements of homology groups (compare with the comments at the end of Sec. 4.3). As we mentioned earlier, the fundamental group was defined by Poincaré [Poin 4] in 1895. Although the higher homotopy groups were first defined by Čech [Ce 1] in 1932, it was Hurewicz who first used them and who gave homotopy theory its start. Homotopy groups satisfy many of the same properties as homology groups and can also be used for applications in much the same way. In fact, they are stronger invariants but are harder to compute in general. Other early workers in homotopy theory were Hopf, H. Freudenthal (1905-) (known for the Freudenthal suspension theorem in [Freu]), Pontrjagin, and Whitney.

Some recommended texts for learning more about homology and cohomology theory are [Cai], [Fra 2], [Gre], [Hu 2], [Mau], and [May 2]. Cairns' book [Cai] is a particularly good place to read about Poincaré duality because it is written in the same spirit as this book is. For homotopy theory see [Hil] or [Hu 1]. Massey's book [Mas] provides an excellent introduction to the fundamental group of a space and some of its applications.

Appendix A

THE TOPOLOGY OF $\mathbb{R}^n$

The symbol "$\emptyset$" will always denote the empty set. Let X_i, $1 \leq i \leq n$, be sets. We shall say that the X_i are disjoint if $X_i \cap X_j = \emptyset$ whenever $i \neq j$. The Cartesian product $X_1 \times X_2 \times \cdots \times X_n$ of the X_i is defined to be the set of n-tuples $(x_1, x_2, \cdots, x_n)$, where $x_i \in X_i$. Next, let $f : X \to Y$ be a map between two sets X and Y. In keeping with the modern trend we prefer to use the adjectives "injective," "surjective," or "bijective" to describe f when it is one-to-one [$f(x_1) = f(x_2)$ implies $x_1 = x_2$], onto [$y \in Y$ implies that there is an $x \in X$ such that $f(x) = y$], or both, respectively. If f is bijective, then the inverse of f, $f^{-1} : Y \to X$, is defined by $f^{-1}(y) = x$, where $f(x) = y$. We shall say that f is a constant map if there is a $y_0 \in Y$ and $f(x) = y_0$ for all $x \in X$. An extension of f to a set Z, where $X \subset Z$, is a map $F : Z \to Y$ with the property that $F(x) = f(x)$ for all $x \in X$. In this case, f is called the restriction of F to X and is also denoted by $F|X$. The symbol 1_X refers to the identity map of the set X, that is, $1_X(x) = x$ for all $x \in X$. Finally, for notational simplicity we shall usually identify a set containing only a single element with that element.

Throughout this book the symbols $\mathbb{Z}$, $\mathbb{Q}$, $\mathbb{R}$, and $\mathbb{C}$ denote the integers, the rational numbers, the real numbers, and the complex numbers, respectively. Recall that if $z = x + y\sqrt{-1} \in \mathbb{C}$, where x, $y \in \mathbb{R}$, then $x - y\sqrt{-1}$ is called the complex conjugate of z and is denoted by $\bar{z}$. The n-dimensional Euclidean space $\mathbb{R}^n$ is the n-fold Cartesian product of $\mathbb{R}$, that is,

$$\mathbb{R}^n = \{ x = (x_1, x_2, \cdots, x_n) \mid x_i \in \mathbb{R} \text{ for } 1 \leq i \leq n \}$$

As usual, we shall identify the plane $\mathbb{R}^2$ with $\mathbb{C}$ by identifying $(x,y) \in \mathbb{R}^2$ with the complex number $x + y\sqrt{-1}$. Also, by identifying $(x_1, x_2, \cdots, x_n) \in \mathbb{R}^n$ with $(x_1, x_2, \cdots, x_n, 0) \in \mathbb{R}^{n+1}$ we obtain natural inclusions

$$0 = \mathbb{R}^0 \subset \mathbb{R} = \mathbb{R}^1 \subset \mathbb{C} = \mathbb{R}^2 \subset \cdots \subset \mathbb{R}^k \subset \cdots$$

Let $x, y \in \mathbb{R}^n$ and $c \in \mathbb{R}$. The operations

$$x + y = (x_1 + y_1, x_2 + y_2, \cdots, x_n + y_n)$$

and

$$cx = (cx_1, cx_2, \cdots, cx_n)$$

make $\mathbb{R}^n$ into a vector space over $\mathbb{R}$ and it is often convenient to treat points of $\mathbb{R}^n$ as vectors that can be added, subtracted, and multiplied by scalars. Next, define the dot product $\bullet$ between vectors in $\mathbb{R}^n$ by

$$x \bullet y = \sum_{i=1}^{n} x_i y_i$$

and call

$$|x| = \sqrt{x \bullet x}$$

the length of the vector x. Note that $|x - y|$ is the usual Euclidean distance between x and y. The set

$$B_r^n(x) = \{ y \in \mathbb{R}^n : |x - y| < r \}$$

is called the open ball of radius r around x in $\mathbb{R}^n$. If $z \in \mathbb{R}^2$, then let $\arg z \in (-\pi, \pi]$ denote the angle which z makes with the x axis. Equivalently, $\arg z$ is uniquely specified by the equation $z = |z|e^{i(\arg z)}$, where $i = \sqrt{-1}$ and where we are considering z as a complex number.

Let $A \subset \mathbb{R}^n$. We say that A is an open subset of $\mathbb{R}^n$ if for every $x \in A$ there is some $\epsilon > 0$ such that $B^n_\epsilon(x) \subset A$. The subset A is said to be closed if $\mathbb{R}^n - A$ is open. It is useful also to relativize these notions. If $B \subset A$, then we say that B is open in A if there is an open set U in $\mathbb{R}^n$ such that $B = U \cap A$ and B is closed in A if $A - B$ is open in A. Let $a \in A$. A subset N of A is called a neighborhood of a in A if there is an open set V in A such that $a \in V \subset N$.

LEMMA 1. (a) Arbitrary unions and finite intersections of open sets are open.

(b) Arbitrary intersections and finite unions of closed sets are closed.

Proof: Easy. (b) follows from (a) because of the identity

$$\mathbb{R}^n - \bigcap_{\alpha \in J} C_\alpha = \bigcap_{\alpha \in J} (\mathbb{R}^n - C_\alpha)$$

which holds for any collection of sets $\{C_\alpha\}_{\alpha \in J}$.

Next, define a point $z \in \mathbb{R}^n$ to be an accumulation point of a subset $A \subset \mathbb{R}^n$ if $[B^n_r(z) - z] \cap A \neq \emptyset$ for all $r > 0$. Define the closure of A, $c\ell(A)$, by

$$c\ell(A) = \bigcap \{ C \mid C \text{ is a closed subset of } \mathbb{R}^n \text{ and } A \subset C \}$$

Lemma 1(b) shows that $c\ell(A)$ is closed. Clearly, $c\ell(A)$ is the smallest closed set which contains A.

LEMMA 2. We have that

$$c\ell(A) = A \cup \{ z \mid z \text{ is an accumulation point of } A \}$$

Furthermore, the set A is closed if and only if $A = c\ell(A)$.

Proof: Easy.

As usual, given $X \subset \mathbb{R}$, let "ℓub" and "gℓb" stand for "least upper bound of X" and "greatest lower bound of X," respectively. If X is finite, then we prefer to write $\max X$ or $\min X$ for

$\ell ub\ X$ or $g\ell b\ X$, respectively. Define the diameter, diam A, of a nonempty subset $A \subset \mathbb{R}^n$ by

$$\text{diam } A = \ell ub \{ |x - y| : x,y \in A \}$$

if this least upper bound exists; otherwise, set $\text{diam } A = \infty$. If $z \in \mathbb{R}^n$, then let

$$\text{dist}(z,A) = g\ell b \{ |z - x| : x \in A \}$$

denote the distance from z to A. It is convenient to define $\text{diam } \emptyset = \text{dist}(z,\emptyset) = 0$.

LEMMA 3. Let A be a closed subset of $\mathbb{R}^n$. If $z \in \mathbb{R}^n$, then there is an $x_0 \in A$ such that $\text{dist}(z,A) = |z - x_0|$. Furthermore, $\text{dist}(z, A) = 0$ if and only if $z \in A$.

Proof: Easy.

Let $A \subset \mathbb{R}^n$. An open cover of A is a collection $\{V_\alpha\}_{\alpha \in J}$ of subsets of A which are open in A and such that $A \subset \bigcup_{\alpha \in J} V_\alpha$, where J is some indexing set. We say that A is compact if every open cover $\mathcal{O}$ of A has a finite subcover, that is, there is a finite number of sets $V_1, V_2, \cdots, V_k \in \mathcal{O}$ such that $A \subset \bigcup_{i=1}^{k} V_i$. The set A is said to be bounded if $A \subset B_r^n(0)$ for some $r > 0$.

THEOREM 1 (Heine-Borel-Lebesgue). A subset $A \subset \mathbb{R}^n$ is compact if and only if A is closed and bounded.

Proof: See [Sp] or [Eis].

Let $A \subset \mathbb{R}^n$ and suppose that $f : A \to \mathbb{R}^m$ is an arbitrary function. Then f determines functions $f_i : A \to \mathbb{R}$, $1 \leq i \leq m$, such that $f(x) = (f_1(x), f_2(x), \cdots, f_m(x))$ for all $x \in A$. The function f_i is called the i-th component (function) of f. We shall say that f has a limit $b \in \mathbb{R}^m$ at $a \in \mathbb{R}^n$ and write $\lim_{x \to a} f(x) = b$ provided that for every $\varepsilon > 0$ there is a $\delta > 0$ such that $|f(x) - b| < \varepsilon$ whenever $x \in A$ and $0 < |x - a| < \delta$.

LEMMA 4. If a is an accumulation point of A, then $\lim_{x \to a} f(x)$, if it exists, is unique.

Proof: Easy.

Now let $a \in A \subset \mathbb{R}^n$ and define a function $f : A \to \mathbb{R}^m$ to be continuous at a if $\lim_{x \to a} f(x) = f(a)$. The function f is said to be continuous (on A) if it is continuous at every point $a \in A$. Because all the functions considered in this book will be continuous we may sometimes omit the adjective "continuous."

THEOREM 2. Let $A \subset \mathbb{R}^n$. A function $f : A \to \mathbb{R}^m$ is continuous if and only if $f^{-1}(V)$ is an open subset of A for every open set V in $\mathbb{R}^m$.

Proof: This theorem merely rephrases the $(\epsilon - \delta)$-definition of continuity from analysis in more topological terms and its proof is straightforward. See [Sp].

THEOREM 3. Let A be a compact subset of $\mathbb{R}^n$.

(a) If $f : A \to \mathbb{R}^m$ is continuous, then $f(A)$ is compact.

(b) If $f : A \to \mathbb{R}$ is continuous, then f assumes its maximum and minimum on A, that is, there are points $a_1, a_2 \in A$ such that $f(a_1) \leq f(x) \leq f(a_2)$ for all $x \in A$.

Proof: See [Sp].

THEOREM 4. Let A be a compact subset of $\mathbb{R}^n$ and let $\mathcal{O}$ be an open cover of A. Then there is a $\delta > 0$ with the property that for each $x \in A$, $[B^n_\delta(x) \cap A] \subset V_x$ for some $V_x \in \mathcal{O}$. Such a δ is called a Lebesgue number for $\mathcal{O}$.

Proof: Define a function f on A by

$$f(x) = \ell ub \{ \operatorname{dist}(x, A - V) \mid V \in \mathcal{O} \}$$

Note that by Theorem 1 f never assumes the value ∞. In fact, f is a function from A to the positive real numbers. It is also not hard to show that f is continuous. Therefore, by Theorem 3(b) there is some point $a \in A$ such that $0 < f(a) \leq f(x)$ for all $x \in A$. Any δ satisfying $0 < \delta \leq f(a)$ will be a Lebesgue number for $\mathcal{O}$.

Appendix B

PERMUTATIONS AND ABELIAN GROUPS

We begin with a discussion of permutations.

DEFINITION. A permutation of a set X is a bijective function $\sigma : X \to X$. The set of permutations on X will be denoted by $S(X)$.

Given two permutations $\sigma, \tau \in S(X)$, their composition $\sigma \circ \tau$ belongs to $S(X)$. This operation $\circ$ actually makes $S(X)$ into a (noncommutative) group, but we shall not go into this aspect here. Instead, we shall collect a few basic facts about a particular set of permutations, namely, $S_n = S(\{1,2,\cdots,n\})$. (The set S_n is usually called the symmetric group of degree n.) Of course, whatever we say about S_n would hold equally well for the set of permutations on any other finite set of n elements.

DEFINITION. A transposition is a permutation $\tau \in S_n$ with the property that it interchanges two numbers and leaves all others fixed, that is, there are $i,j \in \{1,2,\cdots,n\}$ with $i \neq j$ such that $\tau(i) = j$, $\tau(j) = i$, and $\tau(k) = k$ for $k \neq i$ or j.

LEMMA 1. Every permutation in S_n, $n \geq 2$, can be written as a product of transpositions, that is, if $\sigma \in S_n$, then $\sigma = \tau_1 \circ \tau_2 \circ \cdots \circ \tau_k$, where $\tau_i \in S_n$ and τ_i is a transposition.

Proof: We may assume that $\sigma(n) = i$ with $i \neq n$, because if σ is the identity map on $\{1,2,\cdots,n\}$, then $\sigma = \tau \circ \tau$ for any transposition $\tau \in S_n$. Let τ_1 be the transposition in S_n defined by $\tau_1(n) = i$ and $\tau_1(i) = n$. Then $(\tau_1 \circ \sigma)(n) = \tau_1(\sigma(n)) = \tau_1(i) = n$, so that we may consider $\tau_1 \circ \sigma$ as a permutation in S_{n-1}. (There are natural inclusions $S_1 \subset S_2 \subset \cdots$.) Using induction on n, there are transpositions $\tau_2,\tau_3,\cdots,\tau_k \in S_{n-1} \subset S_n$ such that $\tau_1 \circ \sigma = \tau_2 \circ \tau_3 \circ \cdots \circ \tau_k$. Since $\sigma = \tau_1 \circ \tau_1 \circ \sigma = \tau_1 \circ \tau_2 \circ \tau_3 \circ \cdots \circ \tau_k$, the lemma is proved.

The way in which a permutation can be written as a product of transpositions is not unique, but we have

THEOREM 1. Let $\sigma \in S_n$ and suppose that $\sigma = \tau_1 \circ \tau_2 \circ \cdots \circ \tau_k = \eta_1 \circ \eta_2 \circ \cdots \circ \eta_t$, where τ_i and η_j are transpositions in S_n. Then k and t are either both even or both odd.

Proof: For details see [Co]. If $f : \mathbb{R}^n \to \mathbb{R}$ and $\sigma \in S_n$, then define $f_\sigma : \mathbb{R}^n \to \mathbb{R}$ by

$$f_\sigma(x_1,x_2,\cdots,x_n) = f(x_{\sigma(1)},x_{\sigma(2)},\cdots,x_{\sigma(n)})$$

Now consider the function $\Delta : \mathbb{R}^n \to \mathbb{R}$ given by

$$\Delta(x_1,x_2,\cdots,x_n) = \prod_{i<j} (x_j - x_i)$$

One can show that Δ satisfies the following two properties:

(i) If $\sigma,\tau \in S_n$, then $\Delta_{\sigma\circ\tau} = (\Delta_\tau)_\sigma$.

(ii) For each transposition $\tau \in S_n$, $\Delta_\tau = -\Delta$.

Using (i) and (ii) we see that for the σ in the theorem

$$\begin{aligned}\Delta_\sigma = \Delta_{\tau_1\circ\tau_2\circ\cdots\circ\tau_k} &= (-1)^k\Delta \\ = \Delta_{\eta_1\circ\eta_2\circ\cdots\circ\eta_t} &= (-1)^t\Delta\end{aligned}$$

In other words, $(-1)^k = (-1)^t$ and Theorem 1 is proved.

Theorem 1 shows that the next definition is well defined.

DEFINITION. A permutation $\sigma \in S_n$ is said to be even if it can be written as a product of an even number of transpositions. Otherwise, σ is said to be odd.

Clearly, the product of two even permutations is even. Also, $\sigma \in S_n$ is even if and only if σ^{-1} is even. Therefore, if we define $\sigma \sim \tau$ whenever $\sigma \circ \tau^{-1}$ is even, then $\sim$ is an equivalence relation in S_n and S_n gets partitioned into two equivalence classes by $\sim$, namely, the even and the odd permutations. (For a definition of equivalence relations and classes see [Co]. In general, equivalence relations enable one to divide up a set into disjoint sets which are the equivalence classes.)

Next, let G be a set and let $+$ be a binary operation on G, that is, $+$ is a map $+ : G \times G \to G$. As usual in this case, if g_1, $g_2 \in G$, then we shall write $g_1 + g_2$ instead of the functional form $+(g_1,g_2)$.

DEFINITION. The pair $(G,+)$ is called an abelian (or commutative) group provided that $+$ satisfies:

(1) (Associativity) For every $g_1,g_2,g_3 \in G$, $g_1 + (g_2 + g_3) = (g_1 + g_2) + g_3$.

(2) (Identity) There is an element $0 \in G$, called the identity of G, such that $0 + g = g + 0 = g$ for all $g \in G$.

(3) (Inverse) For every $g \in G$ there is an element $-g \in G$, called the inverse of g, such that $g + (-g) = (-g) + g = 0$.

(4) (Commutativity) For every $g_1,g_2 \in G$, $g_1 + g_2 = g_2 + g_1$.

NOTATION. For simplicity, we shall usually omit the adjective "abelian" and from now on in this section and throughout this book the word "group" will always mean "abelian group." Furthermore, when dealing with a group $(G,+)$ for which the operation $+$ is obvious from the context, we shall simply refer to "the group G". The trivial groups consisting of only one element will be denoted by 0.

One can show that the identity and the inverse of each element in a group G is unique. Also, $-(-g) = g$ for all $g \in G$.

EXAMPLES

(1) The standard examples of groups are $\mathbb{Z}$, $\mathbb{Q}$, $\mathbb{R}$, and $\mathbb{C}$, with respect to addition. The sets $\mathbb{Q} - 0$, $\mathbb{R} - 0$, and $\mathbb{C} - 0$ become groups under multiplication. Vector addition makes $\mathbb{R}^n$ and $\mathbb{Z}^n = \{ (a_1, a_2, \cdots, a_n) \mid a_i \in \mathbb{Z} \}$ into groups.

(2) For each positive integer n let $\mathbb{Z}_n = \{0, 1, 2, \cdots, n - 1\}$ and define an operation $+_n$ on $\mathbb{Z}_n$ as follows: If $a, b \in \mathbb{Z}_n$ and $a + b = xn + d$, where $x, d \in \mathbb{Z}$ and $0 \leq d < n$, then we set $a +_n b = d$. It is easy to check that $(\mathbb{Z}_n, +_n)$ [or $\mathbb{Z}_n$ for short] is a group called the integers mod n.

DEFINITION. Let $(G,+)$ and $(H,+')$ be groups. We say that $(H,+')$ is a subgroup of $(G,+)$ provided that $H \subset G$ and $h_1 +' h_2 = h_1 + h_2$ for all $h_1, h_2 \in H$, that is, $+' = + | (H \times H)$.

EXAMPLES

(1) Two trivial subgroups of any group are 0 and the whole group itself.

(2) For each $k \in \mathbb{Z}$, $\{ kn \mid n \in \mathbb{Z} \}$ is a subgroup of $\mathbb{Z}$ (with respect to addition) and all subgroups of $\mathbb{Z}$ are of that form.

(3) $\mathbb{Z}^n$ is a subgroup of $\mathbb{R}^n$.

(4) $\{0,3,6\}$ forms a subgroup of $\mathbb{Z}_9$.

(5) $\{ 2n \mid n \in \mathbb{Z} \} \cup \{3\}$ is not a subgroup of $\mathbb{Z}$.

LEMMA 2. (a) A subset H of a group G is a subgroup (under the operation induced from that of G) if and only if $h_1 - h_2 \in H$ for all $h_1, h_2 \in H$.

(b) The intersection of an arbitrary number of subgroups is a subgroup.

Proof: Straightforward.

DEFINITION. Let G and H be groups. A map $f : G \to H$ is called a homomorphism if $f(g_1 + g_2) = f(g_1) + f(g_2)$ for all $g_1, g_2 \in G$. The homomorphism is said to be an isomorphism if it is a bijective map. In that case we say that G is isomorphic to H and write $G \approx H$.

It is easy to see that $f(0) = 0$ and $f(-g) = -f(g)$ for a homomorphism f. If f is an isomorphism then so is its inverse f^{-1}.

EXAMPLES

(1) Inclusion maps such as $\mathbb{Z} \subset \mathbb{Q} \subset \mathbb{R}$ are clearly homomorphisms.

(2) Define $\pi_n : \mathbb{Z} \to \mathbb{Z}_n$ as follows: If $k \in \mathbb{Z}$ and $k = an + b$, where $a, b \in \mathbb{Z}$ and $0 \leq b < n$, then $\pi_n(k) = b$. The map π_n is a homomorphism.

Next, let $f : G \to H$ be a homomorphism of groups.

DEFINITION. The kernel of f, $\operatorname{Ker} f$, and the image of f, $\operatorname{Im} f$, are defined by

$$\operatorname{Ker} f = \{ g \in G \mid f(g) = 0 \}$$

and

$$\operatorname{Im} f = \{ f(g) \mid g \in G \}$$

LEMMA 3. (a) $\operatorname{Ker} f$ and $\operatorname{Im} f$ are subgroups of G and H, respectively.

(b) The homomorphism f is injective if and only if $\operatorname{Ker} f = 0$.

Proof: Easy.

Now let H be a subgroup of a group G.

DEFINITION. Let $g \in G$. The set

$$g + H = \{ g + h \mid h \in H \}$$

is called the coset of H in G generated by g.

It can be shown that two cosets of H in G are either identical or they are totally disjoint. In fact, if we define a relation $\sim$ in G by $g_1 \sim g_2$ if $g_1 - g_2 \in H$, then $\sim$ is an equivalence relation and the cosets of H in G are nothing but the equivalence classes of $\sim$.

EXAMPLES

(1) The cosets of $\{0,3,6\}$ in $\mathbb{Z}_9$ are $\{0,3,6\}$, $\{1,4,7\}$, and $\{2,5,8\}$.

(2) The cosets of $\mathbb{Z} = \mathbb{Z}^1$ in $\mathbb{Z}^2$ are the sets $\{ (n,k) \mid n \in \mathbb{Z} \}$ for $k \in \mathbb{Z}$.

DEFINITION. The factor group (or quotient group) of a group G by a subgroup H, is defined to be the pair $(G/H,+')$, where

$$G/H = \{ g + H \mid g \in G \}$$

and

$$(g_1 + H) +' (g_2 + H) = (g_1 + g_2) + H \quad \text{for all} \quad g_1, g_2 \in G.$$

It is straightforward to check that $(G/H,+')$ is a well-defined group usually denoted simply by G/H. If $H = 0$, one always identifies G/H with G in the natural way.

EXAMPLES

(1) If $G = \mathbb{Z}_9$ and $H = \{0,3.6\}$, then $G/H \approx \mathbb{Z}_3$.

(2) If $G = \mathbb{Z}^2$ and $H = \mathbb{Z}$, then $G/H \approx \mathbb{Z}$.

(3) If $G = \mathbb{Z}$ and $H = \{ 2n \mid n \in \mathbb{Z} \}$, then $G/H \approx \mathbb{Z}_2$.

LEMMA 4. Let G and H be groups. If $f : G \to H$ is a surjective homomorphism, then $H \approx G/(\text{Ker } f)$.

Proof: An isomorphism between H and $G/(\text{Ker } f)$ is given by the map which sends $g + (\text{Ker } f) \in G/(\text{Ker } f)$ onto $f(g) \in H$.

DEFINITION. Let G be a group and let $g_1, g_2, \cdots, g_k \in G$. Then

$$\mathbb{Z}g_1 + \mathbb{Z}g_2 + \cdots + \mathbb{Z}g_k = \{ n_1 g_1 + n_2 g_2 + \cdots + n_k g_k \mid n_i \in \mathbb{Z} \}$$

is called the subgroup of G generated by the $g_1, g_2, \cdots, g_k$.

It is easy to show that $\mathbb{Z}g_1 + \mathbb{Z}g_2 + \cdots + \mathbb{Z}g_k$ is in fact a subgroup of G and also that it is the intersection of all subgroups of G which contain the elements $g_1, g_2, \cdots, g_k$.

DEFINITION. A group G is said to be cyclic if $G = \mathbb{Z}g$ for some $g \in G$.

The standard examples of cyclic groups are $\mathbb{Z} = \mathbb{Z}1$ and $\mathbb{Z}_n = \mathbb{Z}1 = \mathbb{Z}k$, where $k \in \mathbb{Z}_n$ and k is relatively prime to n. It follows from the next lemma that there are no others.

LEMMA 5. A cyclic group G is isomorphic to either $\mathbb{Z}$ or $\mathbb{Z}_n$ for some $n \in \mathbb{Z}$.

Proof: Let $G = \mathbb{Z}g$ and define a homomorphism $\varphi : \mathbb{Z} \to G$ by $\varphi(k) = kg$. If $\text{Ker } \varphi = 0$, then $G \approx \mathbb{Z}$; otherwise, $\text{Ker } \varphi = \{ kn \mid k \in \mathbb{Z} \}$ for some $n \in \mathbb{Z}$ and $G \approx \mathbb{Z}_n$.

DEFINITION. Let G be a group. If $k \in \mathbb{Z}$ and $g \in G$, let

$$kg = \underbrace{g + g + \cdots + g}_{k \text{ times}}$$

Define the order of g, $o(g)$, to be the smallest of the integers $k > 0$ such that $kg = 0$ if such integers exist; otherwise, define the order of g to be ∞.

EXAMPLE: Let $G = \mathbb{Z}_6 = \{0,1,2,3,4,5\}$. Then $o(1) = 6 = o(5)$, $o(2) = 3 = o(4)$, $o(3) = 2$, and $o(0) = 1$.

LEMMA 6. For any group G the elements of finite order form a unique subgroup $T(G)$ called the torsion subgroup of G.

Proof: The proof is clear since $o(0) = 1$, $o(-g) = o(g)$, and $o(g + h) \mid o(g)o(h)$ for all $g,h \in G$.

DEFINITION. A group G is said to be torsion-free if it has no element of finite order other than 0, that is, if $T(G) = 0$.

Clearly, $G/T(G)$ is a torsion-free group for every group G.

DEFINITION. The number of elements in a group G, denoted by $o(G)$, is called the order of G. If G has only a finite number of elements, then G is called a group of finite order, or simply a finite group.

NOTE. Two finite groups of the same order need not be isomorphic.

THEOREM 2 (Lagrange). Let G be a finite group.

(a) If H is a subgroup of G, then $o(H) \mid o(G)$. In fact, $o(G) = o(H)o(G/H)$.

(b) If $g \in G$, then $o(g) \mid o(G)$.

Proof: (a) follows from the fact that G is a union of cosets of H, each of which has the same number of elements. To prove (b) one only has to observe that $o(g) = o(\mathbb{Z}g)$ and apply (a).

DEFINITION. A group G is said to be finitely generated if $G = \mathbb{Z}g_1 + \mathbb{Z}g_2 + \cdots + \mathbb{Z}g_n$ for some $g_1,g_2,\cdots,g_n \in G$. In that case, the g_i are called generators for G.

EXAMPLES

(1) All cyclic groups are finitely generated.

(2) The group $\mathbb{Z}^2$ is not cyclic, but it is finitely generated since $\mathbb{Z}^2 = \mathbb{Z}(1,0) + \mathbb{Z}(0,1)$. A similar statement holds for $\mathbb{Z}^n$, $n \geq 2$.

(3) The groups $\mathbb{Q}$ and $\mathbb{R}$ are not finitely generated.

LEMMA 7. Subgroups of finitely generated groups are finitely generated.

Proof: See [Co].

DEFINITION. Let G_1, G_2, $\cdots$, and G_n be groups. We define the abelian group $G_1 \hat{\oplus} G_2 \hat{\oplus} \cdots \hat{\oplus} G_n$, called the external direct sum of the G_i, to be the group $(G_1 \times G_2 \times \cdots \times G_n, +)$, where the operation $+$ is given by

$$(g_1, g_2, \cdots, g_n) + (g_1', g_2', \cdots, g_n') = (g_1 + g_1', g_2 + g_2', \cdots, g_n + g_n')$$

EXAMPLE: The groups $\mathbb{Z}^n$ and $\mathbb{R}^n$ are the external direct sums of n copies of $\mathbb{Z}$ and $\mathbb{R}$, respectively.

DEFINITION. If G_1, G_2, $\cdots$, and G_n are subgroups of a given group G and if each element $g \in G$ can be written uniquely in the form $g_1 + g_2 + \cdots + g_n$ with $g_i \in G_i$, then G is called the internal direct sum of the G_i and we write $G = G_1 \oplus G_2 \oplus \cdots \oplus G_n$.

It is easy to show that if $G = G_1 \oplus G_2 \oplus \cdots \oplus G_n$, then G is isomorphic to $G_1 \hat{\oplus} G_2 \hat{\oplus} \cdots \hat{\oplus} G_n$. For this reason one usually drops the adjectives "external" and "internal" and uses the same symbol $\oplus$ to denote either direct sum. The context will decide which is being used.

THEOREM 3 (The Fundamental Theorem of Finitely Generated Abelian Groups). Let G be a finitely generated abelian group. Then

$$G \approx \underbrace{\mathbb{Z} \oplus \mathbb{Z} \oplus \cdots \oplus \mathbb{Z}}_{r} \oplus \mathbb{Z}_{n_1} \oplus \mathbb{Z}_{n_2} \oplus \cdots \oplus \mathbb{Z}_{n_t}$$

where $1 < n_i \in \mathbb{Z}$ and $n_i \mid n_{i+1}$. The integer r is called the rank of G and is denoted by rk G. The integers $n_1, n_2, \cdots$, and n_t are called the torsion coefficients of G. Both the rank and the torsion coefficients are uniquely determined by G.

Proof: See [Co].

NOTE. The example $\mathbb{Z}_2 \oplus \mathbb{Z}_3 \approx \mathbb{Z}_6$ shows that the condition $n_i \mid n_{i+1}$ is important for the uniqueness part of Theorem 3.

THEOREM 4. Suppose that G, H, and K are finitely generated groups.

(a) If $\varphi : K \to G$ and $\psi : G \to H$ are homomorphisms satisfying

(1) φ is injective,

(2) $\text{Im } \varphi = \text{Ker } \psi$, and

(3) ψ is surjective,

then

$$\text{rk } G = \text{rk } K + \text{rk } H$$

(b) $\text{rk } (H \oplus K) = \text{rk } H + \text{rk } K$, that is, the function rk is "additive."

Proof: We first introduce some more notation.

DEFINITION. The subset $\{g_1, g_2, \cdots, g_n\}$ of a group G is said to form a basis for G provided that

(a) $G = \mathbb{Z}g_1 + \mathbb{Z}g_2 + \cdots + \mathbb{Z}g_n$, and

(b) if $k_1 g_1 + k_2 g_2 + \cdots + k_n g_n = 0$ for $k_i \in \mathbb{Z}$, then $k_i g_i = 0$.

It is easy to show that $\{g_1, g_2, \cdots, g_n\}$ is a basis for the group G if and only if $G = \mathbb{Z}g_1 \oplus \mathbb{Z}g_2 \oplus \cdots \oplus \mathbb{Z}g_n$. Thus by Theorem 3 each finitely generated group has a basis.

DEFINITION. Any group which is isomorphic to $G_1 \oplus G_2 \oplus \cdots \oplus G_n$, where $G_i \approx \mathbb{Z}$, is called a free group.

It follows easily from Theorem 3 and the definitions that the following conditions on a finitely generated group are equivalent:

(1) G is free.

(2) G is torsion-free.

(3) G has a basis consisting of elements of infinite order.

To prove part (a) of Theorem 4 we may assume without loss of generality that all three groups, G, H, and K, are torsion-free. Let $\{h_1,h_2,\cdots,h_s\}$ and $\{k_1,k_2,\cdots,k_t\}$ be bases for H and K, respectively, where $s = \text{rk } H$ and $t = \text{rk } K$. Choose elements $g_1, g_2,\cdots,g_s \in G$ such that $\psi(g_i) = h_i$. It is now easy to show that $\{\varphi(k_1),\varphi(k_2),\cdots,\varphi(k_t),g_1,g_2,\cdots,g_s\}$ forms a basis for G. In other words, $\text{rk } G = t + s = \text{rk } K + \text{rk } H$. Part (b) follows from (a) by letting $G = H \oplus K$ and letting φ and ψ be the natural inclusion and projection, respectively, and Theorem 4 is proved.

THEOREM 5. Let G be a free group with basis $\{g_1,g_2,\cdots,g_n\}$. If H is any group and $h_1,h_2,\cdots,h_n \in H$, then there exists a unique homomorphism $\varphi : G \to H$ such that $\varphi(g_i) = h_i$.

Proof: One shows that any element g of G has a unique representation of the form $k_1g_1 + k_2g_2 + \cdots + k_ng_n$, $k_i \in \mathbb{Z}$, and then defines $\varphi(g) = k_1h_1 + k_2h_2 + \cdots + k_nh_n$.

Theorem 5 can be paraphrased by saying that given a free group G, and any other group H, in order to define a homomorphism $\varphi : G \to H$ it suffices to define it on a basis of G, because any map from a basis of G to H extends uniquely to a homomorphism φ.

EXAMPLE: Consider $G = \mathbb{Z}_3 = \{0,1,2\}$ and define $\psi(1) = 1 \in \mathbb{Z}$. The map ψ does not extend to a homomorphism $\varphi : \mathbb{Z}_3 \to \mathbb{Z}$. Of course, although 1 is a basis for $\mathbb{Z}_3$, the group $\mathbb{Z}_3$ is not a free group.

Next, let G and H be groups.

DEFINITION. Let

$$\text{Hom}(G,H) = \{ h \mid h : G \to H \text{ is a homomorphism} \}$$

and if $h_i \in \text{Hom}(G,H)$, then define $h_1 + h_2 : G \to H$ by $(h_1 + h_2)(g) = h_1(g) + h_2(g)$ for $g \in G$.

LEMMA 8. $(\text{Hom}(G,H),+)$ is an abelian group.

Proof: Straightforward.

LEMMA 9. Let G be any group which is isomorphic to $\mathbb{Z}$. If $h \in \mathrm{Hom}(G,G)$, then there is a unique integer k such that $h(g) = kg$ for all $g \in G$.

Proof: Since $G \approx \mathbb{Z}$, there is some $g_0 \in G$ such that $G = \mathbb{Z}g_0$. It follows that $h(g_0) = kg_0$ for some $k \in \mathbb{Z}$, because $\{g_0\}$ is a basis for G. The fact that $o(g_0) = \infty$ implies that the integer k is unique. Now let $g \in G$. Again, there is some $t \in \mathbb{Z}$ with $g = tg_0$. Thus, $h(g) = h(tg_0) = th(g_0) = t(kg_0) = k(tg_0) = kg$, and the lemma is proved.

Note that an easy consequence of Lemma 9 is that $\mathrm{Hom}(G,G) \approx \mathbb{Z}$ if $G \approx \mathbb{Z}$. In fact, $\mathrm{Hom}(G,G) = \mathbb{Z}1_G$ in that case.

Appendix C

THE INCIDENCE MATRICES

In this appendix we describe briefly the so-called incidence matrices which are associated to every simplicial complex and which played an historically important role. An excellent detailed account can be found in [Cai].

Let K be a simplicial complex of dimension n and let n_q denote the number of q-simplices in K. For each q, let $S_q^+ = \{[\sigma_1^q], [\sigma_2^q], \cdots, [\sigma_{n_q}^q]\}$ be a collection of oriented q-simplices of K as in Sec. 4.3. We have seen that S_q^+ forms a basis for the free abelian group $C_q(K)$. Define integers ϵ_{ij}^q by the equation

$$\partial_{q+1}([\sigma_j^{q+1}]) = \sum_{i=1}^{n_q} \epsilon_{ij}^q [\sigma_i^q]$$

Observe that $\epsilon_{ij}^q = \pm 1$ if σ_i^q is a face of σ_j^{q+1} and that $\epsilon_{ij}^q = 0$ otherwise.

DEFINITION. The integer ϵ_{ij}^q is called the incidence number of $[\sigma_i^q]$ and $[\sigma_j^{q+1}]$. The q-th incidence matrix E^q, $0 \leq q < \dim K$, of K is defined to be the $(n_q \times n_{q+1})$-matrix

$$E^q = (\epsilon^q_{ij}) = \begin{array}{c|cccc} & [\sigma^{q+1}_1] & [\sigma^{q+1}_2] & \cdots & [\sigma^{q+1}_{n_{q+1}}] \\ \hline [\sigma^q_1] & \epsilon^q_{11} & \epsilon^q_{12} & \cdots & \epsilon^q_{1n_{q+1}} \\ [\sigma^q_2] & \epsilon^q_{21} & \epsilon^q_{22} & \cdots & \epsilon^q_{2n_{q+1}} \\ \vdots & \vdots & \vdots & \ddots & \vdots \\ [\sigma^q_{n_q}] & \epsilon^q_{n_q1} & \epsilon^q_{n_q2} & \cdots & \epsilon^q_{n_qn_{q+1}} \end{array}$$

whose rows and columns are indexed by the elements of S^+_q and S^+_{q+1}, respectively.

EXAMPLE. Suppose that $K = \overline{v_0v_1v_2}$ and that we have chosen the S^+_q as follows: $S^+_0 = \{v_0,v_1,v_2\}$, $S^+_1 = \{[v_0v_1],[v_1v_2],[v_0v_2]\}$, and $S^+_2 = \{[v_0v_1v_2]\}$. The incidence matrices of K are then given by

$$E^1 = \begin{array}{c|c} & [v_0v_1v_2] \\ \hline [v_0v_1] & +1 \\ [v_1v_2] & +1 \\ [v_0v_2] & -1 \end{array} \qquad \text{and} \qquad E^0 = \begin{array}{c|ccc} & [v_0v_1] & [v_1v_2] & [v_0v_2] \\ \hline v_0 & +1 & 0 & -1 \\ v_1 & -1 & -1 & 0 \\ v_2 & 0 & +1 & +1 \end{array}$$

Next, let us define incidence matrices with respect to arbitrary bases for the free abelian groups $C_q(K)$. Suppose that for each q, $\{c^q\} = \{c^q_1,c^q_2,\cdots,c^q_{n_q}\}$ is a basis for $C_q(K)$. It follows that each $\partial_{q+1}(c^{q+1}_j)$ can be expressed as a unique linear combination of the c^q_i, that is, there are unique integers η^q_{ij} such that

$$\partial_{q+1}(c^{q+1}_j) = \sum_{i=1}^{n_q} \eta^q_{ij}c^q_i$$

DEFINITION. The q-th incidence matrix of K with respect to the bases $\{c^{q+1}\}$ and $\{c^q\}$ is defined to be the $(n_q \times n_{q+1})$-matrix

$$(\eta^q_{ij}) = \begin{array}{c|ccc} & \cdots & c^{q+1}_j & \cdots \\ \hline \vdots & & \vdots & \\ c^q_i & \cdots & \eta^q_{ij} & \cdots \\ \vdots & & \vdots & \end{array}$$

Let $c \in C_{q+1}(K)$. Since $c = \sum_{j=1}^{n_{q+1}} a_j c_j^{q+1}$ for some unique integers a_j and

$$\partial_{q+1}(c) = \sum_{j=1}^{n_{q+1}} a_j \partial_{q+1}(c_j^{q+1}) = \sum_{i=1}^{n_q} \left(\sum_{j=1}^{n_{q+1}} a_j \eta^q_{ij} \right) c^q_i$$

it is clear that the boundary homomorphisms ∂_q are completely determined once the incidence matrices are known with respect to some bases. It should not be surprising therefore if the homology groups of K are computable from knowledge of the incidence matrices E^q of K alone. We shall now indicate how this can be done.

LEMMA 1. Choose a basis for each group $C_q(K)$. The matrix product of any two successive incidence matrices with respect to any such choice of bases is the zero matrix. In the notation above,

$$(\eta^{q-1}_{ij}) \cdot (\eta^q_{ij}) = 0_q \quad \text{for all } q$$

where 0_q is the $(n_{q-1} \times n_{q+1})$-matrix all of whose entries are zero.

Proof: Lemma 1 is really equivalent to Lemma 2 in Sec. 4.3 since we have the equations

$$(\partial_q \circ \partial_{q+1})(c_j^{q+1}) = \sum_{i=1}^{n_q} \eta^q_{ij} \partial_q (c^q_i)$$

$$= \sum_{i=1}^{n_q} \eta_{ij}^{q} \left(\sum_{t=1}^{n_{q-1}} \eta_{ti}^{q-1} c_t^{q-1} \right)$$

$$= \sum_{t=1}^{n_{q-1}} \left(\sum_{i=1}^{n_q} \eta_{ti}^{q-1} \eta_{ij}^{q} \right) c_t^{q-1}$$

Using Lemma 1 and some mostly algebraic manipulations which have little to do with topology we now prove

THEOREM 1. It is possible to choose bases for all the groups $C_q(K)$ simultaneously with respect to which the q-th incidence matrix has the normalized form

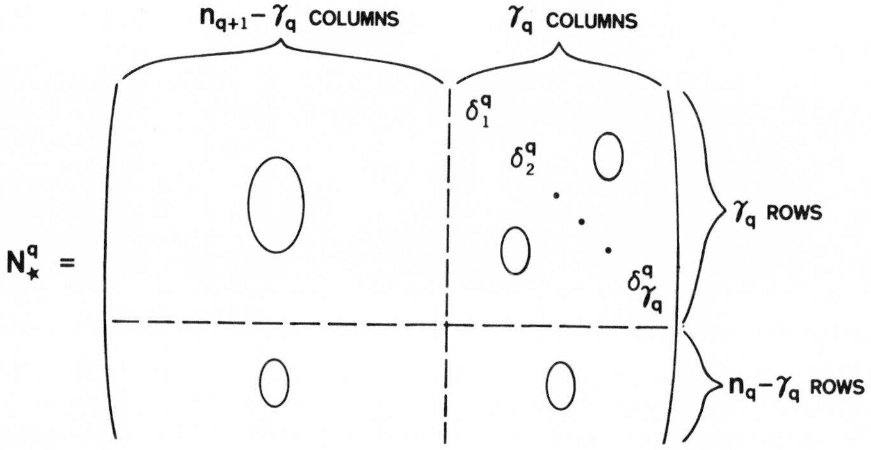

where the δ^q's are positive integers, δ_{i+1}^q divides δ_i^q, and $n_q - \gamma_q \geq \gamma_{q-1}$.

Outline of Proof: Suppose that a basis $\{c^q\} = \{c_1^q, c_2^q, \cdots, c_{n_q}^q\}$ has been chosen for each group $H_q(K)$ and that N^q is the q-th incidence matrix with respect to these bases. We must see what happens to the matrices N^q when we change the bases. It is clear that changing the basis $\{c^q\}$ affects both N^q and N^{q-1}.

CLAIM 1. (a) Replacing $\{c_1^q,\cdots,c_i^q,\cdots,c_{n_q}^q\}$ by $\{c_1^q,\cdots,-c_i^q,\cdots,c_{n_q}^q\}$ corresponds to changing all signs in the i-th column of N^{q-1} and i-th row of N^q.

(b) Replacing $\{c_1^q,\cdots,c_i^q,\cdots,c_j^q,\cdots,c_{n_q}^q\}$ by $\{c_1^q,\cdots,c_j^q,\cdots,c_i^q,\cdots,c_{n_q}^q\}$ corresponds to interchanging the i-th and j-th columns in N^{q-1} and the i-th and j-th row in N^q.

(c) Replacing $\{c_1^q,\cdots,c_i^q,\cdots,c_{n_q}^q\}$ by $\{c_1^q,\cdots,c_i^q + kc_j^q,\cdots,c_{n_q}^q\}$ for some integer k and $i \neq j$ corresponds to replacing the i-th column of N^{q-1} by [i-th column + k•(j-th column)] and replacing the j-th row of N^q by [j-th row - k•(i-th row)].

Parts (a) and (b) of Claim 1 are easy to verify. Part (c) follows from the fact that $\partial_q(c_i^q + kc_j^q) = \partial_q(c_i^q) + k\partial_q(c_j^q)$ and that if

$$\partial_{q+1}(c_s^{q+1}) = \sum_{t=1}^{n_q} \eta_{ts}^q c_t^q$$

then

$$\partial_{q+1}(c_s^{q+1}) = \eta_{is}^q(c_i^q + kc_j^q) + (\eta_{js}^q - k\eta_{is}^q)c_j^q + \sum_{\substack{t=1 \\ t\neq i,j}}^{n_q} \eta_{ts}^q c_t^q$$

Now consider the 0-th incidence matrix $E^0 = (\epsilon_{ij}^0)$ of K. It has the property that each column has only two nonzero entries, namely, one +1 and one -1. If P is any integer matrix, then consider the following property:

(*) Each nonzero column of P has precisely two nonzero entries, one +1 and one -1.

CLAIM 2. Any matrix $P = (p_{ij})$ satisfying (*) can be transformed into a normalized matrix (as defined by Theorem 1) via a sequence of operations of the type (a)-(c) above.

By interchanging rows we may assume that $p_{11} = +1$. If $p_{1j} = \pm 1$ for some $j > 1$, then we replace the j-th column of P by [j-th column - $p_{1j}\cdot$(1st column)]. The new matrix $P' = (p'_{ij})$ will have $p'_{1j} = 0$ and either the j-th column consists of zeros or there are again just two nonzero entries, one +1 and one -1. Therefore, by a sequence of such operations we arrive at a matrix $P'' = (p''_{ij})$ such that $p''_{11} = 1$, $p''_{1j} = 0$ for $j > 1$, and P'' satisfies (*). If $p''_{i1} = -1$, then by adding $(-1)\cdot$(1st row) to the i-th row, we may assume that $p''_{i1} = 0$ for $i > 1$. Induction applied to $(p''_{ij})_{2\leq i,2\leq j}$ finishes the proof of Claim 2.

It follows from Claim 2 that by changing the basis of $C_0(K)$ and $C_1(K)$ appropriately we can transform E^0 into the normalized form N_*^0. In fact, the proof shows that $\delta_i^0 = 1$ for $1 \leq i \leq \gamma_0$. Note also that the i-th incidence matrix for $i \geq 2$ is still E^i but that the 1st incidence matrix will have changed. Assume inductively that for some $k > 0$ it is possible to choose bases for the groups $C_q(K)$ so that with respect to these bases the incidence matrices are $N_*^0,\ldots,N_*^{k-1},N^k,E^{k+1},\ldots,E^{n-1}$. (Clearly, the matrices in dimension i for $i < 0$ or $i \geq n$ are of no interest since they always consist of zeros.)

By Lemma 1, the product $N_*^{k-1} \cdot N^k$ is the zero matrix. From this we see easily that

CLAIM 3. The matrix N^k has only zeros in its last γ_{k-1} rows.

It is a purely algebraic fact, which is rather technical and will not be proved here (see [Co]), that every matrix can be transformed into a normalized form such as is required for Theorem 1 by a sequence of operations of the type (a)-(c) above. Because of Claim 3 we may assume that when applying this procedure to N^k only the first $n_k - \gamma_{k-1}$ rows of N^k will be affected. This means that if we translate these changes into the corresponding changes in the basis for $C_k(K)$ only the first $n_k - \gamma_{k-1}$ columns of N_*^{k-1} are manipulated. Since they consist entirely of zeros, the matrix N_*^{k-1} is left unchanged. On the other hand, the matrix E^{k+1} will certainly have changed. At any rate, we have completed the inductive step and Theorem 1 is proved.

Next, let us show how Theorem 1 leads to a determination of the homology groups of K. Recall that the rows and columns of the incidence matrices are indexed by the chains in the chosen basis of the appropriate chain groups of K. Therefore, consider the q-th incidence matrix

$$N_*^q = \begin{array}{c|c|c|c:c} & A_1^{q+1}\cdots A_{\gamma_{q+1}}^{q+1} & B_1^{q+1}\cdots B_{\beta_{q+1}}^{q+1} & C_1^{q+1}\cdots C_{\rho_q}^{q+1} & C_{\rho_q+1}^{q+1}\cdots C_{\gamma_q}^{q+1} \\ \hline \begin{matrix} A_1^q \\ \vdots \\ A_{\rho_q}^q \end{matrix} & 0 & 0 & \begin{matrix} \delta_1^q & & 0 \\ & \ddots & \\ 0 & & \delta_{\rho_q}^q \end{matrix} & 0 \\ \hdashline \begin{matrix} A_{\rho_q+1}^q \\ \vdots \\ A_{\gamma_q}^q \end{matrix} & 0 & 0 & 0 & \begin{matrix} 1 & & 0 \\ & \ddots & \\ 0 & & 1 \end{matrix} \\ \hline \begin{matrix} B_1^q \\ \vdots \\ B_{\beta_q}^q \end{matrix} & 0 & 0 & 0 & 0 \\ \hline \begin{matrix} C_1^q \\ \vdots \\ C_{\gamma_{q-1}}^q \end{matrix} & 0 & 0 & 0 & 0 \end{array}$$

in the conclusion of Theorem 1, where $0 \leq q \leq n$ and the basis elements of $C_q(K)$ corresponding to the rows of N_*^q have been labeled as follows: The first γ_q basis elements are labeled as A_i^q's; the last γ_{q-1} are labeled as C_i^q's; and the remaining $\beta_q = n_q - \gamma_q - \gamma_{q-1}$ basis elements, if there are any, are labeled as B_i^q's. (Note that $\gamma_{-1} = \gamma_n = 0$.) The integer ρ_q is defined to equal the maximum of $\{0\} \cup \{ i \mid \delta_i^q > 1 \}$. All the information we might want to have about the groups $C_q(K)$ and homomorphisms ∂_q are now at our fingertips. In particular, it is easy to see that $\{[A_1^q],\cdots,[A_{\rho_q}^q],$ $[B_1^q],\cdots,[B_{\beta_q}^q]\}$ is a basis for $H_q(K)$. Also, $o([A_i^q]) = \delta_i^q$, the δ_i^q's are the torsion coefficients of $H_q(K)$, $o([B_i^q]) = \infty$, and $\mathrm{rk}\, H_q(K) = \beta_q$.

Finally, let us indicate how a computer can use the method described in this appendix to compute the homology groups of a simplicial complex K. First, one labels all the vertices of K with distinct numbers and then one feeds K into the computer by telling it all collections of numbers which comprise the highest dimensional simplices of K. From this information the computer can determine the incidence matrices of K. Given the incidence matrices the computer can then inductively, starting with the 0-th, transform them into the normalized form described in Theorem 1. This is possible because the operations (a)-(c) which have to be performed on the matrices to achieve this are operations of a type that computers are able to do easily. The discussion in the previous paragraph now shows how the computer can read off the homology groups from these normalized matrices.

Problem

1. Let $\varkappa_k$ and β_k denote as usual the k-th connectivity and Betti number, respectively, of a simplicial complex K. Show that

$$\varkappa_k = \beta_k + g_k + g_{k-1}$$

where g_i denotes the number of even invariant factors in the i-th incidence matrix of K.

LIST OF SYMBOLS

Only the most important and most frequently used symbols are listed here. In general, symbols appearing solely within a single discussion or proof are not included. The list is divided into three categories: symbols involving Roman letters, those involving Greek letters, and finally some common operations and relations.

A or A^+, fixed infinite set, 66

$A_d^{\pm}(a)$, closed halfspace, 242

$A(\varphi,y)$ or $A(\varphi,y,P)$, angle swept out by curve φ around y, 254-255

arg z, angle z makes with x axis, 306

$B_q(K)$ or $B_q(K;G)$, group of q-boundaries of K, 120, 192

$\mathbb{C}$, complex numbers

$|C|$, underlying space of cell complex C, 209

CX, cone of X, 131

$c_{(L,\mathcal{L})}$, 48

(C,h), cellular decomposition, 210

$C_q(K)$, $C_q(K;G)$, $C_q(C)$, or $C_q^S(X)$, group of q-chains, 115, 190, 193, 210, 298

$C_{\#}(K)$, chain complex of K, 115

$C^q(K)$, q-th cochain group of K, 300

chr(S), chromatic number of surface S, 10

$c\ell$, closure, 307

D^n, n-dimensional unit disk, 3
$D_r(y)$, disk in S^n of radius r around y, 279
DK or DX, double of K or X, 200
deg f, degree of f, 225, 235-236, 266, 271
deg(f;y), local degree of f at y, 253
deg(f|γ;w), local degree of f along γ at w, 258
diam X, diameter of X, 308
dim K, dim X, dim(K,L), dim(X,Y), or dim C, dimension, 37, 184, 185, 210
dist(x,A), distance from x to A, 308
$e_j(k)$, j-th edge of regular k-gon Q_k, 61
|f|, map on underlying spaces, 42
f_* or f_{*q}, map between q-th homology or homotopy groups, 144, 165, 172, 192, 210, 304
$f^s_{\#q}$, map between singular q-chains, 299
f^s_{*q}, map between q-th singular homology groups, 299
$f^{\#q}$, map between q-cochains, 300
f^{*q}, map between q-th cohomology groups, 300
172, 192, 210, 304
$f^s_{\#q}$, map between singular q-chains, 299
f^s_{*q}, map between q-th singular homology groups, 299
$f^{\#q}$, map between q-cochains, 300
f^{*q}, map between q-th cohomology groups, 300
[f], homotopy class of f, 147
$[f]_*$, base point preserving homotopy class of f, 302
f|A, restriction of f to A, 305
G/H, quotient group, 316
gℓb, greatest lower bound
$H_q(K)$, $H_q(X)$, $H_q(K;G)$, $H_q(X;G)$, $H_q(c)$, or $H^s_q(X)$, q-th homology group, 120, 133, 192, 193, 194, 210, 299
$H^q(K)$, q-th cohomology group of K, 300
$H_q(K,L)$, q-th relative homology group, 297
Hom(G,H), group of homomorphisms from G to H, 321
I(σ,x), index of vector field σ at x, 274
Im f, image of f, 315
int X, interior of X, 36, 44
K, abstract simplicial complex, 47

$\mathcal{K}_K$, abstract simplicial complex determined by K, 47

$\mathcal{K}_{(L,\mathcal{L})}$, abstract simplicial complex determined by the labeled complex $(L,\mathcal{L})$, 48

$K_{(L,\mathcal{L})}$, simplicial complex determined by the labeled complex $(L,\mathcal{L})$, 48

K^*, 199

$|K|$, underlying space of K, 37

(K,φ) or $((K,L),\varphi)$, triangulation, 42, 185

Ker f, kernel of f, 315

$(L,\mathcal{L})$, labeled simplicial complex, 48

$\ell(w)$, length of w, 69

$LH_k(x;X)$, k-th local homology group of x in X, 175

$\lim_{x \to a} f(x)$, limit of f at x, 308

$\ell k(x)$, link of x, 175

ℓub, least upper bound

$m_f(z,w)$ or $m_f(z,w,\mu_1,\mu_2)$, multiplicity of z in $f(z) = w$, 259, 268-269

$(M,(K,\varphi),\mu)$, closed oriented pseudomanifold, 266

max, maximum, 307

mesh K, 155

min, minimum, 307

n_v, $n_v(S)$, n_e, $n_e(S)$, n_f, $n_f(S)$, number of vertices, edges, or faces, 2, 10

$n_w(a)$, number of times that a or a^{-1} appears in w, 69

$n_q(K)$ or $n_q(C)$, number of q-simplices or q-cells in K or C, respectively, 92, 205, 211

$n(\tau,\lambda)$, 235

$o(g)$ or $o(G)$, order of g or G, 317, 318

P^2, projective plane, 58-60, 75, 77

P_n, stereographic projection, 183

$P_d(a)$, hyperplane, 242

$p(\tau,\lambda)$, 235

$P_{(L,\mathcal{L})}$, 48

$\mathbb{Q}$, rational numbers

Q_k, regular k-gon, 61

$\mathbb{R}$, real numbers

$\mathbb{R}^n$, n-dimensional Euclidean space, 306
rk, rank, 319
S^n, n-dimensional unit sphere, 4
$S^n_{\pm}$, upper and lower hemisphere of S^n, 61-62
S_q, set of all oriented q-simplices, 115
S^+_q, special subset of oriented q-simplices, 116
S_w, surface associated to the string w, 69-75
S^*, surface associated to the bordered surface S, 97
$\mathcal{S}_n$, symmetric group of degree n, 311
SX, suspension of X, 131
sd K or sd^n K, first or n-th barycentric subdivision of K, 155
$sd_\#$, $sd_{\#q}$ or $sd^n_{\#q}$, subdivision map on q-chains, 158, 159, 172
sd_*, sd_{*q} or sd^n_{*q}, subdivision map on q-th homology groups, 159, 172
st(v), star of v, 150
$st^c(x)$, closed star of x, 175
T(G), torsion subgroup of G, 317-318
v^*, 199
vK, cone of K from v, 128
vKw, suspension of K from v and w, 131
$v_0v_1\ldots v_k$, k-dimensional simplex with vertices v_0, v_1,..., and v_k, 32
$[v_0v_1\ldots v_k]$, oriented k-simplex, 114
$\mathcal{W}$, set of nonempty strings of symbols from $\mathcal{A}$, 66
$\mathcal{W}^*$, special subset of $\mathcal{W}$, 70
w_S, symbol for surface S, 66
$w_j(k)$, j-th vertex of regular k-gon, 61
$W(\varphi,y)$, winding number of φ around y, 255
[X,Y], set of homotopy classes of maps from X to Y, 147
$[X,Y]_*$, set of base point preserving homotopy classes of maps from X to Y, 302
$\bar{z}$, complex conjugate of z, 306
$\mathbb{Z}$, integers
$\mathbb{Z}_n$, integers mod n, 314
$\mathbb{Z}g_1 + \mathbb{Z}g_2 + \cdots + \mathbb{Z}g_k$, subgroup generated by the g_i, 317
$Z_q(K)$ or $Z_q(K;G)$, group of q-cycles of K, 120, 192

$\beta_q(K)$ or $\beta_q(X)$, q-th Betti number of K or X, 126, 133

Δ^q, standard q-simplex, 298

∂_q, ∂_q^r, or ∂_q^s, boundary map, 117-118, 192, 194, 210, 297, 299

∂K or X, boundary of K or X, 40, 44

δ^q, coboundary map, 300

$\delta_s^r K$, 186

$\eta_s K$, 186

$\varkappa_q(K)$ or $\varkappa_q(X)$, q-th connectivity number of K or X, 194

$\lambda^r K$, 188

$\mu^r K$, 187

$\pi_n(Y,y_0)$, n-th homotopy group of Y, 303

$\hat{\sigma}$, barycenter of simplex σ, 155

$\bar{\sigma}$, simplicial complex generated by simplex σ, 40

$[\sigma]$, oriented simplex, 114

ϕ, empty set, 305

(φ,K), geometric realization, 47

$\chi(S)$, $\chi(K)$, $\chi(X)$, or $\chi(C)$, Euler characteristic, 11, 92, 205, 206, 211

$\chi_f(K)$, 209

ψ_S, natural map from some polygon Q_k onto surface S, 62

1_X, identity map of X, 305

$\cdot$, dot product between vectors, 306

$\sim$, equivalence relation such as homologous, 41, 121, 313

$\approx$, homeomorphism, isomorphism of simplicial complexes, or isomorphism of groups, 28, 42, 315

$\simeq$, homotopic, chain homotopic, or homotopy type, 147, 160, 171

$\wedge$, delete symbol, 118

$\vee$, wedge, 181

$\prec$, face relation for simplices, 33

$\oplus$, direct sum, 319

#, connected sum, 78

$\hat{}$, barycenter, 155

$\bar{}$, complex generated by simplex or simplices or complex conjugate, 40, 185, 306

| |, underlying space, map of underlying spaces, or length of vector, 37, 42, 209, 306

|| ||, number of points, 30

[ ], greatest integer or equivalence class with respect to some equivalence relation, 13, 114, 121, 147

BIBLIOGRAPHY

The numbers in square brackets following the entries in this bibliography indicate the pages where the work is mentioned.

[Ad] Adams, J. F., Vector fields on spheres, Ann. Math. 75 (1962), 603-632. [291]

[Ag] Agoston, M. K., Generalized antipodes and the Borsuk antipode theorem, Comment. Math. Helv. 46(4) (1971), 457-466. [247]

[Ahl-S] Ahlfors, L. V., and Sario, L., Riemann Surfaces, Princeton Univ. Press, Princeton, 1960. [296]

[Alex 1] Alexander, J. W., A proof of the invariance of certain constants in analysis situs, Trans. Amer. Math. Soc. 16(1915), 148-154. [138,220,298]

[Alex 2] _________, Note on two three-dimensional manifolds with the same group, Trans. Amer. Math. Soc. 20 (1919), 339-342. [137]

[Alex 3] _________, A proof of Jordan's theorem about a simple closed curve, Ann. of Math. 21(1920), 180-184. [294]

[Alex 4] _________, A proof and generalization of the Jordan-Brouwer Theorem, Trans. Amer. Math. Soc. 23(1922), 333-349. [295]

[Alex 5] _________, An example of a simply connected surface bounding a region which is not simply connected, Proc. Nat. Acad. Sci. USA 10(1924), 8-10. [295]

[Alex 6] ________, Combinatorial analysis situs, Trans. Amer. Math. Soc. 28(1926), 301-329. [53,222]

[Alex 7] ________, Topological invariants of knots and links, Trans. Amer. Math. Soc. 30(1928), 275-306. [22]

[Alex 8] ________, A matrix knot invariant, Proc. Nat. Acad. Sci. USA 19(1933), 272-275. [22]

[Alex 9] ________, On the chains of a complex and their duals, Proc. Nat. Acad. Sci. USA 21(1935), 509-511. [299]

[Alex—V] ________, and Veblen, O., Manifolds of n dimensions, Ann. of Math. 2(14) (1912-1913), 163-178. [222]

[Alexf 1] Alexandroff, P. S., Zur Begründung der n-dimensionalen mengentheoretischen Topologie, Math. Ann. 94(1925), 296-308. [53]

[Alexf 2] ________, Simpliziale Approximationen in der allgemeinen Topologie, Math. Ann. 96(1927), 489-511; ibid. 101 (1929), 452-456. [299]

[Alexf 3] ________, Une definition des nombres de Betti pour un ensemble fermé quelconque, Comptes Rendus 184 (1927), 317-319. [299]

[Alexf 4] ________, Untersuchungen über Gestalt und Lage abgeschlossener Mengen beliebiger Dimension, Ann. Math. 2(30) (1928), 101-187. [299]

[Alexf 5] ________, Dimensiontheorie. Ein Beitrag zur Geometrie der abgeschlossenen Mengen, Math. Ann. 106(1932), 161-238. [293]

[Alexf-H] ________, and Hopf, H., Topologie, Springer-Verlag, Berlin, 1935. [ix,22,136,168,265,293]

[Am] Ames, L. D., On the theorem of analysis situs relating to the division of the plane or of space by a closed curve or surface, Bull. Amer. Math. Soc. 10(2) (1904), 301-305. [294]

[Ant] Antoine, L., Sur l'homéomorphie de deux figures et de leurs voisinages, J. Math. Pures Appl. (8)4(1921), 221-325. [295]

[As] Aschkinuse, W. G., Vielecke und Vielflache, Enzyklopädie der Elementarmathematik, Band IV: Geometrie, VEB Deutscher Verlag der Wissenschaften, Berlin, 1969. [8]

[Bai] Baire, M. R., Sur la non-applicabilité de deux continus a n et n+p dimensions, Bull. des Sc. Math. 2(31) (1907), 94-99. [221]

[Bal-C] Ball, W. W. R., and Coxeter, H. S. M., Mathematical Recreations and Essays, Macmillan, New York, 1962. [18,20,275]

[Bar] Barr, S., Experiments in Topology, John Murray, London, 1965. [17]

[Ben] Bendixson, I., Sur les courbes définies par des equations différentiélles, Acta Math. 24(1901), 1-88. [290]

[Ber] Berge, C., The Theory of Graphs and its Applications, Methuen, London, 1962. [15]

[Bern] Bernstein, F., Über einen Schoenfliesschen Satz der Theorie der stetigen Funktionen zweier reeller Veränderlichen, Nachr. Ges. Wiss. Göttingen (1900), 98-102. [221]

[Bet] Betti, E., Sopra gli spazi di un numero qualunque di dimensioni, Ann. Math. Pura Appl. 2(4)(1871), 140-158. [135,137]

[Bin] Bing, R. H., The elusive fixed point property, Amer. Math. Monthly 76(1969), 119-132. [224]

[Bir 1] Birkhoff, G. D., Proof of Poincaré's geometric theorem, Trans. Amer. Math. Soc. 14(1913), 14-22. [224]

[Bir 2] ________, Dynamical Systems, Amer. Math. Soc. Coll. Publ., No. 9, New York, 1927. [224]

[Bir-K] ________, and Kellogg, O. D., Invariant points in function spaces, Trans. Amer. Math. Soc. 23(1922), 96-115. [224]

[Bla] Blackett, D. W., Elementary Topology: A Combinatorial and Algebraic Approach, Academic Press, New York, 1967. [258,265,275]

[Bli] Bliss, G. A., The exterior and interior of a plane curve, Bull. Amer. Math. Soc. 10(2)(1904), 398-404. [294]

[Bo] Bohl, P., Über die Bewegung eines mechanischen Systems in der Nähe einer Gleichgewichtslage, J. Math. 127 (1904), 179-276. [293]

[Bor 1] Borsuk, K., Sur les rétractes, Fund. Math. 17(1931), 152-170. [223]

[Bor 2] ________, Über Schnitte der n-dimensionalen Euklidischen Räume, Math. Ann. 106(1932), 239-248. [293]

[Bor 3] ________, Drei Sätze über die n-dimensionale Euklidische Sphäre, Fund. Math. 20(1933), 177-190. [291]

[Boy] Boy, W., Über die Curvatura integra und die Topologie geschlossener Flächen, Math. Ann. 57(1903), 151-184 (=Diss. Göttingen, 1901). [60]

[Bra] Brahana, H. R., Systems of circuits on two-dimensional manifolds, Ann. of Math. 23(1922), 144-168. [103]

[Brou 1] Brouwer, L. E. J., Zur Analysis Situs, Math. Ann. 68 (1910), 422-434. [294]

[Brou 2] ________, Beweis des Jordanschen Kurvensatzes, Math. Ann. 69(1910), 169-175. [294]

[Brou 3] ________, Beweis der Invarianz der Dimensionzahl, Math. Ann. 70(1911), 161-165. [54,136,168,221,289]

[Brou 4] ________, On looping coefficients, Proc. K. Akad. Wet. Amsterdam 15(1912), 113-122. [22]

[Brou 5] ________, Über Abbildung von Mannigfaltigkeiten, Math. Ann. 71(1912), 97-115. [223,289,291,293]

[Brou 6] ________, Beweis der Invarianz des n-dimensionalen Gebiets, Math. Ann. 71(1912), 305-313; Zur Invarianz des n-dimensionalen Gebiets, ibid. 72(1912), 55-56. [221]

[Brou 7] ________, Beweis des Jordanschen Satzes für den n-dimensionalen Raum, Math. Ann. 71(1912), 314-319 [295]

[Brou 8] ________, "Beweis der Invarianz der geschlossenen Kurve," Math. Ann. 72(1912), 422-425. [294]

[Brou 9] ________, Über den Natürlichen Dimensionbegriff, J. Math. 142(1913), 146-152. [222]

[Brou 10] ________, Aufzählungen der Abbildungsklassen endlichfach zusammenhängender Flächen, Math. Ann. 82(1921), 280-286. [292]

[Brw] Brown, M., A proof of the generalized Schoenflies theorem, Bull. Amer. Math. Soc. 66(1960), 74-76. [295]

[Cai] Cairns, S. S., Introductory Topology, Ronald Press, New York, 1961. [304,323]

[Can] Cantor, G., Ein Beitrag zur Mannigfaltigkeitslehre, J. Math. 84(1878), 242-258. [222]

[Cat] Catalan, E. C., Sur la théorie des polyèdres, J. Éc. Polyt. 24(1865), 1-71. [23]

[Cau] Cauchy, A. L., Recherches sur les polyèdres, J. Éc. Polyt. 16(1812-1813). [23]

[Cay 1] Cayley, A., Chapters in the analytical geometry of n dimensions, Cambridge Math. J. 4(1845), 119-127 (=Coll. Math. Papers, Vol. 1, 55-62). [104,221]

[Cay 2] ________, Contour and slope lines, Philosophical Magazine, London, 4(18)(1859), 264-268(=Coll. Math. Papers, Vol. 4, 108-111). [293]

[Cay 3] ________, On Δ-faced polyacrons, in reference to the problem of the enumeration of polyhedra, Manchester Phil. Soc. Mem. 3(1)(1862), 248-256 (=Coll. Math. Papers, Vol. 5, 38-43). [23]

[Cay 4] ________, A memoir on abstract geometry, London Trans. 160(1869), 51-63(=Coll. Math. Papers, Vol. 6, 456-469). [104,221]

[Ce 1] Čech, E., Höherdimensionale Homotopiegruppen, Proc. Int. Congr. Mathematicians (Zürich, 1932), Vol. 3, 203. [304]

[Ce 2] ________, Théorie générale de l'homologie dans une espace quelconque, Fund. Math. 19(1932), 149-183. [220,299]

[Ce 3] ________, Sur les nombres de Betti locaux, Ann. of Math. 35(1934), 678-701. [221]

[Ch-S] Chinn, W. G., and Steenrod, N. E., First Concepts of Topology, Random House, New York, 1966. [254,265]

[Cli] Clifford, W. K., On the canonical form and dissection of a Riemann surface, Proc. Lond. Math. Soc. 8(1877), 292-304. [103]

[Co] Cohn, P. M., Algebra, Vol. 1, Wiley, New York, 1974. [312,313,319,328]

[Cr-F] Crowell, R., and Fox, R. H., An Introduction to Knot Theory, Blaisdell, New York, 1965. [22]

[Da] D'Alembert, J., Recherches sur le calcul intégral, Histoire de l'Acad. de Sc. de Berlin 2(1746), publ. 1748, Mémoires, 182. [292]

[De] Dehn, M., Topologie des dreidimensionalen Raumes, Math. Ann. 69(1910), 137-168. [22]

[De-H] ________, and Heegaard, P., Analysis Situs, Enzykl. Math. Wiss. III A B 3, Leipzig, 1907, 153-220. [ix,53,103,168]

[Des] Descartes, R., De solidorum elementis, Oeuvres de Descartes, vol. 10, Publ. by C. Adam and P. Tannery, Paris, 1908, 256-276. [2,5]

[Dyc 1] Dyck, W. von, Beiträge zur analysis Situs. I., Math. Ann. 32(1888), 457-512. [21,103,293]

[Dyc 2] ________, Beiträge zur Analysis Situs. II, Math. Ann. 37(1890), 273-316. [135,293]

[Dyc 3] ________, Kronecker's Charakteristiken eines Functionssystem, München Ber. 25(1895), 261, 447. [21,293]

[Dys] Dyson, F. J., Continuous functions defined on spheres, Ann. of Math. 54(2)(1951), 534-536. [291]

[Eb] Eberhard, V., Ein Satz aus der Topologie, Math. Ann. 36(1890), 121-133. [135]

[Ec] Eckmann, B., Gruppentheoretischer Beweis des Satzes von Hurwitz-Radon über die Komposition quadratischer Formen, Comment. Math. Helv. 15(1942), 358-366. [291]

[Eil] Eilenberg, S., Singular homology theory, Ann. Math. 45 (1944), 407-447. [298]

[Eil-M] ________, and Mac Lane, S., Acyclic models, Amer. J. Math. 75(1953), 189-199. [168]

[Eil-S 1] ________, and Steenrod, N. E., Axiomatic approach to homology theory, Proc. Nat. Acad. Sci. USA 31(1945), 117-120. [220]

[Eil-S 2] ________, ________, Foundations of Algebraic Topology, Princeton Univ. Press, Princeton, 1952. [220]

[Eis] Eisenberg, M., Topology, Holt, Rinehart and Winston, New York, 1974. [53,308]

[Eu 1] Euler, L., Solutio problematis ad geometriam situs pertinentis, Comm. Acad. Sci. Petrop. 8(1736), 128-140, publ. 1741. [15]

[Eu 2] ________, Recherches sur les racines imaginaires des equations, 1749(publ. 1751), Opera 6, 1st series, 78-147. [292]

[Eu 3] ________, Elementa doctrinae solidorum, Novi Comm. Acad. Sci. Petrop. 4(1752-1753), 109-140, publ. 1758 (=Opera 26, 1st series, 71-93). [5,22]

[Eu 4] ________, Demonstratio nonnullarum insignium propriet-atum, quibus solida hedris planis inclusa sunt praedita, Novi Comm. Acad. Sci. Petrop. 4(1752-1753), 140-160, publ. 1758(=Opera 26, 1st series, 94-108). [5,22]

[Fon] Foncenex, F. D., Sur les logarithmes des quantités imaginaires, Misc. Soc. Taurin. I., 1759. [292]

[Fox] Fox, R. H., A quick trip through knot theory, Topology of 3-manifolds, edited by M. K. Fort, Jr., Prentice-Hall, 1962, 120-167. [22]

[Fr] Franklin, P., A six colour problem, J. Math. Phys. 13 (1934), 363-369. [13]

[Fra 1] Franz, W., General Topology, Transl. by L. F. Boron, Frederick Ungar, New York, 1965. [53]

[Fra 2] ________, Algebraic Topology, Transl. by L. F. Boron, Frederick Ungar, New York, 1968. [304]

[Fre] Fréchet, M., Sur quelque points du calcul fonctionnel, Rend. Palermo 22(1906), 1-74. [53]

[Fre-F] ________, and Fan, K., Initiation to Combinatorial Topology, translated from the French, with sone notes, by H. W. Eves, Prindle, Weber and Schmidt, Boston, 1967. [5,8,10]

[Freu] Freudenthal, H., Über die Klassen der Sphären-Abbildungen I, Composito Math. 5(1937), 299-314. [304]

[G 1] Gauss, C. F., Demonstratio nova theorematis omnem func-tionem algebraicam rationalem integram unius vari-abilis in factores reales primi vel secundi ordinis resolvi posse, Werke 3, 1-30 (publ. 1799); Demon-stratio nova altera theorematis..., ibid., 31-56, (publ. Dec. 1815); Theorematis de resolubilitate..., ibid., 57-64 (publ. Jan. 1816); Beiträge zur Theorie der algebraischen Gleichungen, Abh. Königl. Ges. Wiss. Göttingen 4(1850), 3-15. [292]

[G 2] ________, Werke 8, 271-286. [21,103]

[G 3] ________, Werke 5, 605 [21]

[Gi] Girard, A., Invention Nouvelle on l'Algebre, Amsterdam, 1629. [292]

[Gra] Grassmann, H., Die Lineale Ausdehnungslehre, Leipzig, 1844. [104,221]

[Gre] Greenberg, M. J., Lectures on Algebraic Topology, W. A. Benjamin, New York, 1967. [304]

[Ha] Hahn, H., Mengentheoretische Characterisierung der stetigen Kurven, Sitz.-Ber. Akad. Wiss. Wien 123 (1914), 24-33; Die allgemeinste ebene Punktmenge, die stetiges Bild einer Strecke ist, Jhber. Deutsch. Math. Vereinig. 23(1914), 318-323. [295]

[Had] Hadamard, J. S., Sur quelques applications de l'indice de Kronecker, Appendix to J. Tannery, Théorie des Fonctions, Paris, 1910. [221,293]

[Har] Harary, F., Graph Theory, Addison-Wesley, Reading, Mass., 1971. [15]

[Hau] Hausdorff, F., Grundzüge der Mengenlehre, First edition, Leipzig (Veit), 1914. [53]

[He] Heath, T. L., The Thirteen Books of Euclid's Elements, 2nd ed., 3 vols., Cambridge University Press, New York, 1926 (reprinted by Dover Publ., 1956). [7]

[Hea] Heawood, P. J., Map-colour theorem, Quart. J. Math. 24 (1890), 332-339. [9,13,18]

[Hil] Hilton, P. J., An Introduction to Homotopy Theory, Cambridge Tracts in Math. and Math. Phys., No. 43, Cambridge Univ. Press, 1961. [304]

[Hoc-Y] Hocking, J. G., and Young, G. S., Topology, Addison-Wesley, Reading, Mass., 1961. [53,190]

[Hop 1] Hopf, H., Abbildungsklassen n-dimensionalen Mannigfaltigkeiten, Math. Ann. 96(1927), 209-224. [220,291]

[Hop 2] ________, Vektorfelder in n-dimensionalen Mannigfaltigkeiten, Math. Ann. 96(1927), 225-250. [290]

[Hop 3] ________, Topologie der Abbildungen von Mannigfaltigkeiten, Math. Ann. 100(1928), 579-608. [220,292,293]

[Hop 4] ________, Zur Topologie der Abbildungen von Mannigfaltigkeiten, II., Math. Ann. 102 (1930), 562-623. [220, 293]

[Hop 5] ________, Über wesentliche und unwesentliche Abbildungen von Komplexen, Recueil. Math. Soc. Math. Moscow 37(1930). [220,293]

[Hop 6] ________, Über die Abbildungen der dreidimensionalen Sphäre auf die Kugelfläche, Math. Ann. 104(1931), 637-665. [220]

[Hop 7] ________, Die Klassen der Abbildungen der n-dimensionalen Polyeder auf die n-dimensionale Sphare, Comm. Math. Helv. 5(1933), 39-54. [220,293]

[Hu 1] Hu, S.-T., Homotopy Theory, Academic Press, New York, 1959. [304]

[Hu 2] ________, Homology Theory: A First Course in Algebraic Topology, Holden-Day, San Francisco, Calif., 1966. [304]

[Hure] Hurewicz, W., Beiträge zur Topologie der Deformationen. I. Höherdimensionale Homotopiegruppen, Proc. K. Akad. Wet. Amsterdam 38(1935), 112-119; II. Homotopie und Homologiegruppen, ibid. 521-528; III. Klassen und Homologietypen von Abbildungen, ibid. 39(1936), 117-126; IV. Asphärische Räume, ibid. 215-224. [220,293]

[Hure-W] ________, and Wallman, H., Dimension Theory, Princeton University Press, Princeton, N.J., 1948. [222]

[Hurw 1] Hurwitz, A. Über die Komposition quadratischer Formen von beliebig vielen Variablen, Nachr. Ges. Wiss. Göttingen 1898, 309-316(=Math. Werke II, 565-571). [291]

[Hurw 2] ________, Über die Komposition der quadratischen Formen, Math. Ann. 88(1923), 1-25. [291]

[J 1] Jordan, C., Sur la deformation des surfaces, J. Math. 2 (11)(1866), 105-109. [103]

[J 2] ________, Des contours tracés sur les surfaces, J. Math. Pures Appl. 2(11)(1866), 110-130. [168]

[J 3] ________, Cours d'Analyse, Paris, 1893. [20,294]

[Kak] Kakutani, S., A proof that there exists a circumscribing cube around any bounded closed convex set in $\mathbb{R}^3$, Ann. of Math. 43(4)(1942), 739-741. [291]

[Kam] Kampen, E. R. van, Die Kombinatorische Topologie und die Dualitätssätze, dissertation, Leiden, 1929. [22,221]

[Ki-S] Kirby, R. C., and Siebenmann, L. C., On the triangulation of manifolds and the Hauptvermutung, Bull. Amer. Math. Soc. 75(1969), 742-749. [57,138,175]

[Kirc] Kirchoff, G., Über die Auflösung der Gleichungen, auf welche man bei der Untersuchung der linearen Verteilung galvanischer Ströme geführt wird, Ann. Phys. Chem. 72(1847), 497-508. [16]

[Kirk] Kirkman, T. P., The k-partitions of N, Manchester Phil. Soc. Mem. 2(12)(1855), 47-70. [21,23]

[Kle 1] Klein, F., Bemerkungen über den Zusammenhang der Flächen, Math. Ann. 7(1874), 549-557. [223]

[Kle 2] ________, Über den Zusammenhang der Flächen, Math. Ann. 9(1876), 476-482. [12]

[Kle 3] ________, Über Riemanns Theorie der algebraischen Funktionen und ihrer Integrale, eine Ergänzung der gewöhnlichen Darstellungen, B. G. Teubner, Leipzig, 1882. [88,103]

[Kn 1] Kneser, H., Glättungen von Flächenabbildungen, Math. Ann. 100(1928), 609-617. [293]

[Kn 2] ________, Geschlossene Flächen in dreidimensionale Mannigfaltigkeiten, Jhber. Deutsch. Math. Vereinig. 38(1929), 248-260. [104]

[Kn 3] ________, Die kleinste Bedeckungszahl innerhalb einer Klasse von Flächenabbildungen, Math. Ann. 103(1930), 347-358. [293]

[Kr] Kronecker, L., Über Systeme von Funktionen mehrer Variablen, Monatsberichte der Berliner Akad. 1869, 159, 688; Über die verschiedenen Sturm'schen Reihen und ihre gegenseitigen Beziehungen, ibid. 1873, 117; Über Sturm'sche Funktionen, ibid. 1878, 95; Über die Charakteristik von Funktionensystemen, ibid. 1878, 145. [292]

[Ku] Kuratowski, K., Sur le problème des courbes gauches en topologie, Fund. Math. 15(1930), 271-283. [16]

[Lag] Lagrange, J. L., Sur la forme des racines imaginaires des équations, Nouv. Mém. de l'Acad. de Berlin 1772, 222. [292]

[Lap] Laplace, P. S., Leçons de Mathématique à l'École Normale en 1795, Oeuvres 14. [292]

[Leb 1] Lebesgue, M. H., Sur la non applicabilité de deux domaines appartenant à des espaces de n et n+p dimensions, Math. Ann. 70(1911), 166-168. [221]

[Leb 2] ________, Sur l'invariance du nombre de dimensions d'un espace, et sur le théoremè de M. Jordan relatif aux variétés fermées, Comptes Rendus 152(1911), 841-843. [22,221,295]

[Leb 3] ________, Quelques conséquences simples de la formule d'Euler, J. Math. Pures Appl., 9e série 19(1940), 27-43. [8]

[Lef 1] Lefschetz, S., Intersections and transformations of complexes and manifolds, Trans. Amer. Math. Soc. 28 (1926), 1-49. [22,217]

[Lef 2] ________, Closed point sets on a manifold, Ann. Math. 29(1928), 232-254. [222]

[Lef 3] ________, Topology, Amer. Math. Soc. Coll. Publ. No. 12, New York, 1930. [22,298,299]

[Lef 4] ________, On singular chains and cycles, Bull. Amer. Math. Soc. 39(1933), 124-129. [298]

[Lef 5] ________, Abstract complexes, Lectures in Topology, edited by R. L. Wilder and W. L. Ayres, Univ. of Michigan Press, Ann Arbor, Mich., 1941, 1-28. [54]

[Lef 6] ________, Algebraic Topology, Amer. Math. Soc. Coll. Publ. No. 27, New York, 1942. [299]

[Leg] Legendre, A. M., Éléments de Géométrie, 7th ed., Paris, 1808. [23]

[Leib 1] Leibniz, G. W., Math. Schriften, 1 Abt., Vol. 2, 1850, 19-20(=Gerhardt, Der Briefwechsel von Leibniz mit Mathematikern 1 (1899), 568 =Chr. Huygens, Oeuv. Comp., 8, No. 2192). [2]

[Leib 2] ________, Characteristica Geometrica, Math. Schriften, 2 Abt., Vol. 1, 1858, 141-168. [2]

[Len] Lennes, N. J., Theorems on the simple polygon and polyhedron, master's thesis, 1903. [294]

[Lh] L'Huilier, S., Memoires sur les solides regulaires, Gergonnes Ann. de Math. 3(1812), 169. [23]

[Lia] Liapunov, A. M., Problème Général de la Stabilité du Mouvement, Ann. of Math. Studies 17, Princeton Univ. Press, Princeton, 1947. [290]

[Lie 1] Lietzmann, W., Lustiges und Merkwürdiges von Zahlen und Formen, Ferdinand Hirt in Breslau, Konigsplatz 1, 1930. [18]

[Lie 2] ________, Visual Topology, American Elsevier, New York, 1965. [6,8,22]

[Lis 1] Listing, J. B., Vorstudien zur Topologie, Göttinger Studien, 1847. [2,21]

[Lis 2] ________, Der Census räumlicher Komplexe, Abh. Ges. Wiss. Göttingen 10(1861), 97-180, and Nachr. Ges. Wiss. Göttingen 1861, 352-358. [11,103]

[Lit] Little, C. N., Transactions of the Connecticut Academy, 1885. [21]

[Lu] Lüroth, J., Über Abbildungen von Mannigfaltigkeiten, Math. Ann. 63(1907), 222-238. [221]

[Man] Mannheim, J. H., The Genesis of Point Set Topology, Macmillan, New York, 1964. [53]

[Mar] Markov, A. A., The unsolvability of the problem of homomorphism, Doklady Akad. Nauk. SSSR 121(1958), 218-220. [220]

[Mas] Massey, W. S., Algebraic Topology: An Introduction, Harcourt, Brace, Jovanovich, New York, 1967. [55,96, 97,100,101,304]

[Mau] Maunder, C. R. F., Algebraic Topology, Van Nostrand-Reinhold, London, 1970. [304]

[Max] Maxwell, J. C., Hills and dales, Philosophical Magazine, London, 4(40)(1870), 421-427 (=Works, 2, 233-240). [293]

[May 1] Mayer, J., Über abstrakte Topologie, Monatsh. Math. Phys. 36(1929), 1-42; Abstrakte Topologie II, ibid., 219-258. [54]

[May 2] ________, Algebraic Topology, Prentice Hall, Englewood Cliffs, N.J., 1972. [304]

[Maz] Mazur, B., On imbeddings of spheres, Bull. Amer. Math. Soc. 65(1959), 59-65. [295]

[Mazu] Mazurkiewicz, S., Sur les lignes de Jordan, Fund. Math. 1(1920), 166-209. [295]

[Meng] Menger, K., Dimensionstheorie, B. G. Teubner, Leipzig, 1928. [222]

[Mey 1] Meyer, F., Anwendungen der Topologie auf die Gestalten der algebraischen Kurven, Thesis, 1878. [21]

[Mey 2] ________, Über algebraische Knoten, Proc. Royal Soc. Edinburgh, 1885-1886. [21]

[Mi 1] Milnor, J., On manifolds homeomorphic to the 7-sphere, Ann. of Math. 64(1956), 399-405. [296]

[Mi 2] ________, Two complexes which are homeomorphic but combinatorially distinct, Ann. of Math. 74(1961), 575-590. [137]

[Mi 3] ________, Topology from the Differentiable Viewpoint, The Univ. Press of Virginia, Charlottesville, 1965. [233,290]

[Möb 1] Möbius, A. F., Theorie der elementaren Verwandtschaft, Königl. Sächs. Ges. der Wiss. Leipzig 15(1863), 18-57 (=Werke, 2, 433-472). [23,52,103,107]

[Möb 2] ________, Werke 2, 513-560. [11,103]

[Möb 3] ________, Über die Bestimmung des Inhaltes eines Polyeders, Königl. Sächs. Ges. Wiss. Leipzig 17(1865), 31-68 (=Werke 2, 473-512). [11,23,103]

[Moi] Moise, E. E., Affine Structures in 3-manifolds: V, Ann. Math. 2(56)(1952), 96-114. [57]

[O] Osgood, W. F., Satz von Schoenflies aus der Theorie der Funktionen zweier reeller Veranderlichen, Nachr. Ges. Wiss. Gottingen 1900, 94-97. [221]

[Pea] Peano, G., Sur une courbe qui remplit toute une aire plane, Math. Ann. 36(1890), 157-160. [19,222]

[Ph] Phillips, A., Turning a surface inside out, Scientific American 214(1966), 112-120. [60]

[Poin 1] Poincaré, H., Mémoire sur les courbes définies par une equation différentielle, J. Math. Pures Appl.(3)7 (1881), 375-422; ibid. (3)8(1882), 251-296; ibid. (4)1(1885), 167-244; ibid. (4)2(1886), 151-217. [289,290,293]

[Poin 2] ________, Sur le problème des trois corps et les équations de la dynamique, Acta Math. 13(1890), 1-270. [290]

[Poin 3] ________, Les Methodes Nouvelles de la Mécanique Céleste, 3 vol., Paris, Gauthiers-Villars, 1892-1899. [290]

[Poin 4] ________, Analysis Situs, J. Éc. Polyt. (2)1(1895), 1-121. [135,136,137,168,223,301,304]

[Poin 5] ________, Complément à l'analysis situs, Rend. Circ. Mat. Palermo 13(1899), 285-343. [135,136]

[Poin 6] ________, Second complément à l'analysis situs, Proc. Lond. Math. Soc. 32(1900), 277-308. [135,136,301]

[Poin 7] ________, Sur certaines surfaces algébriques; troisième complément à l'analysis situs, Bull. Soc. Math. France 30(1902), 49-70. [135]

[Poin 8] ________, Sur les cycles des surfaces algébriques; quatrième complément à l'analysis situs, J. Math. Pures Appl. (5)8(1902), 169-214. [135]

[Poin 9] ________, Cinquième complément à l'analysis situs, Rend. Circ. Mat. Palermo 18(1904), 45-110. [135,137]

[Poin 10] ________, Pourquoi l'espace à trois dimension, Revue de Metaphysique et de Morale 20(1912), 483-504. [222]

[Poin 11] ________, Sur un théorème de geométrie, Rend. Circ. Mat. Palermo 33(1912), 375-407. [223]

[Poins] Poinsot, L., Memoire sur les polygones et les polyèdres, J. Éc. Polyt. 10(1809). [23]

[Pon] Poncelet, J. V., Traite des Proprietes Projectives des Figures, Paris, 1822. [103]

[Pont 1] Pontrjagin, L., Über den algebraischen Inhalt topologischer Dualitätssätze, Math. Ann. 105(1931), 165-205. [22,295]

[Pont 2] ________, The general topological theorem of duality for closed sets, Ann. of Math. 35(1934), 904-914. [222]

[R-T] Rademacher, H., and Toeplitz, O., The Enjoyment of Mathematics, Princeton Univ. Press, Princeton, N.J., 1957. [8]

[Ra 1] Radó, T., Bemerkung zur Arbeit des Herrn Bieberbach: ..., Math. Ann. 90(1923), 30-37. [56]

[Ra 2] ________, Über des Begriff der Riemannschen Fläche, Acta. Sci. Math. Szeged. 2(1925), 101-121. [56]

[Rad] Radon, J., Lineare Scharen orthogonaler Matrizen, Abh. Sem. Hamburg I (1923), 1-14. [291]

[Ree] Reech, F., Propríetées des surfaces fermées, J. Ec. Polyt. 21(1858). [293]

[Rei 1] Reidemeister, K., Knoten und Gruppen, Abh. Hamburg Univ. 1926, 7-23. [22]

[Rei 2] ________, Knoten und Verkettungen, Math. Zeit. 29(1929), 713-729. [22]

[Riem 1] Riemann, B., Grundlagen für eine allgemeine Theorie der Funktionen einer veranderlichen complexen Grosse, Diss. Gött. 1851, Nr. 6 (=Werke, 2. Aufl., 1-48). [23,102]

[Riem 2] ________, Über die Hypothesen, welche der Geometrie zur Grunde liegen, Habilitationschrift, Göttingen, 1854 (=Werke, 272-287). [104,221]

[Riem 3] ________, Theorie der Abel'schen Funktionen, J. für Math. 54(1857), 115-155, (=Werke, 2. Aufl., 88-144). [23,102]

[Ries] Riesz, F., Stetigkeitsbegriff und abstrakte Mengenlehre, Atti del IV Congresso Intern. dei Matem., Vol. 2, Bologna, 1908, 18-24. [53]

[Rin 1] Ringel, G., Färbungsprobleme, Mathematische Monographien, VEB Deutscher Verlag der Wissenschaften, Berlin, 1959. [13]

[Rin 2] ________, Map Color Theorem, Springer Verlag, New York, 1974. [10]

[Rin-Y] ________, and Youngs, J. W. T., Solution of the Heawood map coloring problem, Proc. Nat. Acad. Sci. USA, 60 (1968), 438-445. [13]

[Rot] Rothe, P., Arithmetica Philosophica, Nuremberg, 1600. [292]

[Scha 1] Schauder, J. P., Zur Theorie stetiger Abbildungen in Funktionalraumen, Math. Zeit. 26(1927), 47-65, 417-431. [224]

[Scha 2] ________, Der Fixpunktsatz in Funktionalräumen, Studia Math. 2(1930), 171-180. [224]

[Scha-L] ________, and Leray, J., Topologie et equations fonctionelles, Ann. Sci. Éc. Norm. Sup. 51(1934), 45-78. [224]

[Schl 1] Schläfli, L., Theorie der vielfachen Kontinuität, 1850-1852 (=Schweiz. Naturf. Ges. Denkschr. 190). [104]

[Schl 2] ________, Über die linearen Relationen zwischen den 2p Kreiswegen erster Art und den 2p zweiter Art in der Theorie der Abelschen Funktionen..., J. Math. 76 (1873), 149-155. [103]

[Scho 1] Schoenflies, A., Satz der Analysis Situs, Nachr. Ges. Wiss. Göttingen 1899, 282-290. [221]

[Scho 2] ________, Über einen grundlegenden Satz der Analysis Situs, Nachr. Ges. Wiss. Göttingen 1902. [294]

[Scho 3] ________, Beitrage zur Theorie der Punktmengen. I, Math. Ann. 58(1904), 195-234;..., II., ibid. 59(1904), 129-160;..., III., ibid. 62(1906), 286-328. [294]

[Se-T 1] Seifert, H., and Threlfall, W., Lehrbuch der Topologie, Teubner, Leipzig, 1934. [ix,22,136,220]

[Se-T 2] ________, ________, Old and new results on knots, Can. J. Math. 2(1950), 1-15. [22]

[Sim 1] Simony, O., Über eine Reihe neuer Thatsachen aus dem Gebiete der Topologie, Math. Ann. 19(1882), 110-120; ibid. 24(1884), 253-280. [21]

[Sim 2] ________, Sitz.-Ber. Akad. Wiss. Wien, Vol. 82, 84, 85, 87, 88, and 96, 1880-1883, 1887. [21]

[Sp] Spivak, M., Calculus on Manifolds, W. A. Benjamin, New York, 1965. [308,309]

[Stnr] Steiner, J., Sur le maximum et le minimum des figures dans le plan, sur la sphère et dans l'espace en général, J. Math. 24(1842), 93-152. [23]

[Stnz] Steinitz, E., Polyeder und Raumeinteilungen, Enzykl. Math. Wiss. III AB 12, Leipzig, 1916, 1-139. [ix,8]

[T] Tait, P. G., On knots, Trans. Royal Soc. Edinburgh, 1879, 1884, 1886, and Proc. Royal Soc. Edinburgh, 1876-1879. [21]

[Ti 1] Tietze, H., Über die topologischen Invarianten mehrdimensionaler Mannigfaltigkeiten, Monatsh. Math. Phys. 19(1908), 1-118. [222]

[Ti 2] ________, Über Funktionen, die auf einer abgeschlossenen Menge stetig sind, J. Math. 145(1915), 9-14. [294]

[Ti-V] ________, and Vietoris, L., Beziehungen zwischen den verschiedenen Zweigen der Topologie, Enzyk. Math. Wiss. III AB 13, Leipzig, 1929, 141-237. [ix,53]

[Tu] Tucker, A. W., An abstract approach to manifolds, Ann. of Math. 34(1933), 191-243. [54]

[U 1] Urysohn, P., Les Multiplicités Cantoriennes, Comptes Rendus 175(1922), 440-442. [222]

[U 2] ________, Über die Mächtigkeit der zusammenhängenden Mengen, Math. Ann. 94(1925), 262-295. [294]

[U 3] ________, Memoire sur les multiplicités Cantoriennes, Fund. Math. 8(1926), 225-359. [222]

[Veb 1] Veblen, O., A system of axioms for geometry, Trans. Amer. Math. Soc. 5(1904), 343-384. [294]

[Veb 2] ________, Theory of plane curves in non-metrical analysis situs, Trans. Amer. Math. Soc. 6(1905), 83-99. [294]

[Veb 3] ________, Analysis Situs, Amer. Math. Soc. Coll. Lect., Vol. 5, New York, 1922. [298]

[Veb 4] ________, The intersection numbers, Trans. Amer. Math. Soc. 25(1923), 540-550. [222]

[Vi 1] Vietoris, L., Über den höheren Zusammenhang kompakten Räumen und eine Klasse von zusammenhangstreuen Abbildungen, Math. Ann. 97(1927), 454-472. [220,299]

[Vi 2] ________, Zum höheren Zusammenhang der kompakten Raume, Math. Ann. 101(1929), 219-225. [299]

[Wey 1] Weyl, H., Die Idee der Riemannschen Flache, 2nd edition, Leipzig, 1923. [53]

[Wey 2] ________, Análisis situs combinatorio, Rev. Mat.-Hisp.-Am. 5(1923), 209-218. [53]

[Wh] Whitehead, J. H. C., Combinatorial homotopy: I. Bull. Amer. Math. Soc. 55(1949), 213-245. [223]

[Whi] Whitney, H., On products in a complex, Ann. Math. 39 (1938), 397-432. [299]

[Wil] Willard, S., General Topology, Addison-Wesley, Reading, Mass., 1968. [53]

[Y-Y] Yamabe, H., and Yujobô, Z., On the continuous functions defined on the sphere, Osaka Math. J. 2(1950), 19-22. [291]

[Ya] Yang, C. T., On Theorems of Borsuk-Ulam, Kakutani-Yamabe-Yujobô and Dyson, I., Ann. Math. 60(2)1954, 262-282. [291]

Supplementary Reading

Alexandroff, P. S., Einfachste Grundbegriffe der Topologie, Springer-Verlag, Berlin, 1932.

Bers, L., Topology, Notes from New York University, 1956-1957.

Dugundji, J., Topology, Allyn and Bacon, Boston, 1966.

Gerretsen, J., and Vredenduin, P., Polygone und Polyeder, Grundzüge der Mathematik, Band II: Geometrie, Vandenhoeck und Ruprecht, Göttingen, 1960, 174-225.

Hilbert, D., and Cohn-Vossen, S., Geometry and the Imagination, translation by P. Nemenyi, Chelsea Publ., New York, 1952.

Kerékjartó, B. v., Vorlesungen über Topologie, Springer-Verlag, Berlin, 1923.

Kline, M., Mathematical Thought from Ancient to Modern Times, Oxford Univ. Press, 1972.

Lefschetz, S., Introduction to Topology, Princeton Univ. Press, Princeton, N.J., 1949.

Liusternik, L. A., Convex Figures and Polyhedra, Dover Publ., New York, 1963.

Rademacher, H., and Steinitz, E., Vorlesungen über die Theorie der Polyeder, Springer-Verlag, Berlin, 1934.

Reidemeister, K., Topologie der Polyeder, Akademischer Verlag, Leipzig, 1938.

Roman, T., Reguläre und halbreguläre Polyeder, VEB Deutscher Verlag der Wissenschaften, Berlin, 1968.

Smith, D. E., A Source Book of Mathematics, Dover Publ., New York, 1959.

Tietze, H., Ein Kapitel Topologie, Hamburg. Math. Einzelschriften, 36, Teubner, Leipzig, 1942.

Wall, C. T. C., A Geometric Introduction to Topology, Addison-Wesley, Reading, Mass., 1972.

Weise, K. H., and Noack, H., Ausgewählte Fragen der Topologie, Grundzüge der Mathematik, Band II: Geometrie, Vandenhoeck und Ruprecht, Göttingen, 1960, 530-615.

INDEX

A

Abelian group, 313
Abstract simplicial complex, 47
Accumulation point, 307
Acyclic models, method of, 162-163
Adams-Hurwitz-Radon-Eckmann theorem, 231-232
Alexander horned sphere, 295
Algebraic variety, 135
Analysis situs, 2
Angle swept out by curve, 254-255
Antipodal map, 229
Antipodal points, 229, 247-248
Antipodal preserving, 235, 237

B

Ball, open, 306
Barycenter, 155
Barycentric coordinate, 34
Barycentric subdivision, 155
Base point preserving homotopy, 302
Base point preserving map, 302
Basis for group, 320
Betti group, 136
Betti number
  of polyhedron, 133
  of simplicial complex, 126
Bijective map, 305
Bordered surface, 97
Borsuk-Ulam theorem, 241
Borsuk's separation criterion, 283-284
Boundary
  chain, 119-120
  map, 117-118, 192, 194, 210, 297, 299
  of polyhedron, 44
  of simplicial complex, 40
Boundary point of polyhedron, 44
Bounded, 308
Brouwer fixed-point theorem, 216

C

Cartesian product, 305
Čech homology theory, 299
Cell, 209
Cell complex, 209-210
Cellular decomposition, 210
Chain, 115, 190, 193, 210
  elementary, 116
  singular, 298
Chain complex, 115
Chain equivalence, 163
Chain homotopy, 160

Chain map, 142
Chromatic number
  of graph, 16
  of surface, 10
Classification theorem
  for bordered surfaces, 100
  for surfaces, 82
Closed
  curve, 41
  manifold, 58
  path, 41
  subset, 307
  surface, 58
Closure of set, 307
Coboundary map, 300
Cohomology group
  simplicial, 300
  singular, 300
Cohomology ring, 301
Coloring of maps and graphs, 9-17
Commutative diagram, 140
Commutative group, 313
Compact, 308
Complex, see Simplicial complex
Complex conjugate, 306
Component
  function, 308
  of simplicial complex, 45
  of space, 41
Composite of chain map, 145
Cone, 128, 131
Connected simplicial complex, 40
Connected space, 41
Connected sum, 78
Connected number, 95
Connectivity number, 194, 207-208
Constant map, 305
Continuous map, 309
Convex hull, 31
Convex set, 31
Coset, 315-316
Cup product, 301
Curve, 41, 294-295
  simple closed, 208, 294
Cutting and pasting, 50
Cycle, 119
Cyclic group, 317

D

Degree
  of continuous map, 225, 235-236, 266
  of polynomial, 252
  of rational function, 271
Descartes-Euler theorem, 4, 6-7
Diameter, 308
Dimension, 221-222
  of cell complex, 210
  of polyhedral pair, 185
  of polyhedron, 184
  of simplicial complex, 37
  of simplicial pair, 185
Disjoint sets, 305
Disk, 58
  n-dimensional unit, 3
Distance from point to set, 308
Dot product, 306
Double
  of polyhedron, 200
  of simplicial complex, 200
Dyson-Yang theorem, 245

E

Edge
  of first kind, 83
  of second kind, 83
Equivalence class or relation, 313
Essential map, 276
Euclidean distance, 306
Euclidean space, n-dimensional, 306
Euler characteristic, 11, 92, 99
  of cell complex, 211
  of polyhedron, 206
  of simplicial complex, 205
Euler-Poincaré formula, 205
Euler's theorem, see Descartes-Euler theorem
Even permutation, 313
Extension, 305
External direct sum, 319

F

Face
  of simplex, 33
  of singular simplex, 298
Factor group, 316
Finite group, 318
Finitely generated group, 318
Finite-to-one map, 262
Fixed point, 216
Four color conjecture, 9
Free group, 320
Fundamental group, 137, 304
Fundamental theorem of algebra, 251
Fundamental theorem of finitely generated abelian groups, 319

G

Generators of group, 318
Genus, 93, 103
Geometria situs, 2
Geometric complex, 46
Geometric realization, 47
Gradient, 274
Graph, 15
  planar, 16
Great circle, 9
Group, 313
  of q-boundaries, 119-120
  of q-chains, 115, 190, 193, 210
  of q-cycles, 119
  of singular q-chains, 298
  q'th cochain, 300

H

Hahn-Mazurkiewicz theorem, 295
Hairy billiard ball problem, 231
Hairy circle, 232
Ham sandwich problem, 241-242
  generalized, 242
Hauptvermutung, 137
Heawood conjecture, 13
Heawood number, 13
Heine-Borel-Lebesgue theorem, 308
Hemisphere, upper or lower, 61-62
Hole in space, 125, 304
Homeomorphism, 28
  linear, 42
Homogeneously n-dimensional, 187
Homologous, 121
Homology class, 121
Homology group
  of cell complex, 210
  of polyhedron, 133
  of simplicial complex, 120, 192, 194
  of space, 299
Homomorphism, 315
Homotopic maps, 147
Homotopy, 147
Homotopy class, 147
Homotopy equivalence, 171
Homotopy extension theorem, 281
Homotopy group, 303
Homotopy theory, 149, 220
Homotopy type, 171
Hopf theorem, 291-292
Hyperplane, 242

I

Identity map, 305
Image of homomorphism, 315
Imbedding, 79
Incidence matrix, 323-325
Incidence number, 323
Index of vector field, 274, 289
Induced maps
  on cohomology groups, 300
  on homology groups, 144, 165, 192, 210, 299
  on homotopy groups, 304
  on underlying spaces, 42
Inessential map, 276
Injective map, 305
Integers mod n, 314
Integral curve, 289
Interior or interior point
  of polyhedron, 44
  of simplex, 36
Intermediate value theorem, 253
  generalized, 257

Internal direct sum, 319
Invariance
  of boundary, 189
  of dimension, 184, 185
  of domain, 190
  of homogeneously n-dimensional, 187
  of pseudomanifolds, 198
Inverse of map, 305
Isolated zero of vector field, 274
Isomorphism
  of groups, 315
  of simplicial complexes, 42
Isotopic, 149

J

Jordan curve theorem, 19
  generalized, 286

K

Kakutani-Yamabe-Yujobô theorem, 245
Kernel of homomorphism, 315
Klein bottle, 88
Knot, 20
Knot theory, 20, 149
Königsberg bridge problem, 15
Kronecker characteristic, 292-293

L

Labeled simplicial complex, 48
Lagrange theorem, 318
Lebesgue number, 309
Lefschetz fixed point theorem, 217-218
Length of vector, 306
Limit, 308
Line segment, 31
Linearly independent
  points, 31
  vector fields, 231
Link, 175
Linking number, 21
Local degree, 253, 258
Local homology group, 175

M

Manifold,n-dimensional, 57
Mesh, 155
Möbius strip, 11
Multiplicity
  of root, 259, 268
  of zero, 259

N

Neighborhood, 307
Neighboring regions problem, 18
Nonorientable, see Surface or Pseudomanifold
Norm of partition, 254
Normal form
  for bordered surface, 97-98
  for surface, 82
  of incidence matrix, 326

O

One-sided surface, 11
Open cover, 308
Open subset, 307
Order
  of element, 317
  of group, 318
Orientable, see Surface or Pseudomanifold
Orientation
  at point, 12, 113
  coherent, 201
  induced, 118
  of n-dimensional vector space, 112
  of simplex, 111
Orientation-preserving, 104, 260
Orientation-reversing, 104

P

Partition, 254
Path, 40
Peaks, pits, and passes, 274
Peano curve, 19, 222
Permutation, 311
Poincaré-Bendixson theorem, 290
Poincaré-Birkhoff fixed point theorem, 224
Poincaré-Bohl theorem, 257, 293
Poincaré conjecture, 137
Poincaré duality theorem, 136-137, 301
Poincaré-Hopf theorem, 290
Point at infinity, 183
Polygon, 76
Polyhedral imbedding, 283
Polyhedral pair, 185
Polyhedron, 2, 43
  regular, 7
  simple, 2-3
Projective geometry, 103
Projective plane, 58-60, 75, 77, 103
Pseudomanifold, 195-196
  closed, 196
  closed oriented, 266
  orientable, 201

Q

Querschnitte, 102
Quotient group, 316
Quotient space, 49

R

Radó's theorem, 56
Rank, 319
Rational function, 271
Reflection, 227
Region, 294-295
Relative homology group, 297
Restriction, 305
Retract, 214
Retraction, 214
Riemann surface, 102
Root, 251
Rouché's theorem, 265
Rubber sheet geometry, 26
Rückkehrschnitte, 102-103

S

Schoenflies conjecture, 295
Schoenflies theorem, 294
  generalized, 288
Simplex
  k-dimensional, 32
  oriented, 114
  standard, 298
Simplicial approximation, 152
Simplicial approximation theorem, 164
Simplicial complex, 37
  finite, 37
  infinite, 37
Simplicial map, 42
Simplicial pair, 185
Simply-connected, 137, 295
Singular homology group, 299
Singular simplex, 298
Skeleton of simplicial complex, 127
Solution to equation, 251
Sphere, n-dimensional unit, 4
Star, 150
  closed, 175
Star-shaped, 173
Stereographic projection, 183
Subcomplex of simplicial complex, 39
Subgroup, 314
  generated by elements, 317
Support of chain, 194
Surface, 56
  closed, 58
  non-compact, 100-101
  orientable, 11, 93, 113-114, 201, 203, 204-205
  with boundary, 58, 96
Surjective map, 305
Suspension, 131
Symbol for surface, 66
Symmetric group, 311

T

Tangent vector, 230, 233
Tietze-Urysohn extension theorem, 294
Topological invariance of homology groups, 133
Topologically invariant function, 211
Topology
  algebraic, 29-30
  combinatorial, 1, 126, 219
  definition of, 28
  general or point set, 53, 219
Torsion coefficient
  of finitely generated group, 319
  of polyhedron, 133
  of simplicial complex, 126
Torsion-free group, 318
Torsion subgroup, 317-318
Torus, 11, 27
Transposition, 311
Triangulation, 42, 185
  infinite, 43
  local, 174
  minimal, 96
  proper, 56

U

Underlying space
  of cell complex, 209
  of simplicial complex, 37
Universal coefficient theorem, 193, 302

V

Vector field, 229, 230, 233
  non-zero, 229
  $\mathbb{R}^n$-, 229
  tangent, 230, 233
Vertex, 15, 32
Volume, n-dimensional, 243

W

Wedge of two spaces, 181
Winding number, 254, 255